CELL ADHESION MOLECULES IN CANCER AND INFLAMMATION

Cell Adhesion and Communication

A series of books encompassing monographs on classes of adhesion molecules and monographs giving a broader functional synopsis on adhesion molecules on a particular system.

Edited by Christo Goridis

Volume 1
Cell Adhesion Molecules in Cancer and Inflammation
edited by A. A. Epenetos and M. Pignatelli

Other volumes in preparation

The Laminins
P. Ekblom, R. Timpl

Tenascin and other counteradhesive molecules of the extracellular matrix
K.L. Crossin

This book is part of a series. The publisher will accept continuation orders which may be cancelled at any time and which provide for automatic billing and shipping of each title in the series upon publication. Please write for details.

CELL ADHESION MOLECULES IN CANCER AND INFLAMMATION

Edited by

A. A. Epenetos

Hammersmith Hospital, London, UK

M. Pignatelli

Royal Postgraduate Medical School, London, UK

HARWOOD ACADEMIC PUBLISHERS
Australia • Austria • Belgium • China • France • Germany • India • Japan
Luxembourg • Malaysia • Netherlands • Russia • Singapore • Switzerland
Thailand • United Kingdom • United States

 Printed in Singapore.

Harwood Academic Publishers GmbH
Poststrasse 22
7000 Chur
Switzerland

British Library Cataloguing in Publication Data

Cell Adhesion Molecules in Cancer and
Inflammation. — (Cell Adhesion &
Communication Series, ISSN 1023–7046;
Vol. 1)
I. Epenetos, Agamemnon A.
II. Pignatelli, M. III. Series
616.994

ISBN 3–7186–5516–0

CONTENTS

PREFACE TO THE SERIES

The development and normal functioning of all multicellular organisms is governed to a large part by the interactions cells undergo with neighbouring cells and with their acellular environment. Over the past ten years, it has become clear that many of these interactions are mediated by cell-cell adhesion molecules and extracellular matrix components and their cellular receptors, that is molecules responsible for establishing direct cell-cell or cell-matrix contacts. These molecules are particularly important for determining whether a cell remains where it is or moves elsewhere and, if a cell moves, where it goes and when it stops migrating. These are of course key events during normal development, but they play equally crucial roles in adult physiology and pathology, in the extravasation of white blood cells and lymphocyte traffic, in wound healing and inflammation, in tumour invasiveness and metastasis. It is thus hardly surprising that clinicians are devoting increasing attention to the molecular mechanisms underlying cell adhesion and that cell adhesion molecules are now being considered as suitable targets for drug development. Moreover, recent advances in the field show that at least some of these molecules are also capable of triggering intracellular events, in the same way as diffusible growth or differentiation factors do.

The launching of this new book series has of course to do with the awareness of the importance of cell adhesion for both physiological and pathological processes. It is aimed at graduates starting to do research in this area, at the clinician, who wants to become familiar with a field with many clinical implications, at research workers, both in academics and industry, planning to move into the area and at scientists already working in the field who want a keep abreast with the recent developments outside their own speciality. Hence, each volume should provide a review of the field as well as critical assessment of recent discoveries and future developments. Each volume of the series will have a volume editor who will invite a small number of contributors to cover the different aspects of the topic. By keeping the number of contributors small, we hope to avoid overlaps and redundancies, common pitfalls of multiauthor volumes.

One part of the series will concentrate on the important families of molecules, another one will address the role of cell adhesion in selected physiological and pathological phenomena. The first volume on Cell Adhesion Molecules in Cancer and Inflammation is an excellent example of the second kind; a forthcoming volume will be on Cell-cell Adhesion in Cancer Invasion and Metastasis. Examples of volumes devoted to one class of molecules are the forthcoming volumes on laminins and tenascin. Later volumes will deal with the integrins and the adhesion receptors of the immunoglobulin superfamily and encompass topics such as leucocyte adhesion and the role of cell-cell adhesion in morphogenesis.

I am indebted to the Harwood Academic Publishers staff, who made this enterprise possible, and to the volume editors, each one an authority in his/her field, who agreed to assemble and edit this and the forthcoming volumes.

Christo Goridis (Marseille)

INTRODUCTION

The past decade has been marked by a tremendous advance in the biochemical and functional characterisation of cell surface adhesion molecules mediating cell-matrix and cell-cell interactions. There is no doubt that there has been an exponential increase in interest in this subject since it became obvious that these interactions are key elements in the regulation of morphogenesis, tissue physiology and are disrupted in a variety of disease processes including neoplasia and inflammation. The articles, reviews and progress reports written on adhesion molecules are innumerable. This book, however, gives attention to the basic role of adhesion molecules in human cancer and inflammatory conditions with a view to exploiting them as novel bio-markers in a clinical setting with the longer-term prospect of improved diagnosis and therapy. Thus the ultimate aim of this book is to give an opportunity to the readers to assess whether the increased knowledge in the structural organisation and function of adhesion molecules provided by basic science can now be applied to develop better diagnostic and prognostic markers. Undoubtedly, the understanding we have accumulated about adhesive interactions in normal and transformed cells points the direction for their future incorporation into clinical practice.

M. Pignatelli
A. A. Epenetos

CONTRIBUTORS

A. Albini
Istituto Nazionale per la Ricerca sul Cancro
V. le Benedetto XV 10
16132 Genova
Italy

A. Alexandroff
Department of Surgery
University of Edinburgh
Medical School
Teviot Place
Edinburgh, EH8 9AG
UK

J. Allen
Department of Physiology
University of Cambridge
Downing Street
Cambridge CB2 73G
UK

J. Bartek
Epithelial Cell Biology Laboratory
Imperial Cancer Research Fund
London
UK

P. Brandtzaeg
Laboratory for Immunohistochemistry and Immunopathology
Institute of Pathology
University of Oslo
Oslo
Norway

B. S. Brown
Department of Biochemistry and Molecular biology
University of Manchester
Manchester M13 9PT
UK

M. A. Campbell
National Eye Institute
NIH
Bethesda, MD 20892
USA

G. J. Chader
National Eye Institute
NIH
Bethesda, MD 20892
USA

S. Chang
Royal College of Surgeons
London
UK

D. A. Cheresh
Department of Immunology, IMM 13
The Scripps Research Institute
10666 N. Torrey Pines Road
La Jolla, CA 92037
USA

E. Chignier
INSERM Unité
Institut Pasteur de Lyon
Faculté de médecine Alexis Carrel
Lyon
France

L. Cimino
Laboratorio Oncogenesi Molecolare
Regina Elena Cancer Institute
Via delle Messi D'oro 156
00158 Roma
Italy

M. Coviello
Laboratory of Analysis
Oncology Institute
Bari
Italy

J. Dämmrich
Institute of Pathology
University of Würzburg
Josef Schneiderstrasse 2
8700 Würzburg
Germany

J. Davies
Haemopoiesis Laboratory
Imperial Cancer Research Fund
London
UK

P. Defilippi
Dipartimento di Genetica
Biologia e Chimica Medica
Università di Torino
Via Santena 5bis
10126 Torino
Italy

P. Drillenburg
Department of Pathology
Academic Medical Center
University of Amsterdam
Meibergdreef 9
1105 AZ Amsterdam
The Netherlands

A. J. Edwards
FACS Laboratory
Imperial Cancer Research Fund
London
UK

R. Falcioni
Laboratorio Oncogenesi Molecolare
Regina Elena Cancer Institute
Via delle Messi D'Oro 156
00158 Roma
Italy

J. Fawcett
Cell Adhesion Laboratory
Imperial Cancer Research Fund
Institute of Molecular Medicine
Oxford
UK

B. Felding-Habermann
Department of Molecular and
Experimental Medicine, SBR 8
The Scripps Research Institute
10666 N. Torrey Pines Road
La Jolla, CA 92037
USA

G. Gagliardi
Cell Adhesion Laboratory
Department of Histopathology
Royal Postgraduate Medical School
Du Cane Road
London W12 OHS
UK

S. Gentleman
National Eye Institute
NIH
Bethesda, MD 20892
USA

V. E. Gould
Department of Pathology
Rush Medical College
Chicago, IL
USA

M. M. Griffin
Department of Biochemistry and
Molecular biology
University of Manchester
Manchester M13 9PT
UK

F. Grunert
Institut fur Immunobiologie
Albert-Ludwigs Universitat
Freiburg
Germany

A. M. Hanby
ICRF/RCS Histopathology Unit
35–43 Lincoln's Inn Fields
London WC 2A 3PX
UK

C. Hanski
Universitatsklinikum Stegliz
Freie Universitat Berlin
Hindenburgdamm 30
D-1000 Berlin 45
Germany

J. Heino
Department of Medical
Biochemistry
The MediCity Research Laboratory
University of Turku
Kiinamyllynkatu 10
FIN-20520 Turku
Finland

M. A. Horton
Haemopoiesis Laboratory
Imperial Cancer Research Fund
London
UK

A. M. Jackson
Department of Surgery
University of Edinburgh
Medical School
Teviot Place
Edinburgh, EH8 9AG
UK

K. James
Department of Surgery
University of Edinburgh
Medical School
Teviot Place
Edinburgh, EH8 9AG
UK

J. T. Kemshead
The Imperial Cancer Research
Fund
Paediatric and Neuro-Oncology
Group
Frenchay Hospital
Bristol, BS16 1LE
UK

S. Kern
Institute of Pathology
University of Würzburg
Josef Schneiderstrasse 2
8700 Würzburg
Germany

M. Korhonen
Department of Anatomy
University of Helsinki
P.O. Box 9
University of Helsinki
Finland

D. Kvale
Laboratory for
Immunohistochemistry and
Immunopathology
Institute of Pathology
and Medical Department A
Rikshospitalet
The National Hospital
University of Oslo
Oslo
Norway

E-N. Lalani
Cell Adhesion Laboratory
Department of Histopathology
Royal Postgraduate Medical School
Hammersmith Hospital
London W12 OHS
UK

D. Liu
Cell Adhesion Laboratory
Department of Histopathology
Royal Postgraduate Medical School
Du Cane Road
London W12 ONN
UK

E. Manten-Horst
Department of Pathology
Academic Medical Center
University of Amsterdam
Meibergdreef 9
1105 AZ Amsterdam
The Netherlands

G. Marland
Krebs institute for Biomolecular Research
Department of Molecular Biology and Biochemistry
University of Sheffield
P.O. Box 594
Firth Court
Western Bank
Sheffield S10 2UH
UK

J. L. McGregor
INSERM Unité
Institut Pasteur de Lyon
Faculté de médecine Alexis Carrel
Lyon
France

L. McGregor
INSERM Unité
Institut Pasteur de Lyon
Faculté de médecine Alexis Carrel
Lyon
France

A. Melchiori
Istituto Nazionale per la Ricerca sul Cancro
V. le Benedetto XV 10
16132 Genova
Italy

G. Micelli
Laboratory of Analysis
Oncology Institute
Bari
Italy

B. M. Mueller
Department of Immunology IMM 13
The Scripps Research Institute
10666 N. Torrey Pines Road
La Jolla, CA 92037
USA

G. M. Müller
Department of Dermatology
Medical School
Humboldt University
Berlin
Germany

H. K. Müller-Hermelink
Institute of Pathology
University of Würzburg
Josef Schneiderstrasse 2
8700 Würzburg
Germany

G. Nagel
Institut fur Immunobiologie
Albert-Ludwigs Universitat
Freiburg
Germany

A. K. Nigam
Cell Adhesion Laboratory
Department of Histopathology
Royal Postgraduate Medical School
Du Cane Road
London
UK

A. Nolasco
Surgical Senology Unit
Oncology Institute
Bari
Italy

D. Noonan
Istituto Nazionale per la Ricerca sul Cancro
V. le Benedetto XV 10
16132 Genova
Italy

S. T. Pals
Department of Pathology
Academic Medical Center
University of Amsterdam
Meibergdreef 9
1105 AZ Amsterdam
The Netherlands

T. Papadopoulos
Institute of Pathology
University of Würzburg
Josef Schneiderstrasse 2
8700 Würzburg
Germany

L. Partridge
Krebs Institute for Biomolecular Research
Department of Molecular Biology and Biotechnology
University of Sheffield
P.O. Box 594
Firth Court
Western Bank
Sheffield S10 2UH
UK

K. Patel
The Imperial Cancer Research Fund
Paediatric and Neuro-Oncology Group
Frenchay Hospital
Bristol, BS16 1LE
UK

R. Pellegrini
Istituto Nazionale per lo Studio e la Cura dei Tumori
Via Venezian 1
20133 Milano
Italy

D. Perrotti
Laboratorio Oncogenesi Molecolare
Regina Elena Cancer Institute
Via delle Messi D'Oro 156
00158 Rome
Italy

M. Pignatelli
Cell Adhesion Laboratory
Department of Histopathology
Royal Postgraduate Medical School
Du Cane Road
London W12 0ONN
UK

ICRF/RCS Histopathology Unit
35–43 Lincoln's Inn Fields
London WC 2A 3PX
UK

L. Poggi
Istituto Nazionale per la Ricerca sul Cancro
V. le Benedetto XV 10
16132 Genova
Italy

M. Quaranta
Laboratory of Analysis
Oncology Institute
Bari
Italy

T. Radaszkiewicz
Department of Pathology
University Hospital Vienna
Vienna
Austria

S. Rasbridge
ICRF Clinical Oncology Unit
Guy's Hospital
London SE 2
UK

C. Rella
Laboratory of Analysis
Oncology Institute
Bari
Italy

M. G. Rizzo
Laboratorio Oncogenesi Molecolare
Regina Elena Cancer Institute
Via delle Messi D'Oro 156
00158 Roma
Italy

A. Sacchi
Laboratorio Oncogenesi Molecolare
Regina Elena Cancer Institute
Via delle Messi D'Oro 156
00158 Rome
Italy

G. Sala-Newby
Department of Medical Biochemistry
University of Wales College of Medicine
Cardiff
UK

L. C. Sanders
Department of Immunology, IMM 13
The Scripps Research Institute
10666 N. Torrey Pines Road
La Jolla, CA 92037
USA

F. Schittulli
Surgical Senology Unit
Oncology Institute
Bari
Italy

D. L. Simmons
Cell Adhesion Laboratory
Imperial Cancer Research Fund
Institute of Molecular Medicine
Oxford
UK

K. Sirtl
Institute of Pathology
University of Würzburg
Josef Schneiderstrasse 2
8700 Würzburg
Germany

G. W. H. Stamp
Cell Adhesion Laboratory
Department of Histopathology
Royal Postgraduate Medical School
Du Cane Road
London W12 ONN
UK

ICRF/RCS Histopathology Unit
35–43 Lincoln's Inn Fields
London WC 2A 3PX
UK

F. S. Steven
Department of Biochemistry
and Molecular biology
University of Manchester
Manchester M13 9PT
UK

I. C. Talbot
ICRF Colorectal Unit
Department of Pathology
St. Mark's Hospital
London EC1V 2PS
UK

I. Tausch
Department of Dermatology
Medical School
Humboldt University
Berlin
Germany

E. Tchilian
Tumour Immunology Unit
Imperial Cancer Research Fund
London
UK

A. Teixeira
Medical Oncology Unit
Imperial Cancer Research Fund
London
UK

J. A. Thomas
London School of Hygiene and
Tropical Medicine
London
UK

I. Virtanen
Department of Anatomy
University of Helsinki
P.O. Box 9
University of Helsinki
Finland

H. Waldmann
Department of Pathology
University of Cambridge
Cambridge
UK

S. M. Watt
MRC Molecular Haematology Unit
Institute of Molecular Medicine
The John Radcliffe Hospital
Oxford
UK

1. ADHESION MOLECULES IN NEOPLASIA: AN OVERVIEW

M. PIGNATELLI[1], D. LIU, A. K. NIGAM, G. GAGLIARDI, E-N. LALANI and G. W. H. STAMP

Cell Adhesion Laboratory, Department of Histopathology Royal Postgraduate Medical School, Du Cane Road, London W12 OHS, UK

INTRODUCTION

Cell adhesion is a fundamental process which regulates organogenesis and differentiation of prokaryotic and eukaryotic organisms. Over the last decade a major effort has been made to identify a multitude of transmembrane glycoproteins receptors named Cell Adhesion Molecules (CAMs) and Substratum Adhesion Molecules (SAMs) which mediate adhesion between cells and/or between cells and extracellular matrix proteins, respectively (Edelman and Crossin, 1991). Their function and expression have been elucidated using monoclonal antibodies, cell-cell and cell-substratum adhesion assays and affinity chromatography techniques. It has become apparent that these molecules are expressed early during development on most cells and are subsequently maintained in a number of tissues derived from the three primary germ layers. Different model systems have demonstrated that a limited repertoire of CAMs and SAMs, expressed in a well defined spatiotemporal pattern, is required in tissue morphogenesis and its disruption may, at least in part, explain the loss of tissue architecture and uncontrolled proliferation seen in malignant tumours. Four main families of adhesion molecules (integrins, cadherins, immunoglobulin superfamily and selectins) have been identified (Hynes and Lander, 1992). These receptors are genetically and biochemically distinct although, in some cases, are functionally related (Pignatelli, 1993b). A detailed description of the biochemical and genetic structure of adhesion receptors is beyond the scope of this review. Here, we describe recent data on the role of selected CAMs and SAMs in the control of differentiation, invasion and metastasis in human cancer.

INTEGRINS

Integrins are $\alpha\beta$ heterodimeric transmembrane proteins comprising of at least 14 alpha chains and 8 beta chains which are expressed by epithelial cells

[1] Correspondence. Dr Massimo Pignatelli, Department of Histopathology, Royal Postgraduate Medical School, Du Cane Road, London W12 OHS, UK
Tel: 44 81 743 2030 (ext 4834) Fax: 44 81 740 7417

as well as other cell types (Hynes, 1992). Although the integrins were initially divided into three main subfamilies ($\beta 1$, $\beta 2$, $\beta 3$) based on the sharing of a common β chain, it has become clear that some α chains can associate with more than one β chain, giving further complexity to this classification. The majority of known integrins function as receptors for extracellular matrix proteins (ECM) including collagens, fibronectin and laminin (Hynes, 1992). However some leucocyte integrins ($\beta 2$ subfamily) and at least one member of the $\beta 1$ integrin subfamily ($\alpha 4\beta 1$) mediate heterotypic cell-cell adhesion by binding to adhesion molecule members of the immunoglobulin gene family (ICAM-1, ICAM-2 and V-CAM). Integrins differ from receptors for hormones and soluble factors in that they are expressed at very high density on the cell surface and some of them can bind to the same ligand or may show alternative specificity when expressed on different cell types. (Elices and Hemler, 1990).

Transformed cell lines as well as tumour tissues have been studied. Differences in the integrin repertoire are also evident when transformed cells in culture are compared to human tumours at different stages of progression. Different cell types show different specificities for integrin expression. Fibroblast cell lines show a decreased expression of fibronectin receptors and alterations in their distribution (Chakrabarty *et al.* 1989; Plantefaber and Hynes, 1989; Akiyama *et al.* 1990) when compared to normal fibroblast lines. In addition, tyrosine phosphorylation of some integrins has been described in chicken fibroblasts transformed by the Rous sarcoma virus (Horvath *et al.* 1990). Cells derived from aggressive melanomas tend to have an increased expression of $\alpha 2$ and $\alpha 4$ integrin subunits. Overexpression of the $\alpha 4\beta 1$ receptor may be particularly important as this has been shown to be the counter-receptor for the endothelial cell adhesion molecule, VCAM-1, which mediates melanoma-endothelial cell interactions and thus may play an important role in melanoma metastasis (Rice and Bevilacqua, 1989). Studies on malignant melanoma tissue have also shown an increase in expression of $\alpha 4\beta 1$ in 40% of invasive and metastatic melanomas (Albelda *et al.* 1990) whereas it is not expressed on normal melanocytes. Upregulation of the $\beta 3$ subunit has also been observed and correlated with melanoma progression (Albelda *et al.* 199). This subunit is associated with αv to form the vitronectin receptor ($\alpha v\beta 3$) and several groups have now confirmed its role in the invasive and metastatic behaviour of malignant melanoma (Felding-Habermann *et al.* 1992; Seftor *et al.* 1992). Increased levels of both the secreted and the mRNA for the proteolytic enzyme, 72 kDa gelatinase, have also been observed concomitant with the rise in $\alpha v\beta 3$ integrin expression (Seftor *et al.* 1992; Gehlsen *et al.* 1992). Unlike the consistent upregulation of the vitronectin receptor, Natali and colleagues (1991) have reported a down-regulation of the $\alpha 6$ subunit during melanoma progression.

The integrin expression on normal epithelium and epithelial malignancies has been determined on carcinomas of the lung (Damjanovich *et al.* 1992), breast (Zutter *et al.* 1993; Koukoulis *et al.* 1991; Pignatelli *et al.* 1991,

1992a), pancreas (Hall *et al.* 1991; Weinel *et al.* 1992), prostate (Bonkhoff *et al.* 1993), stomach, kidney (Korhonen *et al.* 1992), skin (Peltonen *et al.* 1989; Stamp and Pignatelli, 1991) and colon (Pignatelli *et al.* 1990b; Koretz *et al.* 1991; Stallmach *et al.* 1992; Nigam *et al.* 1993; Koukoulis *et al.* 1993). The general finding from most of these studies is that each tissue and its cancer appears to have a specific integrin repertoire which to a greater or lesser extent is reduced in malignant tissue. Expression is heterogeneous but down-regulation appears to be consistently related to a decreasing level of differentiation. A relationship between the integrity of the basement membrane and integrin expression has been observed in some studies. Those tumours that were more invasive and lacked intact basement membranes tended not to express basement membrane protein binding integrins, such as $\alpha6$ (D'Ardenne *et al.* 1991). Whether this alteration in expression is a cause or an effect of the changes in the interaction of tumour cells with their surrounding extracellular matrix remains to be determined.

A number of studies have shown that alterations of integrin function at the time of intravenous tumour injection into mice can influence the number of metastases formed. One such study employed a synthetic peptide containing the RGD sequence recognised by many integrins in their ligands and coinjected it with a melanoma tumour cell load into mice. A marked reduction in the number of lung metastases formed was noted at two weeks in the group that had the peptide injection when compared to those that had just the tumour cell load (Humphries *et al.* 1986). It was postulated that the peptide had interfered with the initial binding of the tumour cells during the attachment phase or interfered with cell migration during invasion. This finding has been confirmed *in vitro* using an invasion assay in which RGD containing peptides blocked tumour cell invasion through artificial basement membranes (Gehlsen *et al.* 1988). It has also been reported that cyclic peptides containing the sequence are even more effective (Kumagai *et al.* 1991). Albolabrin, a protein isolated from the venom of the snake *Trimeresurus albolabris*, also contains the RGD sequence and is now part of a family of such peptides known as the disintegrins. These include echistatin, flavoridin and bitistatin and are all obtained from the venom of vipers and have been shown to be potent inhibitors of platelet aggregation and cell-substratum adhesion. Albolabrin has been shown to inhibit murine melanoma cell adhesion and leads to a reduction in experimental metastasis (Soszka *et al.* 1991).

Overexpression of the $\alpha2$ integrin subunit in human rhabdomyosarcoma cells has been shown to increase metastasis *in vivo*. This study employed genetic transfection techniques to overexpress the $\alpha2\beta1$ receptor and then measured its function *in vitro* and *in vivo*. Not only was adhesion increased but experimental metastasis was also increased when compared to non-transfected cells. However, no differences in growth rate or tumorigenicity were noted (Chan *et al.* 1991). An inverse correlation has been noted with overexpression of the $\alpha5\beta1$ receptor and the transformed phenotype.

Chinese hamster ovary cells transfected with the $\alpha 5$ and $\beta 1$ cDNAs have been selected for overexpression of the fibronectin receptor. It was found that the overexpression laid down more fibronectin in their extracellular matrix and migrated less than controls. They also demonstrated reduced tumourigenicity in soft agar and *in vivo* when injected into nude mice (Giancotti and Ruoslahti, 1990). Similarly, selection of Chinese hamster ovary cells that naturally express low levels of this receptor were selected and compared with those that express increased amounts. Tumours produced by subcutaneous injection of the underexpressing clones were larger than those produced by injection of wild-type cells. However, the clones that were overexpressing grew more slowly suggesting an inverse correlation between level of expression of the fibronectin receptor and tumour growth (Schreiner *et al.* 1991). Both of these studies indicated that this integrin played a role in controlling cell proliferation and also suggested that loss of this receptor may be responsible for acquisition of the malignant phenotype *in vivo* and of anchorage independence of transformed cells *in vitro*.

CADHERINS

The cadherins (cad) are Ca^{2+}-dependent cell-cell adhesion molecules that connect cells via homotypic interactions (Takeichi, 1991). They are divided into three main subclasses, E-cad (epithelial cadherin or uvomorulin), P-cad (placental cadherin), N-cad (neural cadherin) which share a basic structure and show selective tissue distribution. Cadherins have three or possibly four regions of internal homology of about 110 aminoacids each (Gallin *et al.* 1987). cDNA analysis of cadherins has shown that these proteins are encoded for by different genes, having close structural relationship and homology. For example, N-cadherin gene has been mapped to chromosome 18q (Walsh *et al.* 1990) whereas E-cadherin gene is on chromosome 16q22.1 (Natt *et al.* 1989). Expression of cadherins is developmentally regulated and the switching on and off of cadherin expression correlates with a variety of morphogenetic events that involve cell aggregation or disaggregation. When cadherins are functionally expressed, the inactivation of other cell-cell adhesion molecules has little effect, indicating that cadherins are the cell-cell adhesion receptors that are most important for the formation of physical cell associations (Duband *et al.* 1987).

E-cad (as well as N-cad) function requires the association with cytoskeletal proteins, named $\alpha-$, $\beta-$ and $\gamma-$catenin, which anchor the E-cad cytoplasmic domain to the actin filaments (Ozawa *et al.* 1989; Ozawa and Kemler, 1992). Loss of heterozygosity on the E-cadherin locus is found frequently in breast (Sato *et al.* 1991), prostate (Carter *et al.* 1990) and liver cancers (Tsuda *et al.* 1990) strongly indicating that E-cadherin may function as a tumour suppressor gene. Indeed, the *fat* tumour suppressor gene in Drosophila has been identified as a novel member of the cadherin gene superfamily

(Mahoney *et al.* 1991). Although malignant transformation does not necessarily affect cadherin activity, it seems that human carcinoma cell lines with a dedifferentiated "fibroblast-like" phenotype are E-cadherin negative and acquire an invasive phenotype in an *in vitro* assay consistent with the differentiating and suppressive invasive role of E-cadherin (Frixen *et al.* 1991; Vleminckx *et al.* 1991; Pignatelli *et al.* 1992; Liu *et al.* 1993). In addition, recent studies have shown that in head and neck (Schipper *et al.* 1991), prostate (Umbas *et al.* 1992), breast (Oka *et al.* 1993), bladder (Bringuier *et al.* 1993), pancreatic (Ansari *et al.* 1994) and colorectal tumours (Pignatelli, 1993a) aberrant E-cadherin expression correlates with high grade and advanced stage of the disease. Large scale clinical studies are in progress to evaluate the use of E-cadherin as a novel prognostic marker in human cancer (M. Pignatelli, personal communication).

IMMUNOGLOBULIN (IG) SUPERFAMILY

The first adhesion molecule of the Ig superfamily to be described was N-CAM (neural cell adhesion molecule) which is expressed by a wide variety of different cell types, predominantly of neural and mesenchymal origin (Chuong and Edelman, 1984). (Edelman and Crossin, 1991). CAMs and SAMs belonging to the Ig family are characterised by a number of Ig-related domains and, in some cases, also contain membrane proximal fibronectin type III repeats (Williams and Barclay, 1988). They mediate either homotypic adhesion between like cells (N-CAM binds to N-CAM on opposite cell) or heterotypic adhesion between two different cell types (V-CAM on endothelial cells binds to $\alpha4\beta1$ on lymphoid cells). Carcinoembryonic antigen (CEA), an oncofoetal protein which is also a member of the Ig family (Williams and Barclay, 1988), has been shown to mediate the Ca^{2+}-independent, homotypic aggregation of colon carcinoma cell lines (Benchimol *et al.* 1989), and acting as an accessory adhesion molecule mediating cell-matrix binding (Pignatelli *et al.* 1990a).

A role in tumour invasion and metastasis has been postulated for this group of adhesion molecules because of their cell-cell and cell-matrix binding specificities. However, because of the divergent nature of the individual members, no clear patterns emerge. N-CAM has been shown to have homology with the gene product encoded by the DCC gene on chromosome 18q (Fearon *et al.* 1990) and it is proposed that loss of this gene, which is a frequent finding in colorectal tumours, could lead to an abnormality of cell-cell adhesion predisposing to the malignant phenotype. Evidence for a possible mechanism of action of the N-CAM molecule was suggested by a report which showed that N-CAM was involved in the contact-dependent inhibition of growth of a mouse fibroblast cell line (Aoki *et al.* 1991). Viral transformation of this cell line resulted in a complete loss of N-CAM from the cell surface and a loss of contact inhibition. It was proposed that cross-linking of N-CAM molecule to N-CAM molecule resulted in intracellular signals that modulated cell growth. Loss of these signals could conceivably contribute to unchecked neoplastic growth. N-CAM also shows homology with MUC-18, a melanoma associated antigen whose expression correlates with tumour progression and metastasis (Lehmann *et al.* 1989). However, the exact role of this molecule is unknown and its possible function in adhesion and tumour progression are, at present, speculative.

N-CAM expression has been evaluated immunohistochemically on many neural and neuroendocrine tumours including Wilm's tumour, pituitary adenomas, phaechromocytomas and small cell carcinomas of the lung (Edelman and Crossin, 1991; Carbone *et al.* 1991; Jin *et al.* 1991). In a large study of 450 human tumours of varying origins, N-CAM immunoreactivity was found to be negligible in tumours of epithelial origin (carcinomas) but expressed to varying degress in tumours of neural and mesenchymal origin (Garin-Chesa *et al.* 1991).

VCAM-1 (vascular cell adhesion molecule) may also play a part in tumour cell interactions. Certain tumour lines show an increased rate of adherence to cytokine stimulated endothelial cells than to resting cells (Rice *et al.* 1988). VCAM-1 is a cytokine-inducible cell adhesion molecule found primarily on vascular endothelial cells and serves as a counter-receptor for the $\alpha 4\beta 1$ integrin receptor. The latter is found largely on leucocytes, but has been shown to be expressed on malignant melanomas, both *in vitro* and in tumour tissue (Albelda *et al.* 1990). It is conceivable, therefore, that VCAM-1 serves as a tumour adhesion receptor for those melanomas that express $\alpha 4\beta 1$ and malignancies of the myeloid line that also express this receptor.

Increased expression of intercellular adhesion molecule-1 (ICAM-1) has also been shown to be correlated with melanoma progression and an increased propensity to metastasis (Johnson *et al.* 1989). Elevated circulating levels of this molecule in the serum also appear to correlate with shortened survival in melanoma (Harning *et al.* 1991). Other malignancies have shown a differential expression of ICAM-1 between primary tumours and metastases (Natali *et al.* 1990) and also differences in serum evaluations for this molecule (Tsujisaki *et al.* 1991).

Interestingly, the candidate tumour suppressor gene, DCC (Deleted in Colorectal Cancer) (Fearon *et al.* 1990) which is located on human chromosome 18q, encodes for a putative Ig-like CAM whose expression and function is lost in almost 50% of large (>1cm) colorectal adenomas and in more than 70% of invasive colorectal carcinomas (Vogelstein *et al.* 1988).

SELECTINS

Selectins also known as LECAMs are transmembrane proteins which mediate heterotypic interactions between blood cells and endothelial cells during lymphocyte homing and leucocyte adhesion (Hynes and Lander, 1992). They contain a single Ca^{2+}-dependent C-type lectin domain which binds to specific carbohydrate groups (such as the sialylated $Lewis_x$ antigen) that are fucosylated and sialylated, a single EGF-like repeat and a number of repeats related to complement-binding proteins. At least two of the selectins, P-selectin (also known as GMP140 or PADGEM) and E-selectin (also known as ELAM-1), are present on activated endothelial cells and, therefore, may provide a binding site for cells expressing the appropriate carbohydrate ligand. Colorectal carcinoma cell lines which express the sialylated $Lewis_x$ antigen (Matsushita *et al.* 1990) are able to adhere to cytokine-activated endothelial cells, and the binding is inhibited by E-selectin antibodies (Lauri *et al.* 1991). E-selectin-mediated adhesion of colorectal carcinoma cell lines to endothelial cells may, therefore, be an important mechanism mediating the vascular adhesion of malignant cells in the metastatic pathway.

CD44

CD44 is a cell surface glycoprotein that participates in lymphocyte homing, T-cell activation and adhesion to hyaluronate and extracellular matrix proteins (Aruffo *et al.* 1990). Up to nine isoforms of the CD44 protein are known to exist. This is due to different messenger RNA (mRNA) transcripts being created through alternative splicing. These then translate into functionally diverse but structurally similar proteins (Matsumura and Tarin, 1992). The 85kDa isoform, CD44H, is broadly distributed in haematopoietic cells, fibroblasts, and numerous tumours of both mesenchymal and neuro-ectodermal origin (Stamenkovic *et al.* 1989). The expression of the 150kDa isoform, CD44E, appears restricted to subsets of epithelial cells (Stamenkovic *et al.* 1991).

However, the interest in this molecule has arisen because of the expression of the gene products in carcinomas. Breast and colonic carcinomas and their metastases showed amplified expression of at least nine alternatively spliced products of the CD44 gene in contrast to non-neoplastic tissues (Matsumura and Tarin, 1992). The ratio of the epithelial to the haemopoietic variant has also been studied in colorectal cancers, and although both are overexpressed, greater amplification is seen in the epithelial variant and may therefore facilitate metastasis (Tanabe *et al.* 1993). This is following an *in vitro* study in which fibrosarcomas transfected to overexpress the CD44 epithelial variant spontaneously metastasised more frequently than did those transfected with the haemopoietic variant (Dougherty *et al.* 1992). The role of CD44 in metastasis has been further elaborated by the isolation of cell lines that overexpress CD44. Melanoma cells which were high expressors of the product were found to give rise to more lung metastases in nude mice than low expressors (Birch *et al.* 1991). Isolation of an antibody to a spliced variant of CD44 from a metastatic rat pancreatic carcinoma cell line has been shown to confer metastatic potential into a non-metastasising cell line after transfection with the cDNA product (Gunthert *et al.* 1991). A second variant, smaller in size, has also recently been identified, this also conferring metastatic behaviour on previously sessile cells (Rudy *et al.* 1993). A human homologue of this metastasis-associated variant has been found to be expressed on all invasive colorectal carcinomas and their metastases. However, focal expression was also observed in adenomatous polyps, this being related to areas of dysplasia. This may represent a selective metastatic advantage to those cells in the polyps expressing the CD44 variant (Heider *et al.* 1993). A link between *ras* oncogene and the CD44 protein has recently been demonstrated. Activation of the oncogene induces surface expression of CD44 both by increasing promoter activity and by affecting splice control mechanisms. Thus, metastatic variants are also overexpressed in cells transfected with *ras* constructs (Hofmann *et al.* 1993).

The exact mechanism by which certain CD44 variants appear to promote metastasis is subject to speculation. It has been proposed that the expression of CD44 products on tumour cells enables them to enter lymph node endothelial venules and adhere more readily than they might otherwise. This is because they adopt a 'molecular disguise' by looking and acting as lymphocytes within the lymph node and thus the normal 'homing' function of CD44 is abused.

CONCLUSION

Expression and/or function of SAMs and CAMs seem to be disturbed in transformed cells *in vitro* as well as *in vivo* and these changes may explain at least in part the phenotype and behaviour of malignant tumours. Further investigation of the molecular control of adhesion molecule function, localisation and level of cell surface expression, will lead to a better understanding of their clinical relevance in the histological assessment of diagnosis, prognosis and likely response to treatment of malignant tumours.

Acknowledgements

The authors are grateful to Ms Jane Campbell for typing the manuscript.

REFERENCES

1. Akiyama, S. K., Larjava, H. and Yamada, K. M. (1990) Differences in the biosynthesis and localisation of the fibronectin receptor in normal and transformed cultured human cells. *Cancer Res.*, **50**, 1601–1607.
2. Albelda, S. M., Mette, S. A., Elder, D. E., Stewart, R., Damjanovich, L., Herlyn, M. and Buck, C. A. (1990) Integrin distribution in malignant melanoma: association of the $\beta 3$ subunit with tumor progression. *Cancer Res.*, **50**, 6757–6764.

3. Ansari, T., Lemoine, N., Gunter, P., Liu, D., Hirano, S., Takeichi, M., Kloppel, G. and Pignatelli, M. (1994) Decreased E-Cadherin expression correlates with lymph node metastasis, high grade and advanced stage in pancreatic cancer. *J. Pathol.*, **172**, A12.
4. Aoki, J., Umeda, M. and Taiko, K. (1991) Neural cell adhesion molecule mediated contact-dependent inhibition of growth of near-diploid mouse fibroblast cell line m5s/1M. *J. Cell Biol.*, **115**, 1751–1761.
5. Aruffo, A., Stamenkovic, I., Melnick, M., Underhill, C. B. and Seed, B. (1990) CD44 is the principal cell surface receptor for hyaluronate. *Cell*, **61**, 1303–1313.
6. Benchimol, S., Fuks, A., Jothy, S., Beauchemin, N., Shirota, K. and Stanners, C.P. (1989) Carcinoembryonic antigen, a human tumor marker, functions as an intercellular adhesion molecule. *Cell*, **57**, 327–334.
7. Birch, M., Mitchell, S. and Hart, I. R. (1991) Isolation and characterisation of human melanoma cell variants expressing high and low levels of CD44. *Cancer Res.*, **51**, 6660–6667.
8. Bonkhoff, H., Stein, U. and Remberger, K, (1993) Differential expression of $\alpha 6$ and $\alpha 2$ very late antigen integrins in the normal, hyperplastic and neoplastic prostate. *Hum. Pathol.*, **24**, 243–248.
9. Bringuier, P. P., Umbas, R., Schaafsma, H. W., Karthaus, H. F. M., Debruyne, F. M. J. and Schalken, J. A. (1993) Decreased E-cadherin immunoreactivity correlates with poor survival in patients with bladder tumours. *Cancer Res.*, **53**, 3241–3245.
10. Carbone D. P., Koros, A. M. C., Linnoila, I., Jewett, P. and Gazdar, A. F. (1991) Neural cell adhesion molecule expression and messenger RNA splicing patterns in lung cancer cell lines are correlated with neuroendocrine phenotype and growth morphology. *Cancer Res.*, **51**, 6142–6149.
11. Carter B. S., Ewing, C. W., Ward, W. S., Treiger, B. F., Aalders, T. W., Schalken, J. A., Epstein, J. I. and Isaacs, W. B. (1990) Allelic loss of chromosome 16q and 10q in human prostate cancer. *Proc. Natl. Acad. Sci. USA*, **87**, 8751–8755.
12. Chakrabarty, S., Jan, Y., Levine, A., McClenic, B. and Varani, J. (1989) Fibronectin/laminin and their receptors in aberrant growth control in FR3T3 cells transformed by Ha-*ras* oncogene and epidermal growth factor gene. *Int. J. Cancer*, **44**, 325–331.
13. Chan, B. M. C., Matsuura, N., Takada, Y., Zetter, B. R. and Hemler, M. E. (1991) *In vitro* and *in vivo* consequences of VLA-2 expression on rhabdomyosarcoma cells. *Science*, **251**, 1600–1602.
14. Chuong, C. M. and Edelman, G. M. (1984) Alterations in neural cell adhesion molecules during development of different regions of the nervous system. *J. Neurosci.*, **4**, 2354–2368.
15. Damjanovich, L., Albelda, S. M., Mette, S. A. and Buck, C. A. (1992) The distribution of cell ahesion receptors in normal and malignant lung tissue. *Am. J. Resp. Cell Molec. Biol.*, **6**, 197–206.
16. D'Ardenne, A. J., Richman, P. I., Horton, M. A., McAulay, A. E. and Jordan, S.(1991) *J. Pathol.*, **165**, 213–220.
17. Dougherty, G. J., Dougherty, S. T., Eaves, C. J. and McBride, W. H. (1992) Expression of human CD44R1 enhances the metastatic capacity of murine fibrosarcoma cells. *Proc. Am. Assoc. Cancer Res.*, **33**, 34.

18. Duband, J. L., Dufour, S., Hatta, K., Takeichi, M., Edelman, G. M. and Theiry, J. P. (1987) Adhesion molecules during somatogenesis in the avian embryo. *J. Cell Biol.*, **104**, 1361–1374.
19. Edelman, G. M. and Crossin, K. L. (1991) Cell adhesion molecules: implications for a molecular histology. *Annu. Rev. Biochem.*, **60**, 155–190.
20. Elices, M. J. and Hemler, M. E. (1990) The human integrin VLA-2 is a collagen receptor on some cells and a collagen/laminin receptor on others. *Proc. Natl. Acad. Sci. USA*, **86**, 9906–9910.
21. Fearon, E. R., Cho, K. R., Nigro, J. M., Kern, S. E., Simmons, J. W., Ruppert, J. M., Hamilton, S. R., Preisinger, A. C., Thomas, G., Kinzler, K. W. and Vogelstein, B. (1990) Identification of a chromosome 18q gene that is altered in colorectal cancers. *Science*, **247**, 49–56.
22. Felding-Habermann, B., Mueller, B. M., Romerdahl, C. A. and Cheresh, D. A. (1992) Involvement of αv gene expression in human melanoma tumorigenicity *J. Clin. Invest.*, **89**, 2018–2022.
23. Frixen, U. H., Behrens, J., Sachs, M., Eberle, G., Voss, B., Warda, A., Lochner, D. and Birchmeier, W. (1991) E-cadherin-mediated cell-cell adhesion prevents invasiveness of human carcinoma cells. *J. Cell Biol.*, **113**, 173–185.
24. Gallin, W. J., Sorkin, B. C1., Edelman, G. M. and Cunningham, B. A. (1987) Sequence analysis of a cDNA clone encoding the liver cell adhesion molecule, L-CAM. *Proc. Natl. Acad. Sci. USA*, **84**, 2808–2812.
25. Garin-Chesa, P., Fellinger, E. J., Huvos, A. G., Beresford, H. R., Melamed, M. R., Triche, T. J. and Rettig, W. J. (1991) Immunohistochemical analysis of neural cell adhesion molecules. *Am. J. Pathol.*, **139**, 275–286.
26. Gehlsen, K. R., Argraves, W. S., Pierschbacher, M. D. and Ruoslahti, E. (1988). Inhibition of *in vitro* tumour cell invasion by Arg-Gly-Asp containing synthetic peptides. *J. Cell Biol.*, **106**, 925–930.
27. Gehlsen, K. R., Davis, G. E. and Sriramarao, P. (1992) Integrin expression in human melanoma cells with differing invasive and metastatic properties. *Clin. Exp. Metastasis*, **10**, 111–120.
28. Giancotti, F. G. and Ruoslahti, E. (1990) Elevated levels of the $\alpha 5\beta 1$ fibronectin receptor suppress the transformed phenotype of Chinese hamster ovary cells. *Cell*, **60**, 849–859.
29. Gunthert, U., Hofmann, M., Rudy, W., Reber, S., Zoller, M., Haubmann, I., Martzku, S., Wenzel, A., Ponta, H. and Herrlich, P. (1991) A new variant of glycoprotein CD44 confers metastatic potential to rat carcinoma cells. *Cell*, **65**, 13–24.
30. Hall, P. A., Coates, P., Lemoine, N. R. and Horton, M. A. (1991) Characterisation of integrin chains in normal and neoplastic human pancreas. *J. Pathol.*, **165**, 33–41.
31. Harning, R., Mainlofi, E., Bystryn, J. C., Henn, M., Merluzzi, V. J. and Rothlein, R. (1991) Serum levels of circulating intercellular adhesion molecule 1 in human malignant melanoma. *Cancer Res.*, **51**, 5003–5005.
32. Heider, K. H., Hofmann, M., Hors, E., Van der Berg, F., Ponta, H., Herrlich, P. and Pals, S. T. (1993) A human homologue of the rat metastasis-associated variant of CD44 is expressed in colorectal carcinomas and adenomatous polyps. *J. Cell Biol.*, **120**, 227–233.
33. Hofmann, M., Rudy, W., Gunthert, U., Zimmer. S. G., Zawadzki, V., Zoller, M., Lichtner, R. B., Herrlich, P. and Ponta, H. (1993) A link between *ras* and

metastatic behaviour of tumour cells: *ras* induces CD44 promoter activity and leads to low-level expression of metastasis-specific variants of CD44 in CREF cells. *Cancer Res.*, **53**, 1516–1521.
34. Horvath, A. R., Elmore, M. A. and Kellie, S. (1990) Differential tyrosine-specific phosphorylation of integrin in Rous sarcoma virus transformed cells with differing transformed phenotypes. *Oncogene*, **5**, 1349–1357.
35. Humphries, M. J., Olden, K. and Yamada, K. M. (1986) A synthetic peptide from fibronectin inhibits experimental metastasis of murine melanoma cells. *Science*, **233**, 467–469.
36. Hynes, R. O. (1992) Integrins: versatility, modulation and signaling in cell adhesion. *Cell*, **69**, 11–25.
37. Hynes, R. O. and Lander, A. D. (1992) Contact and adhesive specificities in the associations, migrations, and targeting of cells and axons. *Cell*, **68**, 303–322.
38. Jin, L., Hemperly, J. J. and Lloyd, R. V. (1991) Expression of neural cell adhesion molecule in normal and neoplastic human neuroendocrine tissues. *Am. J. Pathol.*, **138**, 961–969.
39. Johnson, J. P., Stade, B. G., Holzmann, B., Schwable, W. and Riethmuller, G. (1989) *De novo* expression of intercellular-adhesion molecule in melanoma correlates with increased risk of metastasis. *Proc. Natl. Acad. Sci. USA*, **86**, 641–644.
40. Koretz, K., Schlag, P., Boumsell, L. and Moller, P. (1991) Expression of VLA-α2, VLA-α6 and VLA-β1 chains in normal mucosa and adenomas of the colon and in colon carcinomas and their liver metastases. *Am. J. Pathol.*, **138**, 741–750.
41. Korhomen, M., Laitinen, L., Ylanne, J., Koukoulis, G. K., Quaranta, V., Juusela, H., Gould, V. E. and Virtanen, I. (1992) Integrin distributions in renal cell carcinomas of various grades of malignancy. *Am. J. Pathol.*, **141**, 1161–1171.
42. Koukoulis, G. K., Virtanen, I., Korhonen, M., Laitanen, L., Quaranta, V. and Gould, V. E. (1991) Immunohistochemical localisation of integrins in the normal, hyperplastic and neoplastic breast. *Am. J. Pathol.*, **139**, 787–799.
43. Koukoulis, G. K., Virtanen, I., Moll, R., Quaranta, V. and Gould, V. E. (1993) Immunolocalisation of integrins in the normal and neoplastic colonic epithelium. *Virchows Archiv. B Cell Pathol.*, **63**, 373–383.
44. Kumagai, H., Tajima, M., Ueno, Y., Giga-Hama, Y. and Ohba, M. (1991) Effect of cyclic RGD peptide on cell adhesion and tumour metastasis. *Biochem. Biophys. Res. Commun.*, **177**, 74–82.
45. Lauri, D., Needham, L., Martin-Padura, I. and Dejana, E. (1991) Tumour cell adhesion to endothelial cells: Endothelial leukocyte adhesion molecule-1 as an inducible adhesive receptor specific for colon carcinoma cells. *J. Natl. Cancer Inst.*, **83**, 1321–1324.
46. Lehmann, J. M., Riethmuller, G. and Johnson, J. P. (1989) MUC18 a marker of tumour progression in human melanoma shows sequence similarity to the neural cell adhesiom molecules of the immunoglobulin superfamily. *Proc. Natl. Acad. Sci. USA*, **86**, 9891–9895.
47. Liu, D., Nigam, A. K., Lalani, E. N., Stamp, G. W. H. and Pignatelli, M. (1993) Transfection of E-cadherin into a human colon carcinoma cell line induces differentiation and inhibits growth *in vitro*. *Gut*, **34**, 627.
48. Mahoney, P. A., Weber, U., Onofrechuk, P., Biessmann, H., Bryant, P. J. and Goodman, C. S. (1991) The *fat* tumour suppressor gene in Drosophila encodes a novel member of the cadherin gene superfamily. *Cell*, **67**, 853–868.

49. Matsumura, Y. and Tarin, D. (1992) Significance of CD44 gene products for cancer diagnosis and disease evaluation. *Lancet*, **340**, 1053–1058.
50. Matsushita, Y., Cleary, K.R., Ota, D. M., Hoff, S. D. and Irimura, T. (1990) Sialyldomeric lewis-X antigen expressed on mucin-like glycoproteins in colorectal cancer metastases. *Lab. Invest.*, **63**, 780–791.
51. Natali, P. G., Nicotra, M. R., Cavaliere, R., Giannarelli, D. and Bigotti, A. (1991) Tumour progression in human malignant melanoma is associated with changes in $\alpha 6/\beta 1$ laminin receptor. *Int. J. Cancer*, **49**, 168–172.
52. Natt, E., Magents, R. E., Zimmer, J., Mansouri, A. and Scherer, G. (1989) Regional assignment of the human loci for uvomorulin (UVO) and chymotrypsinogen B (CTRB) with the help of two overlapping deletions on the long arm of chromosome 16. *Cytogenet. Cell Genet.*, **50**, 145–148.
53. Nigam, A. K., Savage, F. J., Boulos, P. B., Stamp, G. W. H., Liu, D. and Pignatelli, M. (1993) Loss of cell-cell and cell-matrix adhesion molecules in colorectal cancer. *Br. J. Cancer*, **68**, 507–514.
54. Oka, H., Shiozaki, H., Kobayashi, K., Inoue, M., Tahara, H., Kobayashi, T., Takatsuka, Y., Matsuyoshi, N., Hirano, S., Takeichi, M. and Mori, T. (1993) Expression of E-cadherin cell adhesion molecules in human breast cancer tissues and its relationship to metastasis. *Cancer Res.*, **53**, 1696–1701.
55. Ozawa, M., Banbault, H. and Kemler, R. (1989) The cytoplasmic domain of the cell adhesion molecule uvomorulin associates with three independent proteins structurally related in different species. *EMBO J.*, **8**, 1711–1717.
56. Ozawa, M. and Kemler, R. (1992) Molecular organisation of the Uvomorulin — Cadherin complex. *J. Cell. Biol.*, **116**, 989–996.
57. Peltonen, J., Larjave, H. and Jaakkola, D. (1989) Localisation of integrin receptors for fibronectin, collagen and laminin in human skin. *J. Clin. Invest.*, **84**, 1916–1923.
58. Picker, L. J., Nakache, M. and Butcher, E. C. (1989) Monoclonal antibodies to human lymphocyte homing receptors define a novel class of adhesion molecules in diverse cell types. *J. Cell Biol.*, **109**, 927–937.
59. Pignatelli, M. (1993a) E-cadherin: a biological marker of tumour differentiation. *J. Pathol.*, **171**, 81–82.
60. Pignatelli, M. (1993b) Models of colorectal tumour differentiation. In The molecular pathology of cancer edited by Lemoine, N. R. and Wright, N. A. *Cancer Surveys*, **16**, 3–13.
61. Pignatelli, M., Durbin, H. and Bodmer, W. F. (1990a) Carcinoembryonic antigen functions as an accessory adhesion molecule mediating colon epithelial cell-collagen interactions. *Proc. Natl. Acad. Sci. USA*, **87**, 1541–1545.
62. Pignatelli, M., Hanby, A. M. and Stamp, G. W. H. (1991) Low expression of $\beta 1$, $\alpha 2$ and $\alpha 3$ subunits of VLA integrins in malignant mammary tumours *J. Pathol.*, **165**, 25–32.
63. Pignatelli, M., Smith, M. E. F. and Bodmer, W. F. (1990b) Low expression of collagen receptors in moderate and poorly differentiated colorectal adenocarcinomas. *Br. J. Cancer*, **61**, 636–638.
64. Pignatelli, M., Cardillo, M. R., Hanby, A. N. and Stamp, G. W. H. (1992a) Integrins and their accessory adhesion molecules in mammary carcinoma — Loss of polarisation in poorly differentiated tumours. *Human Pathol.*, **23**, 1159–1166.
65. Pignatelli, M., Liu, D., Nasim, M. M. Stamp, G. W. H., Hirano, S. and Takeichi,

M. (1992b) Morphoregulatory activities of E-cadherin and beta-1 integrins in colorectal tumour cells. *Br. J. Cancer*, **66**, 629–634.
66. Plantefaber, L. C. and Hynes, R. O. (1989) Changes in integrin receptors on oncogenically tranformed cells. *Cell*, **56**, 281–290.
67. Rice, G. E. Gimbrone, M. A., Jr. and Bevilacqua, M. P. (1988) Tumour cell-endothelial cell interactions. Increased adhesion of human melanoma cells to activated vascular endothelium. *Am. J. Pathol.*, **133**, 204–210.
68. Rice, G. E. and Bevilacqua, M. P. (1989) An inducible cell surface glycoprotein mediates melanoma adhesion. *Science*, **246**, 1303–1306.
69. Rudy, W., Hofmann, M., Schwartz-Albiez, R., Zoller, M., Heider, K. H., Ponta, H. and Herrlich, P. (1993) The two major CD44 proteins expressed on a metastatic rat tumour cell line are derived from different splice variants: each one individually suffices to confer metastatic behaviour. *Cancer Res.*, **53**, 1262–1268.
70. Sato, T., Akiyama, F., Sakamoto, G., Kasumi, F. and Nakamura, Y. (1991) Accumulation of genetic alterations and progression of primary breast cancer. *Cancer Res.*, **51**, 5794–5799.
71. Schipper, J. H., Frixen, U. H., Behrens, J., Unger, A., Jahnke, K. and Birchmeier, W. (1991) E-cadherin expression in squamous cell carcinomas of head and neck: Inverse correlation with tumour dedifferentiation and lymph node metastasis. *Cancer Res.*, **51**, 6328–6337.
72. Schreiner, C., Fisher, M., Hussein, S. and Juliano, R. L. (1991) Increased tumorigenicity of fibronectin receptor deficient Chinese hamster ovary cell variants. *Cancer Res.*, **51**, 6328–6337.
73. Seftor, R. E. B., Seftor, E. A. Gehlsen, K. R., Stetler-Stevenson, W. G., Brown, P. D., Ruoslahti, E. and Hendrix, M. J. C. (1992) Role of the $\alpha v\beta 3$ integrin in human melanoma cell invasion. *Proc. Natl. Acad. Sci. USA*, **89**, 1557–1561.
74. Soszka, T., Kundsen, K. A., Bevila, L., Rossi, C., Poggi, A. and Niewiarowski, S. (1991) Inhibition of murine melanoma cell-matrix adhesion and experimental metastasis by albolabrin, and RGD containing peptide isolated from the venom of *Trimeresurus albolabris*. *Exp. Cell Res.*, **196**, 6–12.
75. Stallmach, A., Von Lampe, B., Matthes, H., Bornhoft, G. and Riecken, E. O. (1992) Diminished expression of integrin adhesion molecules on human colonic epithelial cells during the benign to malign tumour transformation. *Gut*, **33**, 342–346.
76. Stamenkovic, I., Amiot, M., Pesando, J. M. and Seed, B. (1989) A lymphocyte molecule implicated in lymph node homing is a member of the cartilage link protein family. *Cell*, **56**, 1057–1062.
77. Stamenkovic, I., Aruffo, A., Amiot, M. and Seed, B. (1991) The hematopoietic and epithelial forms of CD44 are distinct polypeptides with different adhesion potentials for hyaluronate bearing cells. *EMBO (Eur. Mol. Biol. Organ) J.*, **10**, 343–348.
78. Stamp, G. W. H. and Pignatelli, M. (1991) Distribution of $\beta 1$, $\alpha 1$, $\alpha 2$ and $\alpha 3$ integrin chains in basal cell carcinomas. *J. Pathol.*, **163**, 307–313.
79. Takeichi, M. (1991) Cadherin cell adhesion receptors as a morphogenetic regulator. *Science*, **251**, 1451–1455.
80. Tanabe, K. K., Ellis, L. M. and Saya, H. (1993) Expression of CD44 R1 adhesion molecule in colon carcinoma metastases. *Lancet.*, **341**, 725–726.

81. Tsuda, H., Zhang, Y., Shimosato, Y., Yokota, Y., Terada, M., Sugimura, T., Miyamura, T. and Hiroshashi, S. (1990) Allele loss on chromosome 16 associated with progression of human hepatocellular carcinoma. *Proc. Natl. Acad. Sci. USA*, **87**, 6791–6794.
82. Tsujisaki, M., Imai, K. and Hirata, H. (1991) Detection of circulating intercellular adhesion molecule-1 antigen in malignant diseases. *Clin. Exp. Immunol.*, **85**, 3–8.
83. Umbas, R., Schalken, J. A., Aalders, T. W., Carter, B. S., Karthaus, F. M., Schaafsma, H. E., Debruyne, F. M. J. and Isaacs, W. B. (1992) Expression of the cellular adhesion molecule E-cadherin is reduced or absent in high-grade prostate cancer. *Cancer Res.*, **52**, 5104–5109.
84. Vleminckx, K., Vakaet, L. Jr., Mareel, M., Fiers, W. and Van Roy, F. (1991) Genetic manipulation of E-cadherin expression by epithelial tumour cells reveals an invasion suppressor role. *Cell*, **66**, 107–119.
85. Vogelstein, B., Fearon, E. R., Hamilton, S. R., Kern, S. E., Preisinger, A. C., Leppert, M., Nakamura, Y., White, R., Smits, A. M. M. and Bos, J. L. (1988) Genetic alterations during colorectal-tumour development. *N. Engl. J. Med.*, **319**, 525–532.
86. Walsh, F. S., Barton, C. H., Putt, W., Moore, S., Kelsell, D., Spurn, N. and Goodfellow, P. N. (1990) N-cadherin gene maps to human chromosome 18 and is not linked to the E-cadherin gene. *J. Neurochem.*, **55**, 805–812.
87. Weinel, R. J., Rosendahl, A., Neumann, K., Chaloupka, B., Erb, D., Rothmund, M and Santoso, S. (1992) Expression and function of VLA-$\alpha 2$, $\alpha 3$, $\alpha 5$ and $\alpha 6$ integrin receptors in pancreatic carcinoma. *Int. J. Cancer*, **52**, 827–833.
88. Williams, A. F. and Barclay, A. N. (1988) The immunoglobulin superfamily — domains for cell surface recognition. *Annu. Rev. Immunol.*, **6**, 381–405.
89. Zutter, M. M., Krigman, H. R. and Santoro, S. A. (1993) Altered integrin expression in adenocarcinoma of the breast. *Am. J. Pathol.*, **142**, 1439–1448.

2. CELL ADHESION MOLECULES IN MAMMARY CARCINOMAS

G. W. H. STAMP[1,2,4], A. M. HANBY[2], S. RASBRIDGE[3], E-N. LALANI[1] and M. PIGNATELLI[1,2]

[1]*Department of Histopathology, Royal Postgraduate Medical School, Hammersmith Hospital, London W12 OHS, UK*
[2]*ICRF/RCS Histopathology Unit, 35–43 Lincoln's Inn Fields, London, WC 2A 3 PX, UK*
[3]*ICRF Clinical Oncology Unit, Guy's Hospital, London, SE 2, UK*
[4]*To whom correspondence should be addressed, Tel: 44 81 740 3229 Fax: 44 81 740 7417*

INTRODUCTION

The investigation of the role of integrin receptor molecules in mammary neoplasia requires a knowledge of the *in vivo* distribution of these molecules, both in the normal situation and in benign and malignant disease. From such studies trends emerge which indicate that certain of these receptors are more likely to be of significance, either because of under- or over-expression, and may reflect a significant role in tumour progression. The aim is to define these alterations as closely as possible and then to select an appropriate experimental system in order to verify the functional significance of such observations as in isolation, these can remain of restricted value either in the clinical or scientific context.

The expression of integrins in the adult non-neoplastic mammary gland is assessed prior to reviewing what is currently known about carcinomas. The normal adult non-lactating gland is composed of innumerable terminal duct-lobular units (TDLU), comprising a single terminal duct leading into a branching aggregates of tubular units, which undergo lactational change in pregnancy to produce the secretory unit. The extracellular matrix and fibroblastic mesenchyme appear to dictate normal growth and mesenchyme, and may limit growth, which may centrally involve TGFβ, which not only enhances fibroblastic growth, matrix depostion but is also known to regulate integrin expression in epithelial and mesenchymal cells. Pregnancy associated differentiation and lactation are also dependent upon basement membrane integrity (Howlett and Bissell, 1993). At higher power, the tubules have a bilayered structure, composed of an outer layer of myoepithelial cells which are flattened, sometimes undergoing clearing of the cytoplasm later in the cycle or in pregnancy. In many places they appear to have a radial orientation and the cell bodies span many of the luminal cells, but in three dimensions they form a continuous layer more or less enveloping the basal aspects of the ductal cells. The ductal luminal cells are cuboidal or columnar depending on their state of activation and the nuclei are well polarised. Both cell types are derived from a common precursor, and lie above the basement membrane, which is largely thought to be a myoepithelial product, although the duct luminal cells do show fine cytoplasmic adhesions to the basement membrane at the ultrastructural level.

We have assessed the distribution of integrin chains by a standard Avidin-Biotin peroxidase technique using antibodies obtained from the sources listed in Table 2.1, and the results are summarised in Table 2.2. Membrane staining was assessed to reflect potentially functional receptor expression, and with most of these antibodies, at the correctly determined dilutions, there was relatively little cytoplasmic positivity, except to some degree with $\beta1$ chain–this may represent the cytoplasmic pool of $\beta1$.

We did not often find cytoplasmic positivity using the $\alpha2$ and $\alpha3$ antibodies, which has been reported for P1H5 and J143 ($\alpha3$) (D'Ardenne *et al.* 1992). Myoepithelial cells generally show strong membrane positivity compared to the ductal luminal cells. There is circumferential staining for $\beta1$, $\alpha1$–3 and E Cadherin, but there is basal accentuation of $\alpha6$ and accentuated basal polarisation of $\beta4$ chain, consistent with $\alpha6\beta4$ binding to a hemidesmosomal component. Alpha v chain is expressed at a low level on normal myoepithelial cells. Membrane staining is extremely difficult to assess on cryostat sections because of the less than optimal resolution, and the close apposition of the myoepithelial cells on the basal aspects. However, we have been able to demonstrate $\beta1$ and $\beta4$ chains on methacarn-fixed paraffin-embedded sections, and slightly tangentially orientated ductules clearly show membrane staining for $\beta1$. $\beta4$ chain is again predominantly basal. Alpha 1 chain is apparently expressed on myoepithelial cells according to Koukoulis *et al.* (1991, 1993) but in our hands it is relatively weak on myoepithelial cells, and is very difficult to find on duct luminal cells.

Table 2.1 Antibody specificities and sources

Specificity	*Antibody*	*Source*
$\beta1$	DH 12	B De Strooper
$\beta4$	439.9B	S Kennel
$\alpha1$	TS 2/7.1.1	M Hemler
$\alpha2$	HAS 3	F Watt
	B1 515	K Pischel
$\alpha3$	E1 56	K Pischel
$\alpha5$	P1D6	Chemicon Inc
$\alpha6$	MP 4F10	M Pignatelli
αv	13 C2	M Horton
$\alpha v\beta3$	23 C6	M Horton
E Cad	HECD 1	M Takeichi

Table 2.2 Integrin Expression in Normal Mammary Epithelium

	$\beta 1$	$\alpha 1$	$\alpha 2$	$\alpha 3$	$\alpha 4$	$\alpha 5$	$\alpha 6$	αv	$\alpha v \beta 3$	$\beta 4$	E Cad
Myoepithelial Cells	+ + +	+ +	+ + +	+ + +	−	−	+ + p	+	− / +	+ + p	+ + +
Luminal Cells	+	−	+	+	−	−	+ / −	− / +	+	+	+ / + +

NON-NEOPLASTIC BREAST DISEASE

There are few comprehensive studies of integrin expression in benign breast disease, that of Koukoulis *et al.* (1991) providing the most detail. In general, there is maintenance of the normal phenotype in non-proliferative conditions and in mild epithelial hyperplasias, as the myoepithelial layer remains intact. Epithelial Hyperplasia (Epitheliosis/papillomatosis) shows some enhancement of the membrane staining of the proliferating cohesive cellular aggregates contained within the ductal-lobular system, which are a mixture of luminal cells and some myoepithelial cells (Fig. 2.1). This may be accompanied by membrane-associated $\alpha 2$ and $\alpha 3$ across the bridging hyperplastic cells. As yet,

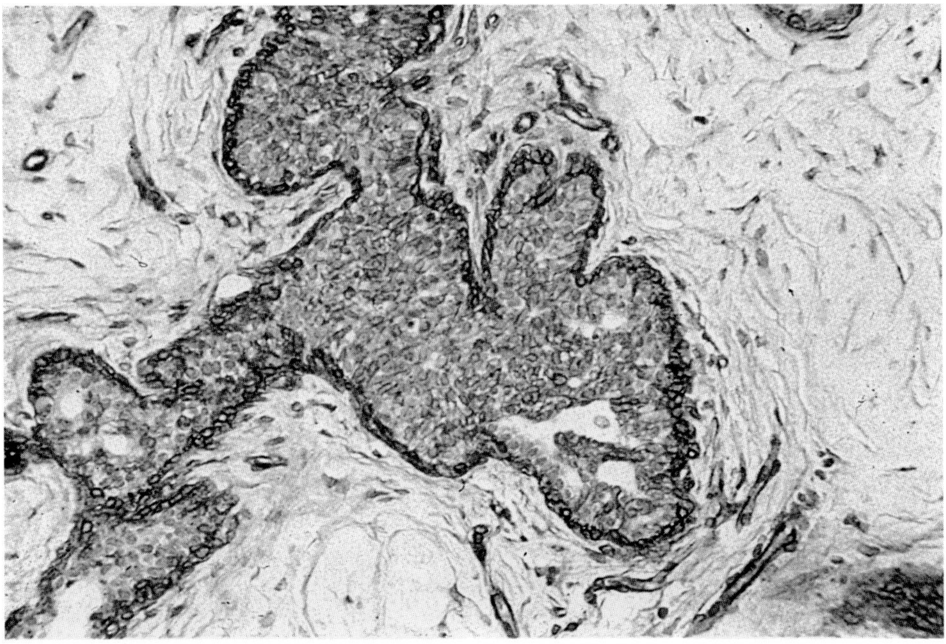

Figure 2.1 Benign epithelial hyperplasia ('epitheliosis') with accentuation of central hyperplastic cell membranes for $\beta 1$ chain. Note the peripheral rim of myoepithelial cells. (See Colour Plate I at the back.)

no study has addressed variations in the stromal expression of $\alpha 1$ and αv integrins in benign sclerosing proliferations in the breast.

MAMMARY CARCINOMAS

Different forms of mammary carcinoma display considerable differences not only in growth patterns and cytological appearances, but also in their biological behaviour. At least some of their properties may be related to cell adhesion molecule expression and function, and to the mechanisms of interaction with the stromal cells. It is important to bear these differences in mind when studying particular cell lines *in vitro* and to relate this to their line of differentiation. It is also worth pointing out that breast carcinomas rarely show differentiation towards a myoepithelial phenotype, but more usually show some features of the luminal cells including secretion. However, they are capable of synthesising extracellular matrix proteins and receptors, and directly interact with ECM without the interposition of a myoepithelial cell layer.

In situ carcinomas of the breast have to be distinguished from benign epithelial hyperplasias on morphological grounds. They are termed *in situ* because the basement membrance is intact, and can be demonstrated immunohistologically if necessary. The two main categories, ductal and lobular, are not reflecting the so-called site of origin, since both arise within the terminal duct lobular units, but is based on their resemblance to the invasive counterparts and their identification as residual foci of *in situ* carcinoma in tumours which have started to invade elsewhere in the ductal lobular system. The reason for identifying the two types is that they are associated with a different risk for subsequent development of invasive carcinoma, which is predominantly ipsilateral in DCIS but confers a bilateral risk in LCIS, and this has implications for the clinical management.

Ductal carcinoma *in situ* tends to show cuboidal/columnar cells with glandular formations and architecturally distinct structures called trabecular bars and bridges. More aggressive variants (ie those more frequently associated with a concurrent invasive carcinoma or which have a greater incidence of subsequent development of invasive cancer after biopsy alone) have more solid proliferations with larger more pleomorphic but cohesive cell populations. Lobular carcinoma *in situ* comprises relatively uniform rounded cells which show reduced cellular cohesion, manifest by spaces between the cells and their spherical shape, a morphology suggesting a lower level of intercellular adhesion. One other characteristic feature is the presence of cytoplasmic spaces lined by apical differentiated plasma membrane with microvilli, in which a small amount of mucin may be found, the intracytoplasmic lumina; it is almost as if each cell acts as an autonomous secretory unit, functioning irrespective to its neighbours.

Invasive ductal carcinomas (IDC) of the breast are the commoner form of invasive cancer. Over 75% fall into the group known as invasive ductal carcinoma NOS (Not Otherwise Specified) which may be further subdivided into Grades I, II and III on the basis of degree of a) tubule or gland formation, b) nuclear morphology and c) number of mitotic figures. Hence Grade I carcinomas show well-formed tubular-glandular structures, with some degree of cytological polarisation, and ordered nuclear features. Poorly differentiated carcinomas are more disorganised with less glandular formation, more solid aggregates and variable less regular neoplastic cells. Some of the distinctive types of IDC have a less aggressive clinical behaviour when in pure form. Of these, the tubular carcinoma is of interest as it is frequently encountered as a small early carcinoma in mammographic screening programmes, and may represent a more indolent type of cancer. It is characterised by distinct well formed tubular glands lying in a well formed cellular stroma. One feature of these tumours is that they are devoid of basement membrane despite their well ordered structure. Mucinous carcinomas have a tenuous vascular supply and exist as clusters apparently floating in a sea of their secreted epithelial mucin, but do not disperse.

Since cytoskeletal organisation may directly or indirectly control nuclear structure, gene expression and proliferation, b) and c) may be critically dependent upon a). Thus it may be that if we can define the series of changes which dictate form and function mammary carcinoma cells we would have the basis for a more comprehensive assessment of cellular behaviour than the current system, which while it is broadly reproducible (in expert hands), is still unable to provide a sophisticated analysis which would identify subtypes with prognostic significance.

Invasive lobular carcinomas (ILC) represent 10–15% of all cancers and show a distinctive invasive pattern, consisting of cords of cells lying between interstitial collagen, so-called ‘indian filing’. In more cellular examples the tumour masses comprise isolated cells with little cooperative architecture and no duct or tubular formations (at least in the classical forms). Invasive lobular carcinomas also tend to invade in a more diffuse pattern and may be more difficult to detect clinically at an early stage. In addition they may show a different pattern of dissemination in the body to IDC, preferentially metastasising to sites such as the skin, gut mucosa, lower genital tract among others as well as diffusely invading lymph nodes and bone marrow. Despite the fact that they often have higher levels of oestrogen receptors, and show lower proliferative indices, ILC is still an aggressive malignancy with only a weak survival advantage over IDC in some of the larger series.

Bearing in mind the diversity of architecture and cytology in carcinomas, we and others have defined the expression of cell surface molecules *in vivo*, and their relationship to differentiation and growth. The following is a summary based on the distributions of the individual subunits, and is based on our own series where the intensity of staining was compared to that of non-neoplastic elements including myoepithelial cells of the lobules (Pignatelli *et al.* 1991, 1992).

Beta 1 Chain

$\beta 1$ is found on a wide range of cells and is detectable on the membranes of a majority of carcinomas, both ductal and lobular, but a greater number of Grade III IDC show decreased or absent expression which achieved statistical significance in our series. Interestingly, most lobular carcinomas showed weak membrane positivity (Fig. 2.2). Stromal cells show $\beta 1$ staining, particularly when neoplastic cells are weak, equivocal or negative in IDC. This staining showed some correspondence to $\alpha 1$ and αv distribution in mammary carcinomas. It has to be said that in these cases the assessment of which membrane is positive (tumour or stromal cell) can be very difficult and often required critical examination of many fields and restaining of these cases at differing dilutions.

Areas of *in situ* carcinoma were noted in a proportion of IDC and ILC, and were highlighted by a strongly stained rim of peripheral cells. Although it was previously believed myoepithelial cells are discontinuous or absent in *in situ* carcinomas, other studies have indicated they remain. However, the resolution of cryostat sections did not allow us to positively identify these as non-neoplastic *residual* myoepithelial cells.

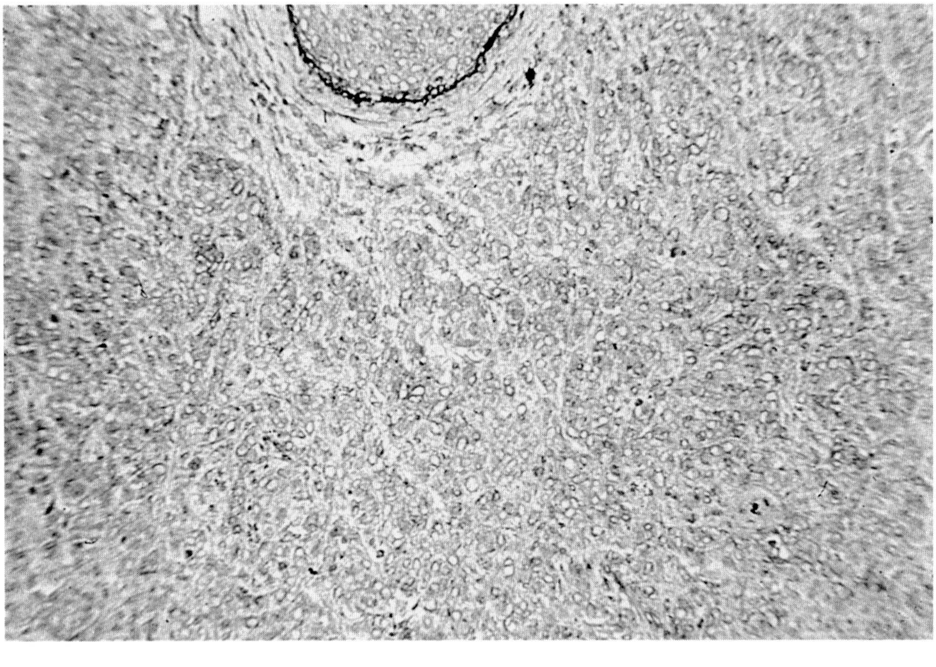

Figure 2.2 Low power view of in situ ductal carcinoma with peripheral residual myoepithelial cells (top centre), but the Grade III invasive ductal carcinoma around it is negative for $\beta 1$ chain. (See Colour Plate II at the back.)

We were able to study a series of *in situ* carcinomas of the breast on methacarn fixed, paraffin embedded material using the antibodies to the $\beta 1$ chain and also $\beta 4$ (Hanby *et al.* 1993). The better morphology allowed us to identify myoepithelial and neoplastic cells. Using this technique, all cases showed positive membrane staining on myoepithelial cells at the periphery of lobules containing DCIS. The neoplastic cells also stained for $\beta 4$ in 14/18 cases of DCIS, on all cells in 10 cases and focally in the remainder (Fig. 2.3). $\beta 4$ chain was only found on the neoplastic cell membranes in 4/18 examples, none of them strongly.

Alpha 1 chain

We have previously demonstrated $\alpha 1$ is abundant in the stroma of colorectal carcinomas, basal cell carcinomas of the skin and mammary carcinomas on desmoplastic fibroblasts, but not convincingly on the tumour cells and D'Ardenne *et al.* 1992 have come to the same conclusion.

Alpha 2 and 3 Chains

Alpha 2 chain is reduced or absent in many poorly differentiated carcinomas, and tends to parallel the reduction in $\beta 1$ chain, indicating VLA 2 is downregulated at least in ductal carcinomas (Fig. 2.4). Comparing Grade III

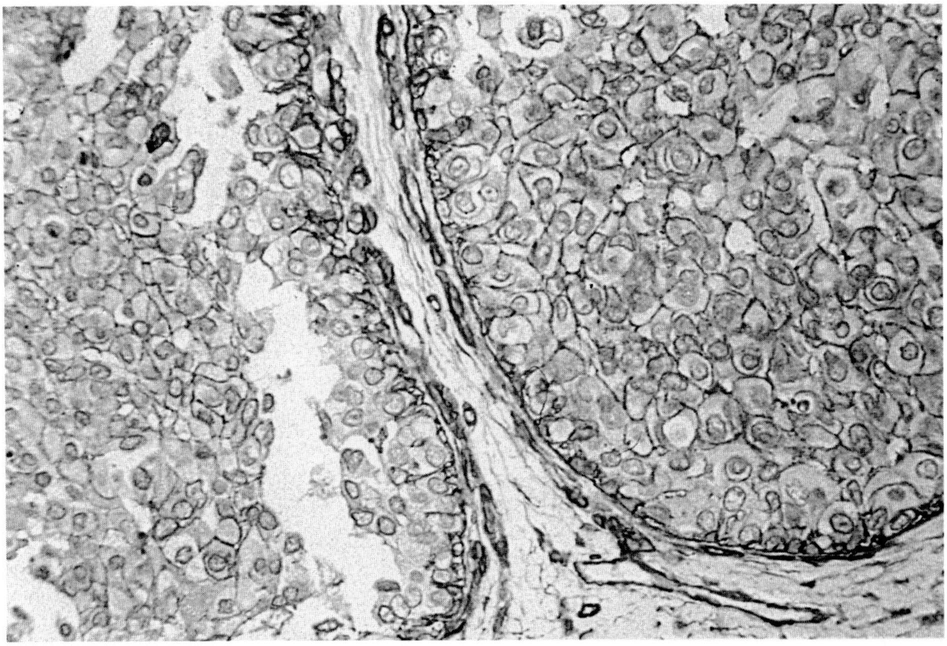

Figure 2.3 Strong membrane staining for $\beta 1$ chain on an in situ ductal carcinoma. (See Colour Plate III at the back.)

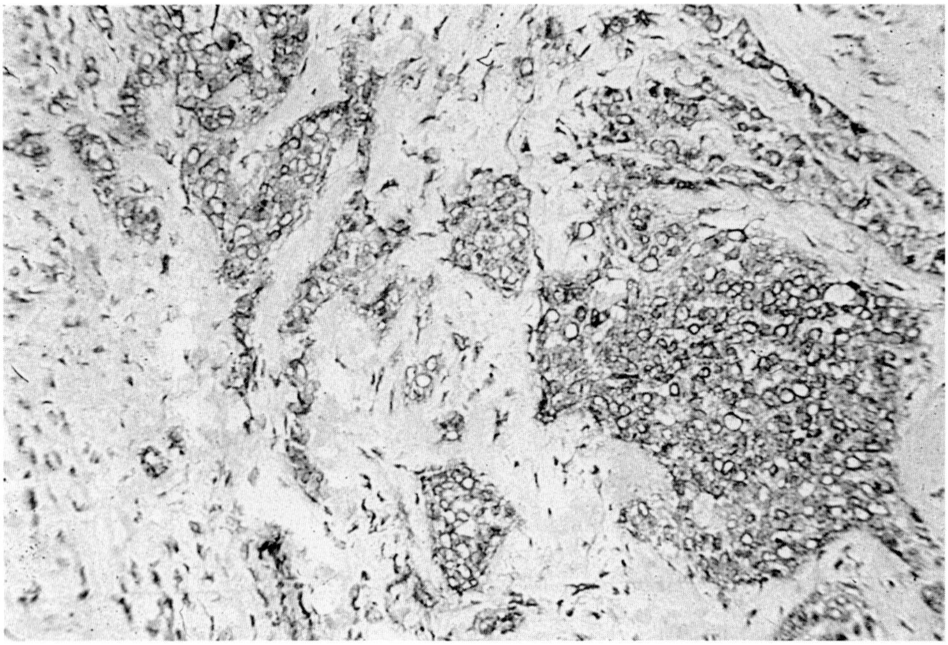

Figure 2.4 Weak $\alpha2$ chain expression on an invasive ductal carcinoma of the breast. (See Colour Plate IV at the back.)

to Grade I & II IDC achieved statistical significance in our series (Pignatelli *et al.* 1992), and confirmed the findings of others (Zutter *et al.* 1990; Koukoulis *et al.* 1991). Jones *et al.* (1992) reported loss of $\alpha2$ chain in 65% of carcinomas but remarkably 90% were unreactive for $\beta1$ chain. Arihuro *et al.* (1993) showed that 46% of their series of IDC were negative for VLA 2, but only 11% of the corresponding DCIS were negative, implying the loss of expression was associated with progression, although the contribution from residual myoepithelial cells in *in situ* carcinomas makes such assessments extremely difficult in our experience. Moreover, loss of $\alpha2$ chain has previously been noted in colorectal carcinomas. ILC showed relatively more $\alpha2$ chain expression, albeit weak in the majority. Interestingly, De Souza *et al.* (1993) showed that overexpression of c erb B2 (Her 2/neu), which is expressed at higher levels in high grade IDC and has prognostic significance, may selectively inhibit expression of $\alpha2$ in transfected immortalised human mammary cell lines, and that expression of c erb B2 may parallel reduction in expression of $\alpha2$ integrin subunit and mRNA. Such a mechanism may also explain observations in pancreatic carcinoma but would not be consistent with the observations in colorectal carcinoma where c erb B2 expression is uncommon, so that the observed reduction in expression may not be entirely due to such a mechanism. ILC showed relatively more $\alpha2$ chain expression in our series, albeit weak in the majority, but the series of Koukoulis *et al.*

(1991) showed relatively little positivity for $\alpha2$ chain in ILC, with weak staining for $\alpha3$ chain.

Alpha 3 chain was reduced in parallel with $\alpha2$ and $\beta1$ but we were unable to study enough cases to achieve statistical significance. Bartolazzi *et al.* (1993) also observed reduced expression of $\alpha3\beta1$ in 75% of IDC and 40% of ILC, but reduction in $\alpha2$ and $\alpha3$ has not been confirmed by other groups, but in these studies cytoplasmic staining was present in all cells, although a reduction in staining intensity was noted in some high grade tumours. However, the role of $\alpha3\beta1$ may be different in tumour progression as there is evidence that cells with enhanced metastatic capacity adhere to substrates and lymphoid stroma significantly and express greater levels of $\alpha3$ (and $\alpha5$) than non-metastatic cells. Although both $\alpha2$ and $\alpha3$ are receptors have collagen as the preferred ligand, it appears that their function may be distinct. Antibodies to VLA 2 block morphogenesis and cell spreading in mammary cell lines whereas antibodies to VLA 3 do not (Berdichevsky *et al.* 1993).

Reduction in $\alpha2$ and $\alpha3$ has not been confirmed by D'Ardenne *et al.* (1992) however, but in this study cytoplasmic staining was present in all cells, (although a reduction in staining intensity was noted in some invasive tumours).

Although $\alpha2$ chain is expressed on fibroblasts, it was only weak or equivocal in our experience and that of D'Ardenne *et al.* (1992) and Koukoulis *et al.* (1991).

Alpha 4 and Alpha 5 Chains

VLA 4 is rarely expressed on human mammary carcinoma cells *in vivo*, although Bao and Tarin (1993) claimed that VLA 4 expression is associated with the metastatic phenotype in the poorly differentiated mammary carcinoma cell line MDAMB 435 which metatstasises to lungs and other sites in nude mice (Price *et al.* 1990). Koukoulis *et al.*(1991) observed expression specifically on the mucinous carcinomas of the breast, which are distinctive because the cells are largely suspended in a sea of extracellular mucin with a delicate fibrovascular supply, but have little contact with conventional ECM and do not form basement membrane.

Alpha 6 and Beta 4 Chains

$\alpha6$ chain is detectable in a basement membrance distribution in a proportion of invasive ductal carcinomas, especially in glandular areas. In more poorly differentiated tumours membrane staining can be seen but it tends to be less polarised. D'Ardenne *et al.* (1992) found a close correlation with laminin distribution and polarised $\alpha6$ expression, and it has been previously documented that basement membrane formation is more defective or absent in high grade IDC and other sites (Gusterson *et al.* 1982). In our series, $\alpha6\beta4$

was present in a proportion of IDC which were weak equivocal or negative for $\beta 1$ chain. Similar observations were also made by Jones *et al.* (1992) who observed absent expression of $\alpha 6$ chain in up to 80% of invasive carcinomas of the breast (greater than $\alpha 2$ chain loss of 65%).

We and others have shown positivity for $\alpha 6$ chain which was weakly to moderately positive in 14 of 18 cases of ILC we examined and in half of these linear staining was seen at the epithelial stromal interface rather than at the cell-cell junctions. In many of these cases there was corresponding pattern for both $\beta 1$ and $\beta 4$ chain, but the association only reached significance for $\alpha 6\beta 4$ in our series (Pignatelli *et al.* 1992). Koukoulis *et al.* (1993) claimed the most intense localisation for $\alpha 6\beta 4$ was on the peripheral cell membranes where they show amoeboid projections (invadopodia) into the connective tissue. This is of interest because these projections also show localisation of type IV collagenase activity in lobular carcinomas (vide infra).

Interestingly, two IDC variants which do not have formed basement membranes, the mucinous and tubular carcinomas, are largely negative for $\alpha 6\beta 4$ at least on the membranes, but these are relatively indolent cancers which rarely metastasise when in pure form. Koukoulis *et al.* (1993) claim to have demonstated $\beta 4$ on the membranes of mucinous carcinomas, a curious finding since there is not an accessible ligand for the tumour cell membrane, (or perhaps more conservatively any *recognised* ligand) in such tumours. Is it possible that mucin itself may act as an ECM in certain circumstances?

Loss of $\alpha 6$ expression at the mRNA level has also been found in some mammary carcinoma cell lines by using differential display techniques in Northern analysis (Sager *et al.* 1993). It was also determined that this was not due to genetic deletion, and suggests that such down regulation may be reversible. No studies have yet correlated this with reduction in Type IV collagen or Laminin expression *in vitro*.

It has been suggested that $\alpha 6\beta 4$ may facilitate metastatic capacity of cells, evidenced from certain mammary carcinoma cells lines including one where there is high expression (Kemperman *et al.* 1993). In this cell line enhanced hepatic metastasis is seen in *in vivo* models. Natali *et al.* (1992) found a relative decrease in expression in metastases to the pleural cavity compared to lymph node metastases suggesting alterations in the microenvironment may be responsible for reduced expression.

ALPHA V EXPRESSION

We and others find $\alpha v\beta 3$ heterodimer is rarely expressed in IDC, and is weak on a proportion of ILC, in contrast to strong localisation to endothelial cells using the antibody 23C6. This contrasts with the situation in malignant melanomas where $\alpha v\beta 3$ is overexpressed in the most aggressive variants (Albelda *et al.* 1990). It is possible that relatively greater amounts of $\alpha v\beta 3$ on ILC may account for some aspects of their invasive behaviour, however. On the other hand the αv chain is identified on a majority of invasive ductal carcinomas of all grades, where it is accentuated at the stromal interface (Fig. 2.5). In this situation it is likely that it is associated with other β chains which bind to multiple ligands, but identifying which ones is unlikely to emerge from straightforward observational studies. The great majority of ILC in our series also have detectable αv single chain which may sometimes be associated with polarisation to the stromal interface. Koukoulis *et al.* (1991) also found that αv was generally weak on the cell processes. Alpha v was sometimes found in tumours negative for $\alpha 2\beta 1$, so it may be that the αv subunit is still expressed when other subunits are downregulated, rather than being frankly upregulated itself.

The stromal localisation was very variable in our series, showing some co-distribution with $\beta 1$ chain, and was positive on desmoplastic fibroblasts which were negative for $\alpha 1$, especially in tubular carcinomas which have a prominent cellular stroma. This is interesting in that basal cell carcinomas of the skin, which have an equally prominent and organised stroma, but negligible metastatic potential, have abundant αv and $\alpha 1$ chains in their peritumoural desmoplastic fibroblasts (Stamp and Pignatelli, 1991). In many invasive ductal carcinomas there was enhancement of αv immunoreactivity at the invasive periphery of the tumours where there is an immature and rather disorganised stromal reaction.

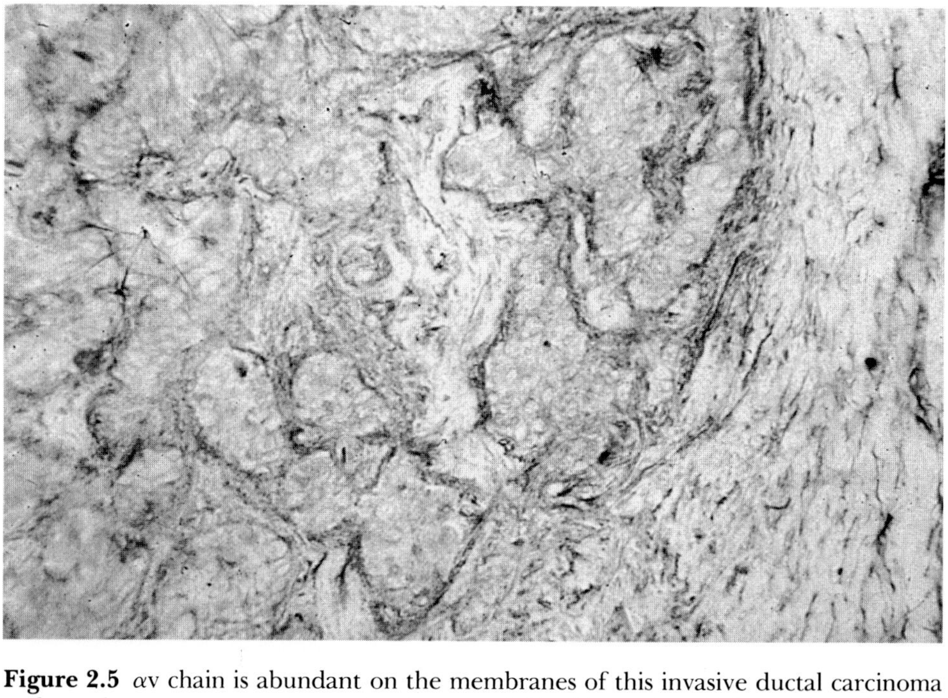

Figure 2.5 αv chain is abundant on the membranes of this invasive ductal carcinoma and to a lesser extent on desmoplastic stromal fibroblasts. (See Colour Plate V at the back.)

E CADHERIN

E Cadherin is strongly expressed on most IDC, although the intensity is variable, and in a series from Guy's Hospital Clinical Oncology Unit there was no significant correlation with pattern of staining and tumour grade, although tubular carcinomas of the breast show high expression of E cd (Fig. 2.6). This has also been confirmed by others (Gammallo *et al.* 1993), but Oka *et al.* (1993) claimed to have demonstrated significant reduction of expression in poorly differentiated ductal carcinomas especially advanced tumours and metastases. However, many ILC are weak or negative for E Cadherin (Fig. 2.7) and where present it sometimes assumed a cytoplasmic rather than membranous localisation. This would conveniently explain why lobular carcinomas display their characteristic morphology and tendency to individual cell invasion. E Cadherin is responisble for cellular aggregation in mammary carcinoma cell lines HS 578 and BT 549 but transfection of the gene does not lead to a reversal of invasive pheotype, nor to aquisition of epithelial morphology (Sommers *et al.* 1993), although many fibroblastoid mammary cell lines are negative for E Cadherin. However, it is conceivable that his results from defective coupling of the intercellular linkage via catenins, as beta actin and plakoglobulin were also absent from the cellular membrances. This might accord with our own and others observations of

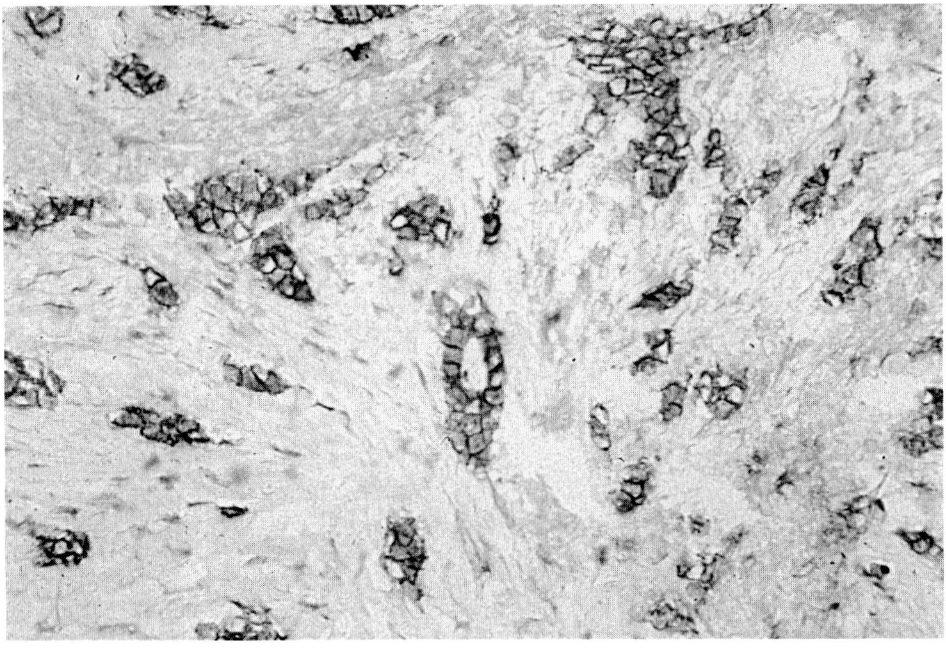

Figure 2.6 Grade I IDC strongly positive for E Cadherin. (See Colour Plate VI at the back.)

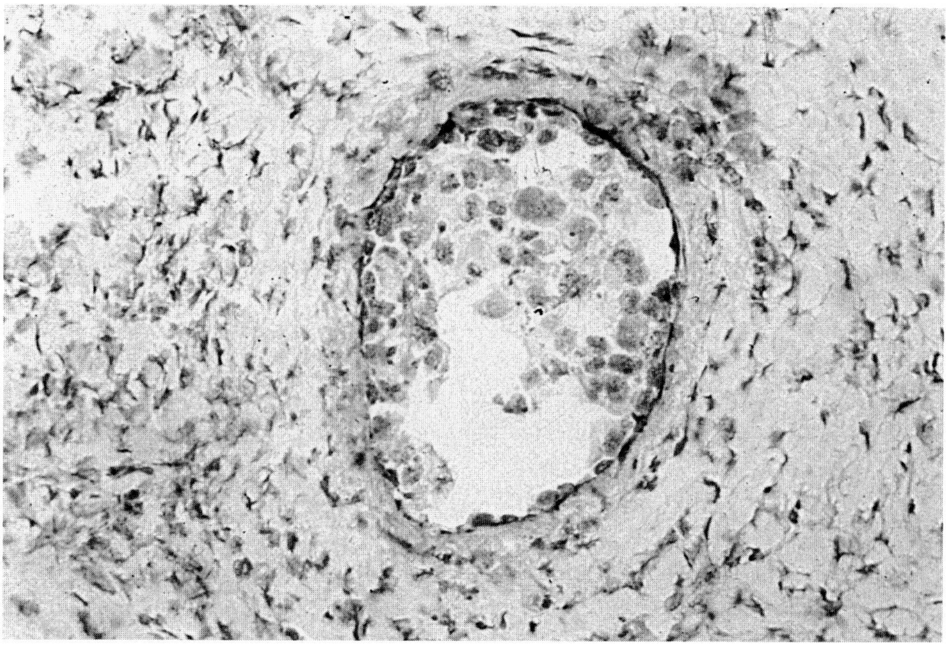

Figure 2.7 In situ and invasive lobular carcinoma negative for E Cadherin. The ring of positive cells (centre) are residual myoepithelial cells in the *in situ* component. (See Colour Plate VII at the back.)

cytoplasmic staining for E Cadherin in some carcinomas (Inoue *et al.* 1992). Others have noted that carcinoma cell lines with an epithelioid cytology are E cd positive, whereas invasive fibroblastoid cell lines did not and that their invasive capacity could be reduced by transfection with the gene for E cd (Frixen *et al.* 1991).

In our series of cases no significant correlations were found between hormone receptor status as determined by cytosol assay, but in the series of Zutter *et al.* (1990) strong positive immunolocalisation for ER appeared to show a close correlation to high levels of $\alpha 2$ expression. We are currently studying the independent effect of oestrogens and progesterones in a number of mammary carcinoma cell lines, our preliminary data suggests that neither oestrogen nor progesterone have any regulatory effect on surface integrin expression when cultured in media depleted of hormones and phenol red (which has an oestrogenic effect). However, integrin cell surface receptors are downregulated in the absence of serum in the medium, which cannot be reversed by exogenous hormones. The mechanisms underlying these changes are clearly complex and indicate that a simple correlation will be unlikely to be definable *in vivo*.

In summary, we and others find that there is a significantly reduced expression of $\alpha 2\beta 1$, which mediated cell-ECM and to a lesser extent cell-cell

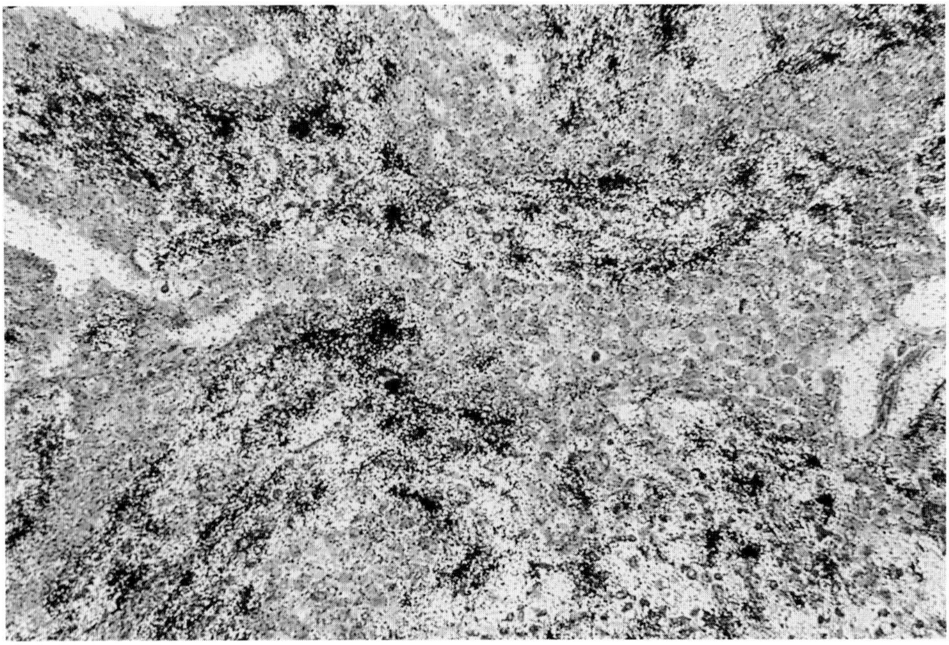

Figure 2.8 Isotopic in situ hybridisation on an IDC showing labelling for 72 kDa Type IV collagenase mRNA in the stromal cells. TIMP 2 localised to the stromal compartment also. (See Colour Plate VIII at the back.)

interactions in epithelial cells in poorly differentiated IDC, although a corresponding decrease has not been convincingly demonstrated for $\alpha 3$ chain and this requires more study. Also in poorly differentiated IDC there is reduced $\alpha 6$ expression and loss of polarity at the interface which may coincide with reduced synthesis of the ligand laminin.

One way of investigating the relevance of this is to transfect a full cDNA for alpha 2 chain into a mammary carcinoma cell line which is $\alpha 2$ neagtive and see if this affects adhesion and differentiation using *in vitro* and *in vivo* models, which we are currently attempting.

E Cadherin, which mediates cell-cell interactions, and present on most IDC but is more variable on the poorly differentiated tumours. In contrast ILC display markedly reduced potential for cell-cell interactions by underexpressing E Cadherin on the surface membrane. They also show reduced $\alpha 2$ expression, but in our series there is a definitely greater incidence of staining for $\alpha 6\beta 4$ ($^{+}/_{-}\alpha 6\beta 1$) on the malignant cells. Similarly $\alpha v\beta 3$ was detected on the majority of ILC. The differences in these three parameters may underlie most of the morphological and behavioural differences in cells showing lobular as opposed to ductal differentiation.

TUMOUR STROMA IN BREAST CARCINOMAS

Little attention has been paid to the role of the integrin expression of stromal cells in modulating epithelial cell behaviour but there are some differences. While there is little $\alpha 4$ except on inflammatory cells, $\alpha 5$ is seen on fibroblasts in some cases especially where there is αv expression. Expression of $\alpha v\beta 1$ may well be important in the growth of the desmoplastic stromal cells in some breast carcinomas, but the significance of this has not been tested *in vitro*. By way of a complementary approach, we had been also looking at the behaviour of the ECM and its modulators, and discovered some changes that we did not anticipate in the stroma of malignant tumours.

STROMAL INFLUENCE ON BASEMENT MEMBRANE STABILITY

It is clear that the dynamic interaction of epithelial and mesenchyme is not solely dependent upon the presence or absence of one or even a number of receptors but also depends on the stability of the ligands in interstitial and basement membrane matrices. This may be manifested by changes both in terms of synthesis (which is increasingly clear that it requires contributions from both epithelium and mesenchyme), and also of their degradation. The influence of growth factors such as the TGFβ family may be highly relevant in modulating both the epithelial cell behaviour, predominantly by modifying the function of the mesenchymal elements in epithelial neoplasms. Other growth factors and cytokines, including TNFα, interleukins and hormones also modify behaviour of neoplastic and desmoplastic cells (especially in hormonodependent tissues such as the mammary gland), and are a part of this complex equilibrium.

Instability of the extracellular matrix would compromise the establishment of the stable phenotype, and moreover require the cell to rely on the primitive interactions which predominate in the embryonic epithelial-mesenchymal environment. Accelerated hydrolytic breakdown of connective tissue barriers is the fundamental process enabling malignant cells to cross organised extracellular matrix (ECM) boundaries and penetrate the lymphatic and vascular system enabling metatstasis (Liotta *et al.* 1991; Hart and Saini, 1992). Degradation of ECM gradually reduces control of growth exerted by cell surface ECM receptors, such as the integrins, leading to the loss of differentiated phenotype which characterises malignant neoplasms. Reduction in level of highly ordered ECM synthesis is paralled by reduced expression of the corresponding cell surface ECM integrin molecules and the upregulation of alternative forms such as fibronectin receptors $\alpha 1\beta 1$ or vitronectin receptors $\alpha v\beta 3$ and $\alpha v\beta 5$, as the malignant cells adapt to an unstable environment which has a composition more reminiscent of the embryonal epithelial-mesenchymal interactive milieu. These alterations underlie the essential features of the invasive malignant cells, *i.e.* loss of cytostructural organisation, enhanced proliferation and migration through connective tissue. If there is accelerated degradation of the basement membrane the neoplastic cells are then interacting with the interstitial matrix rather than the specialised components of the BM. Growth may be proceeding too quickly to enable the formation of an organised BM, and it is possible that transcription of the BM protein genes is downregulated in areas of active growth. However, it seems more likely that there is enhanced degradation of the synthesised BM and other ECM components by specific collagenases produced by the neoplastic cells. There is evidence for this from studies on highly metastatic subclones of transformed cell lines, and from the enhanced metastatic potential of transfected cells containing cDNA coding for Type IV collagenase and Northern analysis reveals that there is increased mRNA for MMP2 in malignant tumours and that antibodies raised to peptides from the procollagenase fragment and to the active enzyme both localise to neoplastic cells (Brown *et al.* 1990; Levy *et al.* 1991).

It is clear that this process must be tightly controlled at the neoplastic cell membrane-ECM interface as uncontrolled pericellular lysis is incompatible with cellular survival nor ECM ligand binding necessary for active migration. While several enzymes are capable of hydrolysing ECM components including Matrix metalloproteinases (MMPs) serine proteases such as plasmin, or cathepsins B and D (Chen, 1992; Duffy, 1992; Liotta *et al.* 1991), early studies identified members of the MMP family to be disproportionately upregulated in malignant tumours compared to their benign counterparts or wound healing (Liotta *et al.* 1991; Hart and Saini, 1992). The MMPs consist of 9 structurally related zinc and calcium dependent endopeptidases that are of particular interest because they have the combined ability to degrade all of the major structural components of the ECM at physiological pH (Matrisian, 1992; Woessner, 1991; Docherty *et al.* 1992). Two MMPs, Gelatinases A and B also termed 72kDa and 92kDa gelatinases (MMP-2 and MMP-9) are particularly relevant because they specifically degrade the basement membrane (as well as denatured collagen) by cleaving the type IV collagen backbone at a pepsin-resistant triple helical domain, generating a characteristic fragmentation pattern. This property is essential for penetration of blood vessel basement membranes and appears to be a prerequisite for the dissemination of carcinoma cells (Zucker *et al.* 1987; Hart and Saini, 1992). Studies have correlated enhanced MMP-2 expression with a metastatic phenotype in simple *in vitro* invasion systems and *in vivo* lung colonisation models (Goldberg and Eisen, 1992; Hendrix *et al.* 1992; McDonnell and Matrisian, 1990; Matrisian *et al.* 1991) and in transfection studies only the 72kDa gelatinase was able to confer the metastatic phenotype to murine cell lines. Furthermore the downregulation of these enzymes by genetic means reduces the metastatic potential of manipulated cell lines. Immunohistochemical studies appeared to localise MMP-2 predominantly to malignant cells, with relatively less in the stroma (Campo *et al.* 1992; Riordan *et al.* 1993). and high levels of mRNA are found in tumour ovary homogenates

(Levy *et al.* 1991). Additional evidence for a role for MMPs came from the discovery that their natural inhibitors, TIMP-1 and TIMP-2, are capable of blocking tumour cell invasion *in vitro*, as do antibodies to MMPs and TIMP-1 or a synthetic analogue prevents metastasis in simple *in vivo* lung colonisation models (de Clerk *et al.* 1991; Albini *et al.* 1991). However, we and others found that the mRNAs for MMP-2, TIMP-2 and Stromelysin 3 localise to desmoplastic fibroblasts in the immediate vicinity of invasive malignant cells Fig 2.8 (Poulsom *et al.* 1992, 1993; Pyke *et al.* 1992; Basset *et al.* 1990), although immunolocalisation is to stromal cells predominantly in the case of Stromelysin 3 (Wolf *et al.* 1993). In the case of MMP-9, most of the mRNA localises to tumour infiltrating macrophages (Naylor *et al.* 1994). The findings in relation to 72kDa gelatinase A suggested a more complex interaction may be occurring in the neoplastic environment than simple upregulation in the neoplastic cells.

There is tight transcriptional control of MMP expression, which may be upregulated in certain cell lines by H ras. TGFβ will upregulate 72kDa gelatinase A in fibroblasts and the cytokines IL-1, TNFα and IFN-γ have also been shown to influence MMP synthesis (Lefebvre *et al.* 1990; Ito *et al.* FEBS 1990) and we have previously shown that ECM-associated TGFβ is abundant in the stroma of malignant tumours which show high stromal expression of 72kDa gelatinase A (Christie-Brown *et al.* 1993). Some highly metastatic melanoma cells overexpress the avβ3 vitronectin receptor, and treatment with antibodies to this integrin resulted in enhancement of invasive activity *in vitro* and increased expression of Gelatinase A (Seftor *et al.* 1992).

All MMPs are secreted as inactive proforms, which appear to fold via a sulphydryl side chain so that the metal atom containing the active site is not exposed, followed by autocatalytic removal of propeptide and once activated these enzymes are susceptible to inhibition by the tight non-covalently binding TIMPs (Docherty *et al.* 1992). The autocatalytic activation event does not occur unless the propeptide is peturbed and it is thought that *in vivo* this may arise from the propeptide being clipped by proteases such as plasmin. Gelatinase A is unlike the other enzymes in that its activation cannot be initiated by other proteases (Hipps *et al.* 1991). The only clue to its *in vivo* activation is the finding that it can occur by some membrane component of cells after stimulation with effectors such as concanavalin A, cytochalasin D, phorbol esters, TGFβ or by plating cells on collagen (Brown *et al.* 1990; Ward *et al.* 1992; Azzam and Thompson, 1992; Overall and Sodek, 1990).

One hypothesis which would explain the apparently paradoxical observations and which would explain controlled local action of 72kDa Gelatinase A at controlled local areas of the tumour membrane ECM boundary, would be cell surface binding and activation of 72kDa Gelatinase, and there is emerging evidence for this concept. There are as yet no reports of any membrane activation mechanism for Stromelysin 3 and no specfic inhibitor has been discovered, so its mode of action may be entirely different. It has been previously demonstrated that tumour cells *in vitro* may take up enzyme fragments *in vitro* (Brown *et al.* 1990), which may explain immunohistochemical observations. At the biochemical level, the purification of Gelatinase A from tumour cell lines has demonstrated its association with the plasma membrane, it has been shown that activation of Gelatinase A by cell membranes requires the enzyme to have an intact C-terminal domain (Murphy *et al.* 1992; Strongin *et al.* 1993). It is also known that TIMP-2 binds to this domain in the proenzyme (Fridman *et al.* 1992; Howard and Banda, 1991), and in doing so blocks activation by cell membranes (Murphy *et al.* 1992; Strongin *et al.* 1993). The binding of progelatinase to concanavalin A stimulated fibroblasts also requires an intact C-terminal domain and that binding can be blocked by TIMP-2 or competition with purified C-terminal domain. It is therefore possible that a role for TIMP-2 *in vivo* may be to prevent progelatinase A having access to sites of activation on the cell surface. We have shown tumour cell surface localisation using a monoclonal antibody to 72kDa gelatinase A (Riordan *et al.* 1993) and we also have preliminary evidence of tumour cell membrane association of 72kDa gelatinase by immunoelectron microscopy. Emonard *et al.* (1992) demonstrated high affinity binding sites for 72kDa gelatinase A on mammary carcinoma cell membranes and binding and activation of exogenous Gelatinase A by the membranes at specific areas of matrix infiltration ('invadopodia') in transformed fibroblasts (Monsky *et al.* 1993). More recently, we have preliminary data that a synthetic inhibitor of activated 72kDa gelatinase localises to the plasma membrane of human mammary carcinomas.

We therefore hypothesise that even if the 72kDa Gelatinase A is recruited from the surrounding stromal cells, its activation at sites on the tumour cell surface may give directionality to tumours using this enzyme in an invasive capacity. The ability to interfere with this process using synthetic inhibitors may offer an interesting alternative therapeutic approach which can potentially inhibit progression of human carcinomas.

REFERENCES

1. Albelda, S. M., Mette, S. A., Elder, D. E., *et al.* (1990) Integrin distribution in malignant melanoma: association of the $\beta 3$ subunit with tumor progression. *Cancer Res.*, **50**, 6757–6764.
2. Azzam, H. S. and Thompson, E. W. (1992) Collagen-induced activation of the Mr 72,000 type IV collagensase in normal and malignant fibroblastoid cells. *Cancer Res.*, 4540–4544.
3. Bao, L. and Tarin, D. (1993) Correlation of VLA-4 integrin expression with metastatic potential in various human tumour cell lines *J. Pathol.*, **167**, 150A.
4. Bartolazzi, A., Cerboni, C., Full, C., Valentini, C., Giorgio-Natali, P., Venturo,

I. and Bigotti, A. (1993) VLA-3 distribution in normal and neoplastic non-lymhoid human tissues. *Pathol. Res. Pract.*, **189**, 387–393.
5. Basset, P., Bellocq, J. P., Wolf, C., Stoll, I., Hutin, P., Limacher, J. M., Podhajcer, O. L., Chenard, M. P., Rio, M. C. and Chambon, P. (1990) A novel metalloproteinase gene specifically expressed in stromal cells of breast carcinomas. *Nature*, **348**, 699–704.
6. Berdichevsky, F., Gilbert, C., Shearer, M. and Taylor-Papadimitriou, J. (1992) Collagen-induced rapid morphogenesis of human mammary epithelial cells: The role of the alpha 2 beta 1 integrin. *J. Cell Sci.*, **102**, 437–446.
7. Brown, P. D., Levy, A. T., Margulies, I. M. K., Liotta, L. A. and Stetler-Stevenson, W. G. (1990) Independent expression and cellular processing of Mr 72, 000 type IV collagenase and interstitial collagenase in human tumorigenic cell lines. *Cancer Res.*, **50**, 6184–6191.
8. Chen, W. T. (1992) Membrane proteases: Roles in tissue remodelling and tumour invasion. *Curr. Op. Cell Biol.*, **4**, 802–809.
9. D'Ardenne, A. J., Richman, P. I., Horton, M. A., McAulay, A. E. and Jordan, S. (1992) Co-ordinate expression of the Alpha-6 Integrin Laminin receptor sub-unit and laminin in breast cancer. *J. Pathol.*, **165**, 213–220.
10. De Souza, B., Berdichevsky, F., Kyprianou, N. and Tayor-Papadimitriou, J. (1993) Collagen-induced morphogenesis and expression of the alpha 2 integrin subunit is inhibited in c-erb B2 transfected human mammary epithelial cells. *Oncogene*, **8**, 1797–1806.
11. Docherty, A. J. P., O'Connell, J., Crabbe, T., Angal, S. and Murphy, G. (1992) The matrix metalloproteinases and their natural inhibitors: Prospects for treating degenerative tissue disease. *Tibtech.*, **10**, 200–207.
12. Duffy, M. J. (1992) The role of proteolytic enzymes in cancer invasion and metastasis. *Clin. Exp. Metastasis*, **10**, 145–155.
13. Emonard, H. P., Remacle, A. G., Noel, A. C., Grimaud, J. A., Stetler-Stevenson, W. G. and Foidart, J. M. (1992) Tumor cell surface-associated receptor for the Mr 72,000 type IV collagenase. *Cancer Res.*, **52**, 5845–5848.
14. Fridman, R., Fuerst, T. R., Bird, R. E., Hoyhtya, M., Oelkuct, M., Kraus, S., Komarek, D., Liotta, L. A., Berman, M. L. and Stetler-Stevenson, W. G. (1992) Doman structure of human 72 kDa gelatinase type IV collagenase; characterisation of proteolytic activity and identification of the tissue inhibitor of metalloproteinase 2 (TIMP-2) binding regions. **267**, 15398–15405.
15. Frixen, U. H., Behrens, J., Sachs, M., Eberle, G., Voss, B., Warda, A., Lochner, D. and Birchmeier, W. (1991) E-cadherin-mediated cell-cell adhesion prevents invasiveness of human carcinoma cells. *J. Cell Biol.*, **113**, 173–185.
16. Goldberg, G. I. and Eisen, A. Z. (1992) Extracellular matrix metalloproteinases in tumor invasion and metastasis. In; Regulatory mechanisms in breast cancer ME Lippman and RB Dickson (Eds) Kluwer Academic Publishers, *Boston*, 421–440.
17. Hales, S. A., Stamp, G., Evans, M. and Fleming, K. A. (1989) Identification of the origin of cells in human basal cell carcinoma xenografts in mice using in situ hybridisation. *Br. J. Dermatol.*, **120**, 351–357.
18. Hanby, A. M., Gillett, C. E., Pignatelli, M. and Stamp, G. W. H. (1992). Demonstration of beta 1 and beta 4 integrin chain expression in ductal carcinoma in situ of the breast. *J. Pathol.*, (in press 1993).
19. Hart, I. R. and Saini, A. (1992) Biology of tumour metastasis. *Lancet.*, **339**, 1453–1457.

20. Hendrix, M. J. C., Seftor, E. A., Grogan, T. M., Seftor, R. E. B., Hersh, E. M., Boyse, E. A., Liotta, L. A., Stetler-Stevenson, W. G. and Ray, C. G. (1992) Expression of type IV collagenase correlates with the invasion of human lymphoblastoid cell lines and pathogenesis in SCID mice. *Mol. Cell Probes,* **6**, 59–65.
21. Hipps, D. S., Hembry, R. M., Docherty, A. J. P., Reynolds, J. J. and Murphy, G. (1991) Purification and characterisation of human 72 kDa gelatinase (Type IV collagenae) — Use of immunolocalisation to demonstrate the non-coordinate regulation of the 72 kDa and 92 kDa gelatinases by human fibroblasts. *Biol. Chem. Hoppe-Seyler*, **372**, 287–296.
22. Howard, E. W. and Banda, M. J. (1991) Binding of tissue inhibitor of metalloproteinases 2 to two distinct sites on human 72 kDa gelatinase. *Identification of a stabilisation site,* **266**, 17972–17977.
23. Jones, J. L., Critchley, D. R. and Walker, R. A. (1992) Alteration of stromal protein and integrin expression in breast – A marker of premalignant change? *J. Pathol.*, **167**, 399–406.
24. Koukoulis, G. K., Virtanen, I., Korhonen, M., Laitinen, L., Quaranta, V. and Gould, V. E. (1991) Immunohistochemical localisation of integrins in the normal, hyperplastic and neoplastic breast. *Am. J. Pathol.*, **139**, 787–799.
25. Levy, A. T., Cioce, V., Sobel, M. E., Garbisa, S., Grigioni, W. F., Litta, L. A. and Stetler-Stevenson, W. G. (1991) Increased expression of the Mr 72,000 type IV collagenase in human colonic adenocarcinoma. *Cancer Res.*, **51**, 439–444.
26. Liotta, L. A., Steeg, P. S. and Stetler-Stevenson, W. G. (1991) Cancer metastasis and angiogenesis; an imbalance of positive and negative regulation. *Cancer Res.*, **64**, 327–336.
27. Matrisian, L. M., McDonnell, S., Millar, D. B., Navre, M., Seftor, E. A. and Hendrix, M. J. C. (1991) *Am. J. Med. Sci.*, **302**, 157–162.
28. Matrisian, L. M. (1992) The matrix-degrading metalloproteinases. *Bioessays,* **14**, 455–463.
29. McDonnell, S. and Matrisian, L. M. (1990) Stromelysin in tumor progression and metastasis. *Cancer Metast. Rev.*, **9**, 305–319.
30. Meshter, C. L., Connolly, J. M. and Rose, D. P. (1992) Influence of regional location of the inoculation site and dietary fat on the pathology of MDA-MB 435 human breast cancer cell-derived tumors grown in nude mice. *Clin. Exp. Metast.*, **10**, 167–173.
31. Monsky, W. L., Kelly, T., Lin, C-Y., Yeh, Y., Stetler-Stevenson, W. G., Mueller, S. C. and Chen, W-T. (1993) Binding and localisation of Mr 72,000 Matrix metalloproteinase at cell surface invadopodia. *Cancer Res.*, **53**, 3159–3164.
32. Monteagudo, C., Merino, M. J., San-Juan, J., Liotta, L. A. and Stetler-Stevenson, W. G. (1990) Immunohistochemical distribution of type IV collagenase in normal, benign and malignant breast tissue. *Am. J. Pathol.*, **136**, 585–592.
33. Muller, D., Wolf, C., Abecassis, J., Millon, R., Engellmann, A., Bronner, G., Rouyer, N., Rio, M-C., Eber, M., Methlin, G., Chambon, P. and Basset, P. (1993) Increased stromelysin 3 gene expression is associated with increased local invasiveness in head and neck squamous carcinomas. *Cancer Res.,* **53**, 165–169.
34. Murphy, G., Houbrechts, A., Cockett, M. I., Williamson, R. A., O'Shea, M. and Docherty, A. J. P. (1991) The N-terminal domain of tissue inhibitor of metalloproteinases retains metalloproteinase inhibitory activity. *Biochemistry,* **30**, 8097–8102.

35. Murphy, G., Willenrock, F., Ward, R. V., Cockett, M. I., Eaton, D. and Docherty, A. J. P. (1992) The C-terminal domain of 72 kDa gelatinase A is not required for catalysis, but is essential for membrane activation and modulates interactions with tissue inhibitors of metalloproteinases. *Biochem. J.*, **283**, 637–641.
36. Murphy, G., Allan, J. A., Willenrock, F., Cockett, M. I., O'Connell, J. P. and Docherty, A. J. P. (1992) The role of the C-terminal domain in collagenase and stromelysin specificity. *J. Biol. Chem.*, **267**, 9612–9618.
37. Natali, P. G., Nicotra, M. R., Botti, C., Mottolese, M., Bigotti, A. and Segatto, O. (1992) Changes in expression of alpha 6 beta 4 integrin heterodimer in primary and metastatic breast cancer. *Br. J. Cancer*, **66**, 318–322.
38. Overall, C. M. and Sodek, J. (1990) Concanavilin A produces a matrix degradative phenotype in human fibroblasts. *J. Biol. Chem.*, **265**, 21141–21151.
39. Pignatelli, M. and Bodmer, W. F. (1988) Genetics and biochemistry of collagen binding-triggered glandular differentiation in a human colon carcinoma cell line. *Proc. Natl. Acad. Sci. USA*, **85**, 5561–5565.
40. Pignatelli, M. and Bodmer, W. F. (1989) Integrin-receptor-mediated differentiation and growth inhibition are enhanced by transforming growth factor-β in colorectal tumour cells grown in collagen gel. *Int. J. Cancer*, **44**, 518–523.
41. Pignatelli, M., Cardillo, M. R., Hanby, A. M. and Stamp, G. W. H. (1992) Expression of integrins and their accessory adhesion molecules in breast carcinomas. Loss of polarisation in poorly differentiated tumours. *Human Pathol.*, (in press).
42. Pignatelli, M., Hanby, A. M. and Stamp, G. W. H. (1991) Low expression of β_1, α_2 and α_3 subunits of VLA integrins in malignant mammary tumours. *J. Pathol.*, **165**, 25–32.
43. Pignatelli, M., Smith, M. E. F. and Bodmer, W. F. (1990) Low expression of collagen receptors in moderate and poorly differentiated colorectal adenocarcinomas. *Br. J. Cancer*, **61**, 636–638.
44. Poulsom, R., Pignatelli, M., Stetler-Stevenson, W., Liotta, L. A., Wright, P. A., Jeffery, R. E., Rogers, L. and Stamp, G. W. H. (1992) Stromal expression of 72 kDa Type IV collagenase (MMP-2) and TIMP2 mRNAs in colorectal neoplasia. *Am. J. Pathol.*, **108**, 81–90.
45. Poulsom, R., Hanby, A. M., Pignatelli, M., Jeffery, R. E., Longcroft, J. M., Rogers, L. M. and Stamp, G. W. H. (1993) Expression of gelatinase A and TIMP-2 mRNAs in desmoplastic fibroblasts in both mammary carcinomas and basal cell carcinomas of the skin. *J. Clin. Pathol.*, **46**, 429–436.
46. Price, J. E., Polyzos, A., Zhang, R. D. and Daniels, L. M. (1990) Tumorigenicity and Metastasis of human breast carcinoma cell lines in nude mice. *Cancer Res.*, **50**, 717–721.
47. Pyke, C., Ralfkiaer, E., Huhtala, P., Hurskainen, T., Dano, K. and Tryggvason, K. (1992) Localisation of messenger RNA for Mr 72,000 and 92,000 type IV collagenases in human skin cancers by in situ hybridisation. *Cancer Res.*, **52**, 1336–1341.
48. Riordan, F., Lalani, E-N., Cardillo, M., Baker, T., Docherty, A., Poulsom, R., Pignatelli, M. and Stamp, G. (1993) Expression of gelatinase A (72 kDa type IV collagenase), TIMP-2 and Stromelysin in ovarian carcinomas. *J. Pathol. Abstrt.*, in press.

49. Stamp, G. W. H. and Pignatelli, M. (1991) Distribution of β_1, α_1, α_2 and α_3 integrin chains in basal cell carcinomas. *J. Pathol.*, **163**, 307–313.
50. Strongin, A. Y., Marmer, B. L., Grant, G. A. and Goldberg, G. I. (1993) Plasma membrane-dependent activation of the 72 kDa type IV collagenase is prevented by complex formation with TIMP-2. *J. Biol. Chem.*, **268**, 14033–14039.
51. Takeichi, M. (1991). Cadherin cell adhesion receptors as a morphogenetic regulator. *Science*, **251**, 1451–1455.
52. Ward, R. V., Atkinson, S. J., Slocombe, P. M., Docherty, A. J. P., Reynolds, J. J. and Murphy, G. (1992) Tissue inhibitor of metalloproteinases 2 inhibits the activation of 72 kDa progelatinase by fibroblast membrances. *Biochim. Biophys. Acta.*, **1079**, 242–246.
53. Woessner, J. F. Jr. (1991) Matrix metalloproteinases and their inhibitors in connective tissue remodelling. *FASEB J.*, **5**, 2145–2154.
54. Wolf, C., Chenard, M-P., Durand de Grossouvre, P., Bellocq, J-P., Chambon, P. and Basset, P. (1993) The breast cancer-associated stromelysin 3 gene is expressed in basal cell carcinoma and during cutaneous wound healing. *Proc. Natl. Acad. Sci.* USA.
55. Zucker, S., Wieman, J. M., Lysik, R. M., Wilkie, D. P., Ramamurthy, N. S., Golub, L. M. and Lane B. (1987) Enrichment of collagen and gelatin degrading activities in the plasma membranes of human cancer cels. *Cancer Res.*, **47**, 1608–1614.
56. Zutter M. M., Mazoujian G. and Santoro S. A. (1990) Decreased expression of integrin adhesive protein receptors in adenocarcinoma of the breast. *Am. J. Pathol.*, **137**, 863–870.

3. FIBRINOLYSIS AND TISSUE DEGRADATION IN BREAST CANCER

MARIA COVIELLO, CARMELA RELLA, MICHELE QUARANTA, GIUSEPPINA MICELLI, *ANGELO NOLASCO and *FRANCESCO SCHITTULLI

*Laboratory of Analysis, *Surgical Senology Unit Oncology Institute, Bari, Italy*

Malignant tissues exhibit fibrinolytic activity, which is known to influence tumour growth and metastasis. The plasma plasminogen activator (PA) and inhibitor (PAI) antigen levels were evaluated in 80 patients with breast cancer, 40 patients with benign disease and 25 healthy controls. The t-PA, u-PA, PAI-1 and PAI-2 levels were determined using immunoenzymatic assays. In plasma of benign disease patients, a significant increase was observed only for PAI-2. The mean t-PA plasma level in primary and metastatic breast cancers was 8.8 ± 3.0 ng/ml and 15.4 ± 7.4 ng/ml, respectively. The mean u-PA plasma level was 0.5 ± 0.6 ng/ml in the former group and 0.7 ± 0.5 ng/ml in the latter. The mean PAI-1 and PAI-2 levels in primary carcinoma patients were 30.8 ± 16.5 ng/ml and 0.6 ± 0.1 ng/ml, respectively, and 40.7 ± 16.2 ng/ml and 0.7 ± 1.2 ng/ml, respectively, in plasma of metastatic patients. An increase of u-PA levels was noted in patients with metastatic lymph nodes. It is *possible* that the patients with elevated plasminogen activator and inhibitor antigen levels represent a group with a particularly dismal prognosis, presuming the involvement of plasminogen activation in tissue degradation and invasion. Therefore, the measurement of these markers in plasma might be of clinical value, but further studies are needed.

INTRODUCTION

Proteolytic degradation of the extracellular matrix is considered an early and crucial feature of the metastatic process (Liotta *et al.* 1981; Tryggvason *et al.* 1987; Schmitt *et al.* 1992). An important and highly regulated cascade of these proteolytic events involves the plasminogen activation system (Duffy, 1992; Quax *et al.* 1992). Plasminogen activators are serine proteases which catalyze the conversion of the zymogen plasminogen to the active plasmin, another serine protease of broad-spectrum specificity. On the basis of their immunological cross-reactivity and fibrin requirement, the plasminogen activators (PAs) can be classified into two forms: tissue-type plasminogen activators (t-PA) and urokinase-type (u-PA). The t-PA is largely dependent on the binding to fibrin for its enzymatic activity to the contrary of u-PA (Camiolo *et al.* 1971). The two forms have been shown to be products of separate genes (Tripputi *et al.* 1986; Tripputi *et al.* 1985) and their synthesis is regulated independently (Rajput *et al.* 1985). Of these two types of plasminogen activators, the u-PA has been implicated in both normal and pathological processes of tissue remodelling, organ involution and invasion whereas the t-PA plays an important role in fibrinolysis and thrombolysis (Dano *et al.* 1985; Saksela and Rifkin, 1988; Duffy *et al.* 1992).

Regulation of these proteases is achieved at many levels, including inhibition by at least two types of fast-acting specific inhibitors, PAI-1 and

PAI-2. These inhibitors are products of different genes and have different biochemical properties (Kruithof, 1988). PAI-1 is secreted by endothelial cells and vascular smooth-muscle cells, and is also detected in plasma and platelets. It forms a 1:1 covalent complex with both t-PA and u-PA, but is thought to be a natural inhibitor of t-PA in plasma (Kruithof *et al.* 1984; Loskutoff *et al.* 1989). PAI-1 is secreted as an active antiprotease and it is often found associated with plasma and extracellular matrix vitronectin which stabilizes it in the active conformation (Seiffert *et al.* 1990). The precise biological role of PAI-2 is unknown; it is present in monocytes, placenta and plasma of prepartum pregnant women and therefore is thought to be involved in the control of local cell-mediated proteolysis during remodelling events (Kopitar *et al.* 1985). In kinetic studies, PAI-2 demonstrated a stronger inhibitory action towards u-PA than for t-PA (Asted *et al.* 1987).

Considerable evidence suggests that PAs are active in the development and diffusion of tumours (Dano *et al.* 1985; Dano *et al.* 1988; Markus, 1988), a mechanism which probably involves the degradation of matrix and basement membrane proteins by plasmin or plasmin-activated proteases (Liotta *et al.* 1981). Plasmin degrades proteins of the extracellular tumour stroma (fibrin-fibronectin matrix) and proteins of the basement membrane (Reich *et al.* 1988). This tumour-associated proteolysis provides the basis for tumour cell invasion and facilitates the release of tumour cells, leading to metastasis. Levels of PA activity are higher in primary tumours, including colon and breast carcinomas, than in corresponding normal tissues (Nagy *et al.* 1977; Kirchleimer *et al.* 1985; De Bruin *et al.* 1988; Evers *et al.* 1982; Duffy *et al.* 1988). Within the circulatory system, tumour cells are often covered with a fibrin coat (Evers *et al.* 1982; Duffy *et al.* 1988) which protects against the immune response, thereby aiding in the process of metastasis. Obviously, PAs and their inhibitors might be involved in the formation and breakdown of this fibrin coat.

In the present study we measured the plasma levels of t-PA, u-PA, PAI-1 and PAI-2 in breast cancer patients, in subjects with benign disease and in healthy female controls, in order to evaluate their clinical use as markers of metastatic activity.

MATERIALS AND METHODS

Patients

Eighty patients with breast cancer (median age: 60 years) and forty patients with benign disease (median age: 37 years) were admitted to the study. Of the neoplastic group, 64 had primary breast cancer and 16 patients had recurrent, diffuse metastases. The latter metastatic subgroup was subjected to radical mastectomy between 1987–1989 and were treated with one or more of the following drugs: cyclophosfamide, methotrexate, 5-fluorouracil, doxorubicin and novantrone.

Surgical tumour samples from the primary breast cancers were examined by a pathologist and evaluated according to the WHO classification: invasive intraductal carcinomas were found in 51/64 (80%) of patients, invasive lobular carcinoma in 7/64 (11%), tubular carcinoma in 6/64 (9%). Lymph node involvement was determined in 47 patients by axillary nodal dissection: 22/47 (47%) patients were lymph node negative whereas 25/47 (53%) patients were lymph node positive.

The t-PA, u-PA, PAI-1 and PAI-2 plasma levels were evaluated in both neoplastic and benign patient subgroups and also in 25 healthy female donors (median age: 40 years) as controls.

Determination of Plasminogen Activator and Inhibitor Antigens

The human plasma was obtained with a Vacutainer system using tubes containing 0.13 M sodium citrate. The tubes were centrifuged at 2000 g for 10 min. The plasma was stored at − 20° C until assays were performed. The t-PA, u-PA, PAI-1 and PAI-2 antigen levels were determined with the commercially available ELISA kits (ELISA for t-PA and ELISA for PAI-1, American Diagnostica Inc., Ortho, Italy; TintElize u-PA and IMULYZE PAI-1, Biopool Sweden, Menarini Italy) which utilize the double antibody principle. Antigen concentrations were expressed as ng/ml of plasma.

Statistical Analysis

The Mann-Whitney test was used for the comparison of mean plasma levels of the proteins studied in benign disease patients and in breast cancer patients compared to controls.

Table 3.1 Plasminogen Activator and Inhibitor Antigen Plasma Levels of Patients and Controls

	*t-PA** *mean ± SD*	*u-PA** *mean ± SD*	*PAI-1** *mean ± SD*	*PAI-2** *mean ± SD*
controls (25**)	4.1 ± 1.2	0.3 ± 0.1	12.8 ± 5.5	0
benign disease (45)	5.7 ± 2.7	0.3 ± 0.2	21.7 ± 9.9	0.4 ± 0.1
primary cancers (80)	10.2 ± 4.7	0.5 ± 0.6	30.8 ± 16.5	0.6 ± 1.0
metastatic cancer (16)	14.5 ± 6.9	0.7 ± 0.5	40.7 ± 16.2	0.7 ± 1.2

*ng/ml
**number of patients in parentheses

RESULTS

Table 3.1 demonstrates the results of plasminogen activator and inhibitor antigen plasma levels. The mean concentration of t-PA, u-PA, and PAI-1 in plasma of patients with benign desease was not significantly increased compared to controls (p = NS). A significant increase was observed only for PAI-2 (0.4 ± 0.1 ng/ml; p = 0.001).

The increase of all plasma antigen levels in the primary breast cancer group and the group with diffuse metastases, however, was statistically significant. In fact, mean t-PA plasma levels in primary and metastatic breast cancers were 8.8 ± 3.0 ng/ml and 15.4 ± 7.4 ng/ml, respectively ($p < 0.0001$); the mean u-PA plasma levels were 0.5 ± 0.6 ng/ml ($p < 0.01$) in the first group and 0.7 ± 0.5 ($p < 0.001$) in the latter group. Moreover, in plasma of patients with primary carcinoma, the mean PAI-1 and PAI-2 levels were 30.8 ± 16.5 ng/ml and 0.6 ± 0.1 ng/ml, respectively, ($p < 0.0001$), and 40.7 ± 16.2 ng/ml and 0.7 ± 1.2, respectively in plasma of patients with metastases ($p < 0.00001$).

When t-PA, u-PA, PAI-1 and PAI-2 concentrations were evaluated with respect to lymph node involvement, no statistically significant correlation was found (p = NS). Nevertheless, we observed an increase in the u-PA plasma levels in lymph node positive patients compared to lymph node negative patients (0.48 ± 0.41 ng/ml versus 0.26 ± 0.2 ng/ml). (Table 3.2).

DISCUSSION

Many studies suggest that the conversion of circulating plasminogen to active plasmin is central to the process of extracellular matrix degradation associated with tissue proliferation and generation of metastases (Dano *et al.* 1985; Mullins and Rohrlich 1983; Markus, 1988). In fact, the formation of metastases is a complex sequence of steps that may involve the presence of fibrin and the processes of blood coagulation and fibrinolysis (Gilbert and Gordon, 1983; Duffy, 1990a). The high PA activity might participate in the destruction of the peritumoral tissue, thus allowing for invasion by tumour cells, whereas its lack in the plasma of cancer patients would increase plasma

Table 3.2 Plasminogen Activator and Inhibitor Plasma Antigen Levels with Reference to Lymph Node Status

	Patients	***t-PA**** *mean ± SD*	***u-PA**** *mean ± SD*	***PAI-1**** *mean ± SD*	***PAI-2**** *mean ± SD*
N –	22	8.3 ± 2.9	0.26 ± 0.2	25.6 ± 14.3	0.5 ± 0.9
N +	24	9.6 ± 3.1	0.48 ± 0.41	27.0 ± 15.3	0.5 ± 1.1

*ng/ml

fibrin and fibrin deposits in the tumour stroma, favouring the formation of metastases.

Tumour presence has been associated with the increase of t-PA plasma levels (Grondahl-Hansen *et al.* 1990); also elevated u-PA plasma levels are found in patients with advanced disease for breast (Grondahl-Hansen *et al.* 1988), prostate (Hienert *et al.* 1988) and gastrointestinal cancers (Kirchleimer *et al.* 1987). Moreover, high PAI-1 levels have often been observed in neoplastic patients (Jahan-Vague, 1985). The cause of these elevated levels has yet to be explained.

The t-PA level can be varied by certain stimuli, such as physical activity, venous occlusion, stress, and various drugs (Kluft, 1985). PAI-1, on the contrary, responds differently to various stimuli; elevated PAI-1 levels have been found in traumatized patients or those with thrombotic disorders. A stimulus for the production of PAI-1 has also been observed in the post-operative period. Furthermore, extracellular u-PA activity may be regulated by some products of tumour cells (growth factors, oncogenes, *etc.*) which can alter its transcription or secretion rate (Testa and Quigley, 1990). PAI-2 is present in plasma only when tissue remodelling events are present (Kopitar *et al.* 1985).

Until now, only limited specific data are available concerning plasminogen activators and plasminogen activator inhibitors in the systemic circulation of neoplastic patients. The present study aims to study both plasminogen activators and inhibitors in order to establish their usefulness as tumour markers.

Our data demonstrate that breast cancer patients show a significant increase of t-PA, u-PA, PAI-1 and PAI-2 plasma levels. Since laboratory plasma tests were carried out only a few days before surgery in our study, it is obvious that this increase cannot be due to a reaction to the acute post-operative phase, but rather to an interaction with the host tumour. The increase observed was higher in patients with diffused metastases, probably due to the result of two phenomena: the presence of the tumour which in itself alters the fibrinolytic components, and the effect of the chemotherapy to which these patients had been subjected. In fact, it is known that chemotherapy influences the fibrinolytic system and exposes patients to the risk of thrombotic disorders (Levine, 1988).

However, significant differences in the t-PA, PAI-1 and PAI-2 plasma levels were not observed in patients with lymph node involvement (N +) with respect to those without lymph node involvement (N –). Only an increase of u-PA levels was observed in patients with axilliary lymph node metastases, in accordance with previous studies (Duffy *et al.* 1990b; Janicke *et al.* 1990) which showed that u-PA antigen levels evaluated in tumour extracts are a strong and independent marker in breast cancer.

In conclusion, this study documented the preoperative alterations of t-PA, u-PA, PAI-1 and PAI-2 plasma levels in breast cancer patients. These alterations may be the result of a tumour-host interaction and therefore must be related to certain biological characteristics of the tumour. Because of the supposed involvement of plasminogen activation in tissue degradation and invasion, it is possible that breast cancer patients with elevated t-PA, u-PA, PAI-1 and PAI-2 plasma levels represent a group of patients with a particularly poor prognosis, and that the measurement of these marker concentrations in plasma might therefore have a real clinical value. Clarification of this point requires larger prospective studies.

Acknowledgements

The authors wish to thank Ms Paulene Maselli Campagna for her help in the preparation of the manuscript.

REFERENCES

1. Astedt, B., Lecander, I. and Ny, T. (1987) The placental type plasminogen activator inhibitor, PAI-2. *Fibrinolysis,* **1**, 203–208.
2. Camiolo, S.M., Thorsen, S. and Astrup, T. (1971) Fibrinogenolysis and fibrinolysis with tissue plasminogen activator, urokinase, streptokinase-activated human globulin, and plasmin. *Proc. Soc. Exp. Biol. Med.*, **138**, 277–280.
3. Dano, K., Andreasen, P. A., Groendahl-Hansen, J., Kristensen, P., Nielsen, L. S. and Skriver, L. (1985) Plasminogen activators, tissue degradation, and cancer. *Adv. Cancer Res.*, **44**, 139–264.
4. Dano, K., Nielsen, L. S., Pyke, C. and Kellerman, G. M. (1988) Plasminogen activators and neoplasia. In Tissue-type plasminogen activator (t-PA): Physiological and Clinical Aspects, (edited by Kluft, C.) CRC Press Inc. Boca Raton, Florida, 19–46.
5. De Bruin, P. A. F., Griffioen, G., Verspaget, H. W., Verheijen, J. H., Dooijewaard, G., van den Ingh, H. F. and Lamers, C. B. H. W. (1988) Plasminogen activator profiles in neoplastic tissues of the human colon. *Cancer Res.*, **48**, 4520–4524.
6. Duffy, M. J., O'Grady, P., Devaney, D., O'Sioroin, L., Fennelly, J. S. and Lijnen, H. R. (1988) Tissue-type plasminogen activator, a new prognostic marker in breast cancer. *Cancer Res.*, **48**, 1348–1349.
7. Duffy, M. J. (1990a) Plasminogen activators and cancer. *Blood Coagul. Fibrin.*, **1**, 681–687.
8. Duffy, M. J., Reilly, D., O'Sullivan, C., O'Higgins, N., Fennelly, J. J. and Andreasen, P. (1990b) Urokinase-plasminogen activator, a new and independent prognostic marker in breast cancer. *Cancer Res.*, **50**, 6827–6829.
9. Duffy, M. J. (1992) The role of proteolityc enzymes in cancer invasion and metastasis. *Clin. Exp. Metastasis,* **10**, 145–155.
10. Evers, J. L., Patel, J., Madeja, J. M., Schneider, S. L., Hobika, G. H., Camiolo, S. M. and Markus, G. (1982) Plasminogen activator activity and composition in human breast cancer. *Cancer Res.*, **42**, 219–226.
11. Gilbert, L. C. and Gordon, S. G. (1983) Relationship between cellular procoagulant activity and metastatic capacity of B16 mouse melanoma variants. *Cancer Res.*, **43**, 536–540.

12. Grondahl-Hansen, J., Agerlin, N. and Munkholm-Larsen, P. (1988) Sensitive and specific enzyme-linked immunosorbent assay for urokinase-type plasminogen activator and its application to plasma from breast cancer patients. *J. Lab. Cli. Med.*, **111**, 42–53.
13. Grondahl-Hansen, J., Bach, F. and Munkholm-Larsen, P. (1990) Tissue-type plasminogen activator in plasma for breast cancer patients determined by enzyme-linked immunosorbent assay. *Br. J. Cancer,* **61**, 412–414.
14. Hienert, G., Kirchleimer, J. C. and Pfluger, H. (1988) Urokinase-type plasminogen activator as marker for the formation of distant metastases in prostatic carcinomas. *J. Urol.*, **140**, 1466–1469.
15. Janicke, F., Schmitt, M., Hafter, R., Hollrieder, A., Babic, R., Ulm, K., Gossner, W. and Graeff, H. (1990) Urokinase-type plasminogen activator (u-PA) antigen is a predictor of early relapse in breast cancer. *Fibrinolysis,* **4**, 69–78.
16. Juhan-Vague, I., Moerman, B., De Cock, F., Aillaud, M. F. and Collen, D. (1985) Plasma levels of a specific inhibitor of tissue-type plasminogen activator (and urokinase) in normal and pathological conditions. *Thromb. Res.*, **33**, 523–530.
17. Kirchleimer, J. C., Pfuger, H., Ritschl, P., Hienert, G. and Binder, B. R. (1985) Plasminogen activator activity in bone metastases of prostatic carcinomas as compared to primary tumors. *Invasion Metastasis,* **5**, 344–355.
18. Kirchleimer, J. C., Hubert, K. and Wagner, O. (1987) Pattern of fibrinolytic parameters in patients with gastrointestinal carcinomas. *Br. J. Haematol.*, **66**, 85–89.
19. Kluft, C., Verheijen, J. H., Jie, A. F. T., Rijken, D. C., Preston, F. E., Sue-Ling, H. M., Jespersen, H. and Aasen, A.O. (1985) The postoperative fibrolytic shutdown: a rapidly reverting acute phase pattern for the fast-acting inhibitor of tissue-type plasminogen activator after trauma. *Scand. J. Clin. Lab. Invest.*, **45**, 605–610.
20. Kopitar, M., Rozman, B., Babnik, J., Turk, V., Mullins, D. E. and Wun, T. C. (1985) Human leukocyte urokinase inhibitor. Purification, characterization and comparative studies against different plasminogen activators. *Thromb. Haemostas.*, **54**, 750–755.
21. Kruithof, E. K. O. (1988) Inhibitors of plasminogen activators. In Tissue-type plasminogen activator (t-PA): Physiological and Clinical Aspects, vol. 1, edited by Kluft, C. CRC Press Inc, Boca Raton, Florida, 190–210.
22. Kruithof, E. K. O., Tran-Thang, C., Ransijn, A. and Bachmann, F. (1984) Demonstration of fast-acting inhibitor of plasminogen activator in plasma. *Blood*, **64**, 907–913.
23. Levine, M. N., Gent, M., Hirsh, J., Arnold, A., Goodyear, M. D., Hrymink, W. and De Pauw, S. (1988) The thrombogenic effect of anticancer drug therapy in women with stage II breast cancer. *New England J. Med.*, **318**, 404–407.
24. Liotta, L. A., Goldfarb, R. H. and Brundage, R. (1981) Effect of plasminogen activator (urokinase), plasmin and thrombin on glycoprotein and collagenous component of basement membrane. *Cancer Res.*, **41**, 4629–4636.
25. Loskutoff D. J., Sawdey, M. and Mimuro, J. (1989). Type 1 plasminogen activator inhibitor. *Plog. Haemostas. Thromb.*, **9**, 87–115
26. Markus, G. (1988) The relevance of plasminogen activators to neoplastic growth. *Enzyme*, **40**, 158–172.
27. Mullins, D. E. and Rohrlich, S. T. (1983) The role of proteinases in cellular invasiveness. *Biochem. Biophys. Acta*, **695**, 177–214.

28. Nagy, B., Ban, J. and Nindar, B. (1977) Fibrinolysis associated with human neoplasia: production of plasminogen activators by human tumors. *Int. J. Cancer*, **19**, 614–620.
29. Ogawa, H., Sekigushi, F., Tanaka, N., Ono, K., Tanaka, K., Kinjo, M., Iwakawa, A. and Naito, S. (1982) Effect of antifibronolysis treatment on human cancer in nude mice. *Anticancer Res.*, **2**, 339–344.
30. Quax, P. H. A., van Muijen, G. N. P., Pedersen, N., de Bart, A. C. W., Schrauwen, Y., Barlovatz-Meimon, G., Blasi, F. and Verheijen, J. H. (1992) The plasminogen activator system in extracellular matrix degradation. *Fibrinolysis*, **6(4)**, 41–44.
31. Rajput, B., Degen, S. F., Reich, E., Waller, E. K., Axelrod, J., Eddy, R. L. and Shows, I. B. (1985) Chromosomal locations of human tissue plasminogen activator and urokinase genes. *Science*, **230**, 672–674.
32. Reich, R., Thompson, E. W., Iwamoto, Y., Martin, G. R., Deason, J. R., Fuller, G. C. and Miskin, R. (1988) Effects of inhibitors of plasminogen activator, serine proteinases, and collagenase IV on the invasion of basement membranes by metastatic cells. *Cancer Res.*, **48**, 3307–3312.
33. Saksela, O. and Rifikin, D. B. (1988) Cell-associated plasminogen activation: regulation and physiological functions. *Annu: Rev.-Cell-Biol.* **4**, 93–126.
34. Seiffert, D., Mimuro, J., Schleef, R.R. and Loskutoff, D. J. (1990) Interactions between type 1 plasminogen activator inhibitor, extracellular matrix and vitronectin. *Cell Differ. Der.*, **32**, 287–292.
35. Sloan, I. G. and Firkin, B. G. (1989) Impaired fibrinolysis in patients with thrombotic or haemostatic defects. *Thromb. Res.*, **55**, 559–567.
36. Schmitt, M., Janicke, F. and Graeff, H. (1992) Tumor associated proteases. *Fibrinolysis*, **6(4)**, 3–26.
37. Sprengers, E. D. and Kluft, C. (1987) Plasminogen activator inhibitors. *Blood*, **69**, 381–387.
38. Testa, J. E. and Quigley, J. P. (1990) The role of urokinase-type plasminogen activator in aggressive tumor cell behavior. *Cancer Metastas. Rev.*, **9**, 353–367.
39. Tryggvason, K., Hoyhtya, M. and Salo, T. (1987) Proteolytic degradation of extracellular matrix in tumor invasion. *Biochim. Biophys. Acta*, **907**, 191–217.
40. Tripputi, P., Blasi, F., Ny, T., Emanuel, B. S., Letofsky, J. and Croce, C. (1986) Tissue type plasminogen activator gene is on chromosome 8. *Cytocenet. Cell Genet.*, **42**, 24–28.
41. Tripputi, P., Blasi, F., Verde, P., Camizzaro, L. A., Emanuel, B. S. and Croce, C. (1985) Human urokinase gene is located on the long arm of chromosome 10. *Proc. Natl. Acad. Sci. USA*, **82**, 4448–4452.
42. Zacharski, L. R., Memoli, V. A. and Rousseau, S. M. (1986) Coagulation-cancer interaction *in situ* in renal cell carcinoma. *Blood*, **68**, 394–399.

4. $\alpha 6/\beta 4$ INTEGRIN IN LUNG CARCINOMA

RITA FALCIONI, LETIZIA CIMINO, DANILO PERROTTI, MARIA GIULIA RIZZO and ADA SACCHI

Laboratorio Oncogenesi Molecolare, Istituto Regina Elena per lo studio e la cura dei tumori, C.R.S. Via delle Messi d'Oro 156, 00158 Roma, Italia

We have studied the expression of $\alpha 6\beta 4$ integrin in human and murine lung carcinomas both *in vitro* and *in vivo* using monoclonal antibodies and cDNA probes. Immunoprecipitation patterns and Northern blot analysis demonstrate that in some murine variants derived from Lewis lung carcinoma the $\beta 4$ subunit is expressed in highly metastatic tumour cells and is absent in low metastatic ones. Previously, by immunochemical studies and immunohistochemistry we demonstrated that the $\beta 4$ subunit is differentially expressed in human carcinomas, being highly expressed in squamous lung carcinomas and undetectable in small cell lung cancers (SCLC). Further studies demonstrated that in SCLC cells the $\alpha 6$ subunit associates only with the $\beta 1$ subunit, while in other tumour cell lines (NSCLC) the $\alpha 6$ associates with both $\beta 1$ and $\beta 4$ subunits. Northern blot analysis confirmed that the mRNA coding for the entire $\beta 4$ subunit (7kb) is undetectable in SCLC cell line but is present in NSCLC cell lines.

The adhesion assays performed on different cell lines demonstrated that lung carcinoma cells which express on their surface the highest levels of $\beta 4$ protein exhibit lower capacity to bind ECM components. Functional studies indicate that laminin adhesion is inhibited at high extent in all lung carcinoma cells (SCLC and NSCLC) by Mabs to the $\alpha 6$ and $\beta 1$ subunits. But, Mabs to the $\beta 4$ subunit inhibit laminin adhesion uniquely in adenocarcinoma cells. In view of the fact that large cell carcinoma cells also express on their surface the $\alpha 6\beta 4$ complex, it is suggested that this receptor has different functions in lung carcinoma cells of different histotype.

In conclusion, data further confirm that the $\alpha 6\beta 4$ integrin might be a good marker of tumour progression in lung carcinomas, and indicate that the evaluation of its expression might be a useful tool in the diagnosis of human lung cancers.

INTRODUCTION

Recently, to define the molecular basis of the alterations in adhesive properties and cytoskeletal organization observed in tumour cells,[1,2] the expression of specific members of the integrin family[3] have been investigated. It has been suggested that the repertoire of integrins on a given cell type may depend on its histological origin and degree of differentiation.[1]

A new member of the integrin superfamily of adhesion receptors was isolated from human epithelial cells.[4] With analogy to other integrins, this molecular complex is a heterodimer composed of structurally unrelated subunits $\alpha 6$ and $\beta 4$. It has been recently reported that the $\alpha 6/\beta 4$ integrin complex is highly expressed in mouse mammary tumour cells and in human carcinoma cell lines.[5–7] Other studies[8] have revealed that Mabs to this complex strongly stained basement membranes from normal epithelia as well as a variety of carcinomas and carcinoma cell lines.[4,5] We

Corresponding. Dr. Ada Sacchi, Laboratorio Oncogenesi Molecolare, Regina Elena Cancer Institute, C.R.S. Via delle Messi D'Oro 156, 00158 Roma, Italia
Tel: 39 6 4985 529; Fax: 39 6 4180473

have also reported that the $\alpha6\beta4$ integrin is expressed in human and murine[9] lung carcinomas (cell lines and solid tumours) and that its expression increases in highly metastatic murine tumour cells.[10] Since human lung carcinomas comprise histotype which may diverge in features such as intermediate filament expression[11] and antigenic phenotype,[12,13] it is conceivable that the expression of cell surface receptors for extracellular matrix proteins in these tumours might vary. In this respect, the expression of the $\alpha6\beta4$ integrin was investigated and, recently, we reported that this integrin is differentially expressed in non-small (NSCLC) versus small cell lung cancers (SCLC).[14] In this paper by using Mabs and cDNA probes to the $\alpha6$ and the $\beta4$ integrin subunits we evaluated the $\alpha6\beta4$ integrin expression in lung carcinoma cells of different origin. The adhesion capacity of tumour cells was studied and correlated with the $\alpha6\beta4$ expression.

MATERIALS AND METHODS

Tumour Lines

In vitro (C87 and BC215) and *in vivo* (M1087 and BM21548) variants derived from Lewis lung carcinoma (3LL) were maintained in culture or in syngeneic animals (C57BI/6J mice) according to previously reported methods.[15,16] Cell suspensions of cultured cells, harvested by trypsinization from log-phase cultures, were injected into syngeneic mice to analyze *in vivo* growth. Spontaneous and artificial metastases were evaluated in C57BI/6J mice injected *i.m.* or *i.v.*, respectively, with 2.5×10^5 viable cells. Tumour bearing mice were sacrified 19 or 24 days after injections and lungs removed and fixed in Bouin's solution. Lung nodules were counted with the aid of a dissecting microscope. *In vivo* tumour lines were passaged in 2–3-month-old male C57BI/6J mice. Each mouse was injected in the leg muscle with 0.1 ml of a suspension containing 2.5×10^5 viable cells.

The human lung adenocarcinoma (acinar pattern) A549,[17] was cultured in McCoy's 5A medium with 10% fetal bovine serum. The large cell carcinoma DG3 and the small cell carcinoma AE2 lines were obtained from Dr. G. Zupi, (Laboratorio Chemioterapia Preclinica, Istituto Regina Elena, Rome, Italy), and were cultured according to previously described methods.[18]

Antibodies and cDNA probes

Mabs 135-13C and 346-11A, respectively to mouse $\alpha6$ and $\beta4$ integrin subunits, and 439-9B to human $\beta4$ integrin subunit were prepared and purified from ascitic fluid as described.[9,10] Mab 135-13C identifies a phylogenetically conserved epitope which is also present in the human $\alpha6$ integrin subunit. Mab GoH3 (CDw49f-VLA6) to human $\alpha6$ integrin subunit[6] was purchased from Immunotech S. A. (Marseille Cedex, France); Mab 4B4

(CDw29) to human β1 integrin subunit[19] was purchased from Coulter Co. (Hialeah, FL., USA); Mab 3E1 to human β4 integrin subunit[20] was purchased from Telios Pharmaceuticals Inc. (San Diego, CA, USA); Mab A1A5 to human β1 integrin subunit[21] was kindly provided by M. E. Hemler (Dana Farber Cancer Institute, Harvard Medical School, Boston, MA, USA).

The BP 9.3 cDNA clone (1.7 kb) coding for the cytoplasmic domain of the human β4 integrin subunit[22] was kindly provided by M. E. Hemler; the pHFBA-1 plasmid containing a cDNA clone coding for chicken β-actin protein was kindly provided by Prof. Renato Baserga (Jefferson Cancer Center, Philadelphia, PA, USA); the cDNA clone coding for the human β1 integrin subunit was purchased from Telios Pharmaceuticals; the cDNA clone 3AB (3 kb) coding for the cytoplasmic and transmembrane domains of the β4 murine integrin subunit was previously described.[23]

Two site assay

The two site assay has been described in detail.[9] Briefly purified Mab 135-13C was incubated (1 μg/ml) in Immulon 2 Removawell strips (Dynatech Laboratories, Inc. Chantilly, France) overnight at 4°C with gentle mixing, and further incubated with 500 μg of BSA diluted in PBS (100 μl/well) for 1 h at 4°C with gentle mixing to saturate binding sites. Unbound material was aspirated and 50 μl of serial dilutions of homogenates from cell lines or tissues to be analyzed, were added for 2 h at 4°C. Homogenates were diluted in PBS with BSA (5 μg/ml) and NP-40 0.05%. Wells were washed in PBS before adding 50 ng of ^{125}I Mab 346-11A diluted in PBS with BSA (5 μg/ml), NP-40 0.05% and 1% normal rat serum for 2 h at 4°C. Cells were washed in PBS, transferred to 12 × 75 mm glass tubes and radioactivity determined in a gamma counter. Binding results were calculated as the average of duplicate samples.

Radiolabelling of the Cells and Immunoprecipitation

Cells were surface labelled with carrier-free ^{125}I using lactoperoxidase, as previously described.[10] After labelling, cells were washed with PBS and solubilized in lysis buffer containing 5 μg/ml BSA, 0.5% NP-40, 1 μM NaN3, 10 μg/ml phenylmethylsulfonylfluoride (Sigma Chemical Co., St. Louis, MO, USA), and 100 μg/ml aprotinin (Sigma). The lysates were clarified by centrifugation (30,000 × g, 1 h, 4°C) and solubilized proteins were directly immunoprecipitated as described.[9]

RNA Extraction and Hybridization

Total RNA was extracted by the guanidinium thyocyanate method[24] from cultured cells or tumours grown under the indicated conditions. The cells were harvested from log-phase cultures and tumours were excised on day 14 postimplant. Total RNA (20 μg/lane) was fractionated on formaldehyde-agarose gels, transferred to nitrocellulose, and hybridized as described[25] with ^{32}P-nick-translated gel purified fragments of DNA probes (2×10^8 c.p.m./μg). Transcript size was determined relative to 18S and 28S markers.

Cell Adhesion and Inhibition Assay

Well in flat-bottom 96-well plates (Constar Corp., Cambridge, MA, USA) were precoated with BSA or one of the four ECM proteins: type I collagen (Sigma), fibronectin (Telios), laminin (Telios). Positive control wells were precoated with PLL (30 μg/ml) in PBS. To each well 6×10^4 – 1×10^5 cells were added as previously described.[26] Following incubation of the cells for 1 h at 37°C, the wells were washed to remove non adherent cells. The remaining adherent cells were stained with crystal violet as previously reported.[26] The stained cells were quantitated following cell lysis with 2% SDS by measuring the absorbance at 600 nm in a microtiter plate reader. To assess the effect of anti-integrin Mabs on cell-substrate adhesion, the cells were pretreated for 1 h at 4°C with saturating concentrations of different Mabs. Following preincubation with antibodies, the cells were added directly to the coated wells and the adhesion assay was carried out as described above.

RESULTS

A tumour associated glycoprotein called TSP-180, was initially identified on the surface of several murine lung carcinomas[27] by using a rat antiserum. Further experiments indicated a preferential expression of this glycoprotein in carcinoma and a correlation was found between the expression of the TSP-180 protein complex and the malignant potential of some tumour types in mice.[10] Recently, the identification of the TSP-180 protein complex with the $\alpha6\beta4$ integrin has been demonstrated.[5] Studying metastatic phenotype of 3LL carcinoma variants we demonstrated that the novel integrin $\alpha6\beta4$ is specifically expressed in highly metastatic tumour cells, whereas the $\alpha6\beta1$ laminin receptor is expressed in all 3LL variants.[23] Since integrins are involved in the metastatic process, because of their role in cell-cell recognition and cell-matrix interactions, we further extended our studies to evaluate the expression of $\alpha6\beta4$ integrin in 3LL metastatic variants. By the use of Mabs to $\alpha6$ (135-13C) and $\beta4$ (346-11A) subunits we quantitated the amount of the $\alpha6\beta4$ integrin in 3LL variants. Using a two site assay, we demonstrated that highly metastatic 3LL cells (M1087 and C87) express the highest amounts of the $\alpha6\beta4$ complex per mg of the total protein as compared with the negative (CCL) control (Table 4.1). These findings have been confirmed by immuno-chemical studies (Table 4.2). Mab 346-11A to the murine $\beta4$ integrin subunit precipitates the band corresponding to the entire $\beta4$ subunit (Mr 204 kd) only from highly metastatic 3LL variants

Table 4.1 Metastatic properties of 3LL variants and $\alpha6\beta4$ quantitation

	*Spontaneous lung metastases**	*ng ^{125}I 346-11A bound mg of protein***
Tumour lines		
3LL parental line	35 ± 4	11 ± 1.5
M1087	110 ± 8	40 ± 5
BM21548	16 ± 3	1.2 ± 0.5
In vitro lines		
C87	20 ± 9.5	29 ± 4
BC215	4 ± 1.5	2 ± 0.5
CCL***	n.e.	< 0.5

Means ± S.E. (number).
* Lung nodules evaluated 24 days after i.m. implantation of 2×10^5 cells.
** Values reported are the average of four experiments ± S.E.
*** CCL is a melanoma murine tumour ad corresponds to the negative control. n.e. = not evaluated.

Table 4.2 Mr* of $\alpha6$ and $\beta4$ integrin subunits immunoprecipitated from cell lysates of 3LL variants.

	α6 (116 kd)	*β4 (204 kd)*
Tumour lines		
3LL parental line	+ +	–
M1087	+ +	+ +
BM21548	+	–
In vitro lines		
C87	+ +	+
BC215	+ +	–

* SDS-PAGE in reducing conditions. Mab 135-13C to α6 and Mab 346-11A to β4 of the integrin subunits. (+ +) high amount; (+) low amount; (–) not detectable.

(M1087 and C87). Total RNA extracted from 3LL metastatic variants was probed with a cDNA clone corresponding to the cytoplasmic domain of the murine $\beta4$ protein[23] and $\beta4$ RNA expression was evaluated. Hybridization blots of 3LL variant tumours (Fig. 4.1) revealed that only highly metastatic tumour cells express relevant amounts of the full length transcript (7kb) coding the entire $\beta4$ protein.[23] By contrast, the RNA bands migrating at lower molecular weights and corresponding to cleaved/spliced forms of the full length transcript were found in all 3LL variants. Similar results were also obtained by analyzing total RNA extracted from *in vitro* variant cell lines (data not shown). The presence of non-functional transcripts and the down regulation of the 7 kb message suggested that, in less metastatic carcinoma cells, post-transcriptional regulation mechanisms were responsible for the absence of the $\beta4$ protein.

The expression of the $\alpha6\beta4$ integrin was also analyzed in human lung carcinomas of different histotype and we demonstrated[14] that the $\beta4$ protein

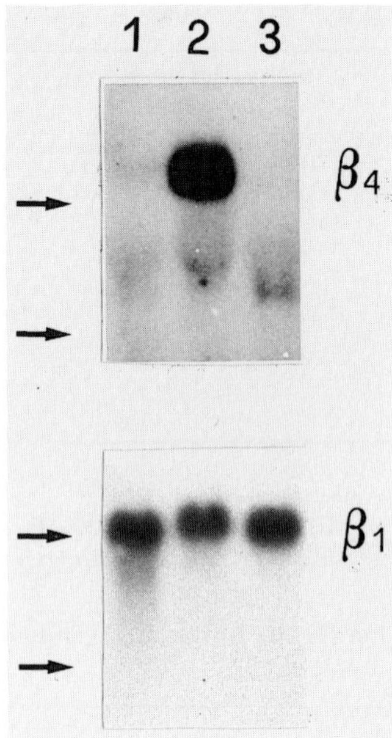

Figure 4.1 $\beta 4$ and $\beta 1$ integrin subunit expression on *in vivo* 3LL variants and on the parental line of 3LL. Total RNA was extracted from primary tumours of different lines (M1087, BM21548 and the parental line of 3LL), was electrophoresed (20 μg/lane) on agarose/formaldehyde gel and blotted onto nitrocellulose. Hybridization was carried out with purified nick-translated inserts, as described in "Materials and Methods". *Lanes* 1: parental line of 3LL; *lanes* 2: M1087 (high metastatic); *lanes* 3: BM21548 (low metastatic). Arrows: 28S and 18S markers.

is expressed in non-small cell lung cancers (NSCLC) but is undetectable in small cell lung cancers (SCLC). To assess $\beta 4$ mRNA expression on lung carcinoma cells of different histotype, total RNA extracted from adenocarcinoma (A549), large cell carcinoma (DG3), small cell carcinoma (AE2) cell lines was probed with a human $\beta 4$ cDNA clone. Northern blot analysis (Fig. 4.2) indicates that $\beta 4$ mRNA (7 kb) is absent in AE2 cells, but is significantly expressed in DG3 and A549 cell lines. The data of figure 4.2 also demonstrates that SCLC and NSCLC cells all express $\beta 1$ and $\alpha 6$ mRNAs as expected by immunoprecipitation analysis. The $\beta 4$ negative blots of SCLC cells matches the lack of protein in cell lysates[14] suggesting that $\beta 4$ expression is regulated at multiple levels in lung carcinomas and that its expression is correlated with the histotype.

The ability of different lung carcinoma cells to adhere to ECM components (laminin, fibronectin, vitronectin, collagen) was studied. The adhesion of 3LL metastatic variants to ECM components is reported in figure 4.3. The data of the adhesion assays demonstrate that more metastatic cells (C87) adhere to ECM components at a lower extent than less metastatic cells

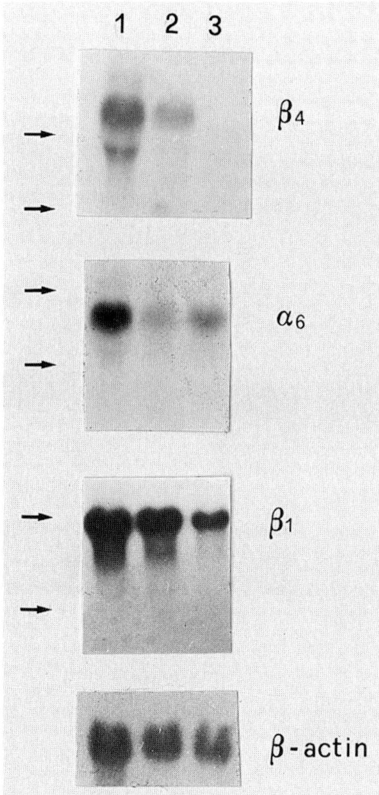

Figure 4.2 $\beta 4$, $\alpha 6$ and $\beta 1$ integrin subunit expression in different human lung carcinoma lines (A549, DG3, AE2). Total RNA was extracted from cultured cells, 20 μg/lane were electroforesed, blotted and hybridized as described in figure 4.1. *Lanes* 1: A549 line (adenocarcinoma); *lanes* 2: DG3 line (large cell carcinoma); *lanes* 3: AE2 line (small cell carcinoma). Arrows: 28S and 18S markers. As a control the β-actin gene expression is reported.

(BC215). These findings are in good agreement with the reported[28] reduced adherence of highly malignant cells. Furthermore, the data reported in figure 4.4 demonstrate that human lung carcinoma cells of different origin adhere with different ability to ECM components. As shown in figure 4.4 DG3 cells (large cell carcinoma) adhere very efficiently to all ECM components, whereas A549 cells (adenocarcinoma) adhere to laminin and fibronectin at a very low extent, and AE2 cells (small cell carcinoma) adhere efficiently only to laminin. Interestingly, the data indicate that lung carcinoma cells of different histotype adhere to laminin to very different extents. Since laminin adhesion is a relevant feature in defining the invasive capacity of tumour cells, we determine whether the adherence of different lung carcinoma cells to laminin is mediated by $\alpha 6\beta 1$ or $\alpha 6\beta 4$ receptors. The

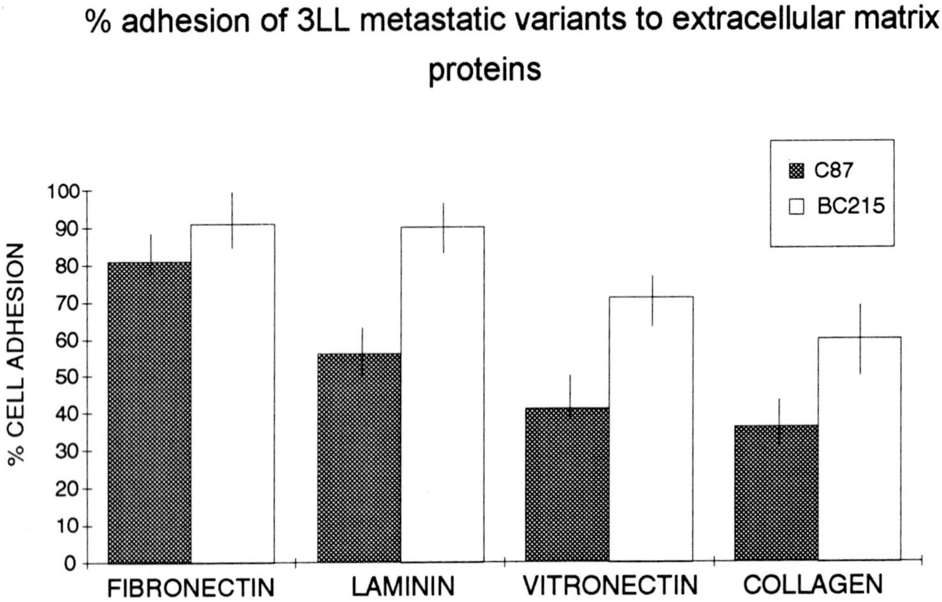

Figure 4.3 Adhesion of 3LL metastatic variants (C87 more metastatic, BC215 less metastatic) to extracellular matrix (ECM) proteins. Different amounts of cells ($3 \times 10^4 - 1 \times 10^5$) were incubated for 60 min at 37°C in quintuplicate microtiter wells coated with each of the indicated ECM proteins or with BSA (bovine serum albumin, negative control) or PLL (poly-L-lysine, positive control). After the remaining cell bounds to the well were washed to remove non-adherent cells, they were quantitated spectrophotometrically after staining as described in "Materials and Methods". *Column*: % adherence with respect to PLL (100%), subtracted of the background (BSA); *bar*: ±S.D.

involvement of different receptors in laminin adhesion was studied by Mab inhibition assays. As shown in figure 4.5, Mabs to the $\beta 1$ and $\alpha 6$ subunits inhibit at a higher extent laminin attachment of A549 cells than Mab to the $\beta 4$ subunit. Because of the absence of $\alpha 6\beta 4$ expression in AE2 cells, Mabs to the $\alpha 6$ and $\beta 1$ subunits and not Mabs to the $\beta 4$ subunit affect laminin adhesion. Interestingly, the adherence of DG3 cells to laminin is affected only by Mabs to the $\alpha 6$ and $\beta 1$ subunits in spite of the fact that these A549 cells express detectable amounts of $\alpha 6\beta 4$ integrin on their cell surface. In conclusion, data reported in figure 4.5 indicate that 1) the $\alpha 6\beta 1$ but not the $\alpha 6\beta 4$ integrin functions as laminin receptor in DG3 (large cell carcinoma) and in AE2 (small cell carcinoma) cells, that 2) both $\alpha 6\beta 1$ and $\alpha 6\beta 4$ integrins function as laminin receptors in A549 (adenocarcinoma) cells, and that, therefore 3) the $\alpha 6\beta 4$ integrin has different ligand specificity in DG3 (large cell carcinoma) or in A549 (adenocarcinoma) cells. Antibody blocking assays

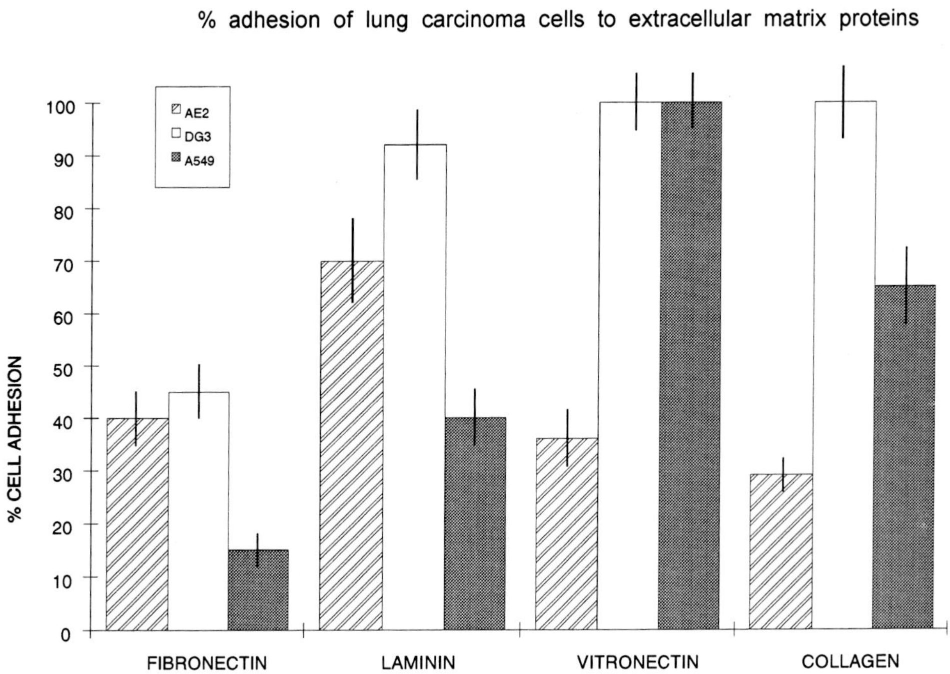

Figure 4.4 Adhesion of different (adenocarcinoma, large cell carcinoma, small cell carcinoma) human lung carcinoma lines (A549, DG3, AE2, respectively) to extracellular matrix (ECM) proteins. The cells (8×10^4) were incubated for 60 min at 37°C in quintuplicate microtiter wells coated with each of the indicated ECM proteins or with BSA (bovine serum albumin, negative control) or PLL (poly-L-lysine, positve control). The assays were performed as in figure 4.3.

were also employed to study the functionality of $\alpha6\beta4$ receptors in murine lung carcinoma cells. In spite of their specificity, however, Mabs 346-11A and 135-13C failed to indicate $\alpha6\beta4$ function (data not shown). The unefficacy of Mabs 135-13C (to the murine and human $\alpha6$ protein) was explained assuming that in these murine cells epitopes different from that reported for the human counterparts are responsible for laminin binding.

DISCUSSION

In previous studies we have demonstrated that in carcinoma cells the $\alpha6\beta4$ complex is involved in several biological processes. In fact, *i*) treatment of tumour-bearing mice with Mab to the $\alpha6$ subunit (135-13C) modulates the metastatic behaviour of Lewis lung carcinoma (3LL);[29,30] *ii*) treatment of Lewis lung carcinoma cells *in vitro* with Mab to $\alpha6$ subunit stimulates cell proliferation;[31] *iii*) treatment of highly malignant carcinoma cells *in vitro* with

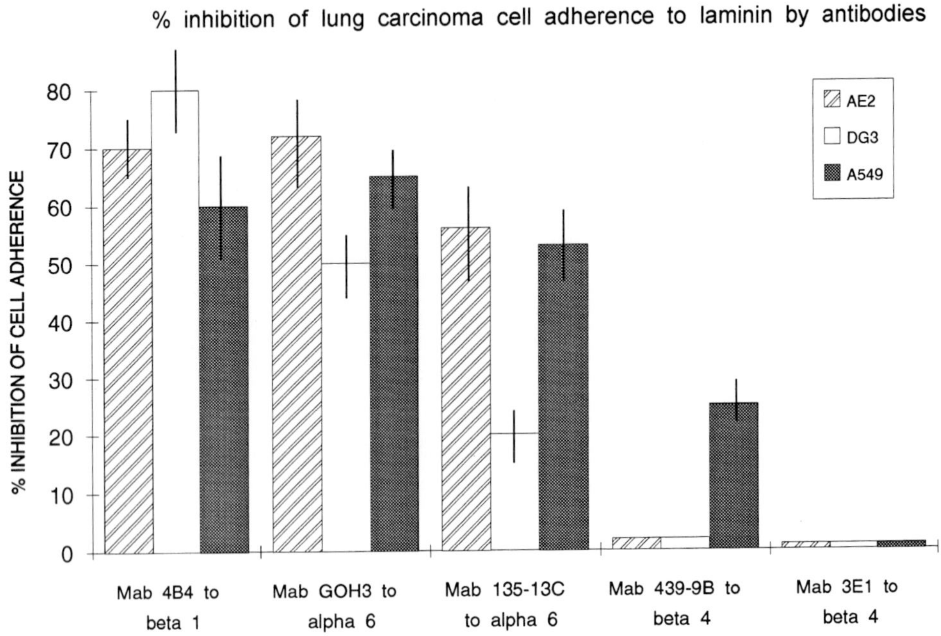

Figure 4.5 Inhibition of cell adherence to laminin by anti-$\beta1$, anti- $\alpha6$, and anti-$\beta4$ Mabs. The cells of different human lung carcinoma lines (A549 adeno-carcinoma, DG3 large cell carcinoma, AE2 small cell carcinoma) were preincubated with saturating concentrations of Mabs specific for the $\beta1$ (Mab 4B4), $\alpha6$ (Mabs GOH3, and 135–13C), and $\beta4$ (Mabs 439–9B, and 3E1) integrin subunits. Without being washed the cells were then added to microtiter wells coated with laminin or BSA. As a control untreated cells were added to microtiter wells coated with laminin, BSA or PLL. Cell adherence was quantitated according to the protocol described in figure 3. *Column*: % inhibition of the adherence with respect to the control (100%), subtracted of the background (BSA); *bar*: ±S.D.

Mab to the $\alpha6$ subunit or with growth factors, such as insulin and EGF, induces the phosphorylation of the β subunit ($\beta4$).[32,33] All these data are consistent with an involvement of the $\alpha6\beta4$ complex in metastatic process and proliferation mechanisms, in good agreement with the suggested role of integrins in cell-cell recognition and cell-matrix interaction. With this view we further extended our studies to define the expression of such a receptor in murine and human lung carcinomas.

Firstly we studied the expression of $\alpha6\beta4$ integrin in metastatic variants derived from Lewis lung carcinoma.[15,16] By a two site assay approach, employing specific Mabs which recognize the α and the β chains of the integrin respectively, we demonstrated that highly metastatic 3LL cells maintained either *in vivo* or *in vitro*, express the highest levels of $\alpha6\beta4$ integrin (Table 1). These results were confirmed by immunoprecipitation of cell line and tumour lysates — (Table 4.2). In fact, the $\beta 4$ subunit was precipitated uniquely from the cell lysates of highly metastatic cells. To evaluate whether the expression of $\beta 4$ protein in 3LL metastatic variants is regulated at transcriptional or post-transcriptional level, carcinomas were probed for $\beta 4$ mRNA. A 7 kb band corresponding to $\beta 4$ full length transcript was revealed uniquely in highly metastatic tumours, whereas other bands migrating to lower molecular weights, and corresponding to cleaved and/or spliced forms of the major transcript (7 kb), were found in all 3LL metastatic variants. These results indicate that post-transcriptional regulation mechanisms of $\beta 4$ expression occur in less metastatic carcinoma cells. Immunochemical and immunohistochemical studies have been performed to assess $\beta 4$ expression in human carcinoma cells. Results reported[14] indicated that all NSCLC tumour tested were positive for $\alpha 6\beta 4$ integrin expression in contrast to SCLC tumours which were found completely negative. Immunoprecipitation experiments using several *in vitro* cell lines confirmed the absence of the $\beta 4$ protein in SCLC cells. Northern blot experiments were performed to assess $\beta 4$ mRNA expression in lung carcinoma cells of different histotype. Here we report that AE2 (small cell carcinoma) cells lack the 7 kb band which corresponds to $\beta 4$ full length transcript. Moreover, we demonstrate that these cells lack also the bands corresponding to lower molecular weight transcripts (cleaved/spliced) present in low metastatic 3LL variants (Fig. 4.1) and observed in other human carcinoma cell lines. Since the $\alpha 6$ and $\beta 1$ transcripts are evident in AE2 cells, the lack of $\beta 4$ reactivity in the blots suggests that in these cells the $\beta 4$ expression is regulated at transcriptional level. These data further confirm that $\beta 4$ integrin expression might be a useful tool to define the subtype of small cell lung cancers.

In order to evaluate the function of $\alpha 6\beta 4$ integrin in lung carcinomas we studied the ability of different lung carcinoma cells to bind extracellular matrix components. The results obtained (Figs 4.3 and 4.4) demonstrate that 1) more metastatic 3LL cells adhere to all the substrata with lower capacity than the lower metastatic counterparts, 2) large cell carcinoma cells adhere to all substrata with high ability, 3) small cell lung carcinoma cells adhere only to the laminin with a good efficiency, 4) adenocarcinoma cells adhere to laminin and fibronectin with the lowest ability. Mabs to different integrin subunits ($\alpha 6$, $\beta 4$, $\beta 1$) were employed to identify the laminin receptors in lung carcinoma cells, but the lack of functional monoclonal antibodies to the murine integrin subunits did not define laminin receptors in 3LL carcinoma cells. Figure 4.5 demonstrates that Mabs to the $\alpha 6$ and $\beta 1$ subunits all inhibit laminin adhesion of human lung carcinoma cells, whereas Mab (439-9B) to the $\beta 4$ subunit inhibits laminin adhesion of adenocarcinoma cells (A549) but not of large cell carcinoma cells (DG3) in spite of the presence of $\alpha 6\beta 4$ receptors on their cell surface.[14] These findings indicate that the $\alpha 6\beta 1$ integrin is a laminin receptor in all lung carcinoma cells, and that the $\alpha6\beta4$ integrin has different functions in lung carcinomas of different histotypes.

In conclusion, further studies on several cell lines and tumours are needed to confirm these results, but data reported indicate that a better knowledge of the structure of the $\alpha6\beta4$ integrin receptor, the identification of the ligand specificity and the availability of adequate reagents could help in elucidating the histotype of lung cancers and their ability to metastasise to distant sites.

Acknowledgements

The work has been partially supported by C.N.R. grant n° 92 02 356 PF 39 and by A.I.R.C.

REFERENCES

1. Cheresh, D. A., Smith, J. W., Cooper, H. M. and Quaranta, V. (1989) A novel integrin vitronectin receptor (αx/βx) is responsible for distinct adhesive properties of carcinoma cells. *Cell*, **57**, 59–69.
2. Hynes, R. O. (1987) Integrins, a family of cell surface receptors. *Cell*, **48**, 549–554.
3. Hynes, R. O. (1981) Relationship between fibronectin and cytoskeleton. *In*: G. Poste and G. L. Nicolson, (eds.). Cytoskeletal Elements and Plasma Membrane Organization. *Cell Surface Review*, **7**, 97–139, Amsterdam, Elsevier Science Publication, BV.
4. Kajiji, S., Tamura, R. N. and Quaranta, V. (1989) A novel integrin (αE/β4) from human epithelial cells suggests a fourth family of integrin adhesion receptors. *EMBO J.*, **8**, 673–680.
5. Kennel, S. J, Foote, L. J., Falcioni, R., Sonnenberg, A., Crouse, C. and Hemler, M. E. (1989) Analysis of the tumor associated antigen TSP-180: identity with $\alpha6/\beta4$ in the integrin superfamily. *J. Biol. Chem.*, **264**, 15515–15521.
6. Sonnenberg, A. F., Hogervorst, F., Osterop, A. and Veltman, F. E. M. (1988) Identification and characterization of a novel antigen complex on mouse mammary tumor cells using monoclonal antibody against platelet glycoprotein Ic. *J. Biol. Chem.*, **263**, 14030–14038.
7. Hemler, M. E., Crouse, C. and Sonnenberg, A. (1989) Association of the VLA of $\alpha6$ subunit with a novel protein: a possible alternative to the common VLA $\beta1$ subunit of certain cells. *J. Biol. Chem.*, **264**, 6529–6535.
8. Sonnenberg, A., Daams, H., van der Valk, M. A., Hilkens, J. and Hilgers, J. (1986) Development of mouse mammary gland: identification of stages in differentiation of luminal and myoepithelial cells using monoclonal antibodies and polyvalent antiserum against keratin. *J. Histochem. Cytochem.*, **34**, 1037–1046.
9. Falcioni, R., Sacchi, A., Reseau, J. and Kennel, S. J. (1988) Monoclonal antibody to human carcinoma-associated protein complex: quantitation in normal and tumor tissue. *Cancer Res.*, **48**, 816–821.
10. Falcioni, R., Kennel, S. J., Giacomini, P., Zupi, G. and Sacchi, A. (1986) Expression of tumor antigen correlated with metastatic potential of Lewis lung carinoma and B16 melanoma clones in mice. *Cancer Res.*, **46**, 5772–5778.

11. Lehto, V. P., Stenman, S., Miettinien, M., Dahl, D. and Virtanen, I. (1983) Expression of a neural type of intermediate filament as a distinguishing feature between oat cell carcinoma and other lung cancers. *Am. J. Pathol.*, **110**, 113–118.
12. Mulshine, J. L., Cutitta, F., Bibro, M., Fedorko, J., Fargion, S., Little, C., Carney, D. N., Gadzar, A. F. and Minna, J. D. (1983) Monoclonal antibodies that distinguish non-small cell from small cell lung cancer. *J. Immunol.*, **131**, 497–502.
13. Ball, E. D., Graziano, R. F., Pettengill, O. S., Sorenson, G. D. and Franger, M. W. (1984) Monoclonal antibodies reactive with cell carcinoma of the lung. *J. Natl. Cancer Inst.*, **72**, 593–598.
14. Mariani-Costantini, R., Falcioni, R., Battista, P., Zupi, G., Kennel, S. J. and Sacchi, A. (1990) Integrin $\alpha6/\beta4$ expression in human lung cancer as monitored by specific monoclonal antibodies. *Cancer Res.*, **50**, 6107–6112.
15. Sacchi, A., Mauro, F. and Zupi, G. (1984) Changes of phenotypic characteristics of variants derived from Lewis lung carcinoma during long-term growth. *Clin. Expl. Metastasis*, **2**, 171–178.
16. Zupi, G., Cavallari, A., Greco, C. and Sacchi, A. (1984) Lewis lung carcinoma lines with different metastatic behavior: response to ICRF-159. *Anticancer Res.*, **4**, 91–95.
17. Giard, D. J., Aaronson, S. A., Todaro, G. J., Arnstein, P., Kersey, J. H., Dosik, H. and Parks, P. (1989) *In vitro* cultivation of human tumors: establishment of cell lines derived from a series of solid tumors. *J. Natl. Cancer Inst.*, **51**, 1417–1423.
18. Laudonio, N., Zupi, G., Storniello, G., Salvati, F. and Greco, C. (1989) Establishment of cell lines from serial samples from two patients with lung tumors before and after chemotherapy. *Cytotechnology*, **2**, 39–47.
19. Morimoto, C., Letvin, N. L., Boyd, A. W., Hagan, M., Brown, H. M., Kornacki, M. M. and Sohlossman, S. F. (1985) The isolation and characterization of the human helper inducer T cell subset. *J. Immunol.*, **134**, 3762.
20. Hessle, H., Sakai, L. Y., Hollister, D. W., Burgeson, R. E. and Engvall, E. (1984) Basement membrane diversity detected by monoclonal antibodies. *Differentiation*, **26**, 49.
21. Hemler, M. E., Huang, C. and Schwartz, L. (1987) The protein family: characterization of five distinct cell surface heterodimers each with a common 130,000 molecular weight β subunit. *J. Biol. Chem.*, **262**, 3300–3309.
22. Ramaswamy, H. and Hemler, M. E. (1990) Cloning primary structure and properties of a novel human β subunit. *EMBO J.*, **9**, 1561–1569.
23. Kennel, S. J., Foote, L., Cimino, L., Rizzo, M. G., Chang, L. Y. and Sacchi, A. Sequence of cDNA for the $\beta4$ subunit of murine integrin. *Gene*, in press.
24. Chomczinsky, P. and Sacchi, N. (1987) Single-step method of RNA isolation by acid guanidinium thyocyanate-phenol extraction. *Anal. Biochem.*, **162**, 156–159.
25. Sambrook, J., Fritsch, E. F. and Maniatis, T. (1989) Molecular cloning: a laboratory manual. Cold Sping Harbor N.Y., Cold Spring Harbor Laboratory.
26. Ruoslahti, E., Hayman, E. G. N., Pierschbacher, M. and Engvall, E. (1982) Fibronectin: purification, immunochemical properties, and biological activities. *Methods in Enzymology*, **82**, 828–831.
27. Kennel, S. J., Lankford, P. F., Foote, L. J., Tsakeres, F. S., Adams, L. M. and Ullrich, R. L. (1980) Distribution of a tumor surface protein common to several murine lung carcinomas. *Cancer Res.*, **40**, 2153–2159.

28. Hynes, R. O. (1992) Integrins: versatility, modulation, and signaling in cell adhesion. *Cell*, **69**, 11–25.
29. Sacchi, A., Kennel, S. J., Apollonj, C. and Natali, P. G. (1985) Treatment with monoclonal antibody to a Lewis lung carcinoma-associated antigen: different effects on a primary tumor and its metastases. *Cancer Treatment Reports*, **69**, 985–991.
30. Sacchi, A., Kennel, S. J., Natali, P. G., Tibursi, G. and Apollonj Ghetti, C. (1987) Metastatic dissemination of 3LL variants after treatment with monoclonal antibody to a tumor-associated antigen. *Clin. Expl. Metastasis*, **5**, 245–257.
31. Sacchi, A., Piaggio, G., Rizzo, M. G., Falcioni, R. and Kennel, S. J. (1989) Stimulation of tumor growth *in vitro* by a monoclonal antibody to a tumor specific protein present on the cell surface of 3LL cells. *Clin. Expl. Metastasis*, **7**, 41–54.
32. Sacchi, A., Falcioni, R., Piaggio, G., Gianfelice, M. A., Perrotti, N. and Kennel, S. J. (1989) Ligand-induced phosphorylation of a tumor surface protein (TSP-180) associated with metastatic phenotype. *Cancer Res.*, **49**, 2615–2620.
33. Falcioni, R., Perrotti, N., Piaggio, G., Kennel, S. J. and Sacchi, A. (1989) Insulin-induced phosphorylation of the $\beta 4$ integrin subunit expressed on a murine metastatic carcinoma cells. *Mol. Carcinogenesis*, **2**, 361–368.

5. LAMININ-INDUCED Y-79 RETINOBLASTOMA CELL DIFFERENTIATION OCCURS IN THE ABSENCE OF "CLASSIC" LAMININ ADHESION MOLECULES

L. POGGI[1], A. MELCHIORI[1], R. PELLEGRINI[2], P. DEFILIPPI[3], D. NOONAN[1], M. A. CAMPBELL[4,5], S. GENTLEMAN[4], G. J. CHADER[4] and A. ALBINI[1]

[1]*Istituto Nazionale per la Ricerca sul Cancro, V. le Benedetto XV10, 16132 Genova, Italy*

[2]*Istituto Nazionale per lo Studio e la Cura dei Tumori, Via Venezian 1, 20133 Milano, Italy*

[3]*Dipartimento di Genetica, Biologia e Chimica Medica, Università di Torino, Via Santena 5 bis, 10126 Torino, Italy*

[4]*National Eye Institute, NIH, Bethesda, MD 20892, USA*

[5]*Present Address: 6 Lovel Rd, Chalfont St. Peter, Bucks, SL9 9NN, UK*

Laminin (LN), a basement membrane glycoprotein, promotes the morphological differentiation of Y-79 retinoblastoma cells grown on a poly-D-lysine (PN) substrate. LN does not act as an attachment substrate for Y-79 cells, and overcoating of the PN substrate with LN actually inhibits Y-79 cell adhesion. Addition of LN to the culture medium increases Y-79 cell adhesion to PN in dose dependent manner after 24 hours. We show here that the effects of LN on Y-79 cell adhesion are calcium dependent. LN also significantly increases Y-79 cell motility in the Boyden chamber assay, but is not chemotactic for these cells. Binding studies using ^{125}I-LN and specific LN peptides suggest that the interaction of LN with Y-79 cells is not mediated by either integrin or high affinity receptors, but by a single, intermediate affinity binding site. Here, by FACS analysis and immunoprecipitation, we confirm the absence of the high affinity 67 kDa LN-binding protein and low levels of LN-binding integrins on these cells; only traces of the β_1 and β_6 integrin subunits were detected. Using LN-Sepharose affinity chromatography of cell-surface iodinated cells, the major cell-surface LN-binding protein was eluted by 20 mM EDTA. This molecule has a molecular mass of 100 kDa which is not affected by reduction. Termed LBM-100, this cell-surface protein could be a "differentiation" LN-binding protein, mediating the effects of LN on Y-79 cells.

INTRODUCTION

Laminin (LN) is a glycoprotein of the basement membrane which is able to promote adhesion, migration and differentiation of many cell types. Generally, the effects induced by LN are thought to be mediated by cell-adhesion to LN.[1–3] Previous studies have shown that the interaction between cells and LN is mediated by many classes of cell-surface receptors, including the high affinity 67 kDa LN-receptor,[4] lectins, galactosyltransferases, sulfatides and integrins.[5] We have investigated the nature of the

Correspondence. Adriana Albini, Servizio di Cancerogenesi Chimica Istituto Nazionale per la Ricerca sul Cancro V.le Benedetto XV 10, 16132 Genova, Italy
Tel: 0039–10–3534640 Fax: 0039–10–352999

molecules that mediate the effects induced by LN on Y-79 retinoblastoma cells. Y-79 are invasive cells,[6] derived from a multipotential stem cell of the neural retina,[7,8] that express an altered mRNA for the Rb gene product.[9] These cells grow in suspension culture but when seeded on poly-D-lysine (PN) attach and show a morphology reminiscent of photoreceptor cells.[10,11] LN as a substrate is anti-adhesive for Y-79 cells, however addition of LN to the culture medium increases Y-79 cell adhesion to PN and induces the cells to differentiate to a neuronal-like morphology.[12] These changes in morphology have been shown to be accompanied by changes in the expression of photoreceptor and neural markers.[12,13] In contrast to the many cell types that differentiate in the presence of LN, Y-79 cells do not adhere directly to LN. YIGSR-and RGD-containing peptides, which are known to be involved in the high affinity 67 kDa receptor binding and in interactions with some integrin receptors, respectively, did not affect Y-79 cell adhesion to PN in the presence of LN. Moreover, the binding between Y-79 cells and LN was not inhibited by YIGSR-containing peptides.[12]

Studies using ^{125}I-LN indicated that Y-79 cells bind LN in a dose dependent manner and with a single intermediate affinity binding site.[12] In this paper we show that the effect of LN and PN on Y-79 cell adhesion is dose and Ca^{++} dependent. We also demonstrate the absence of high affinity 67 kDa LN-binding protein and very low amounts of integrin receptors on Y-79 cells. By affinity chromatography, we have eluted a single major LN-binding protein of 100 kDa (LBM-100)[12] from Y-79 cells which appears to be the only candidate molecule mediating Y-79 cell-LN interactions.

MATERIALS AND METHODS

Cells

Y-79 human retinoblastoma cells (a gift of Dr. P. Russel, NEI, NIH, Bethesda, MD) control A2058 human melanoma cells (donated by Dr. L. Liotta, NCI, NIH, Bethesda, MD) and Me1007 melanoma cells (INT, Milan, Italy) were cultured in RPMI 1640 with 10% heat-inactivated fetal calf serum, antibiotics and glutamine. For certain experiments, cells were maintained in serum-free, chemically-defined medium (ITSS: 5 μg/ml insulin, 5 μ/ml transferrin, 5 ng/l sodium selenite, 0.5% Zn^{++}; Sigma Chem. Co., St. Louis, MO).

Morphology and Attachment Studies

Tissue culture plastic dishes were coated by incubation with LN (Gibco/BRL, Gaithersburg, MD) (12.5 to 50 μg/35 mm dish), PN (Sigma Chem.) (2.5 μg/cm^2) or no substrate for 1 hr followed by saturation with bovine serum albumin (BSA) (0.05%). When indicated, LN was added into the culture medium at the appropriate concentrations. Cell suspensions ($1-2 \times 10^5$/dish) in ITSS were seeded onto the plates and incubated with BSA for 3 or 24 hrs at 37°C. At the end of the incubation period, dishes were rinsed twice with attachment medium and once with PBS to eliminate unattached cells. Attached cells were trypsinized and counted with a hemocytometer. Morphological changes were documented 7–14 days after plating, using phase contrast optics. In some experiments, mouse polyclonal antibodies to laminin (Collaborative Res., Bedford, MA) or fibronectin (Calbiochem., La Jolla, CA) were added to the cells. In Ca^{++} dependent experiments we have used F12 medium supplemented with ITSS, with or without $CaCl_2$ (0.3 mM).

Chemotaxis

Boyden blind well chemotaxis chambers were used for these studies.[14] The chamber consists of two compartments separated by a porous polycarbonate membrane (5 μm pore size) coated with PN (5 $\mu g/cm^2$). The lower compartment contained the stimulatory substances, *i.e.*: fibroblast-conditioned medium (FCM), fibronectin (FN), laminin (LN) or a solution of basement membrane extract, matrigel (MG)[15] in serum-free medium (SFM) containing 0.1% BSA. LN and FN were used at 50 nM, concentration where they elicited maximum migration in preliminary experiments. The upper compartment of the Boyden chambers was filled with $0.5-1 \times 10^6$ cells in SFM containing BSA. Migrated cells were fixed, stained with toluidine blue and counted by a microscope. Triplicate filters were run and 10 fields/filter were counted and averaged.

Immunoprecipitation

Y-79 cells maintained in ITSS medium were seeded at 5×10^6 cells/plate and treated for five days under one of the following conditions: a) growth in suspension culture; b) growth on PN (5 $\mu g/cm^2$) coated dishes; c) growth on PN (5 $\mu g/cm^2$) coated dishes; two days after seeding 12.5 μg/ml of LN was added to the medium for three days; d) growth on PN (5 $\mu g/cm^2$) coated dishes; 12.5 μg/ml of LN was added to the medium for five days. The control A2058 melanoma cells were treated as in d).

At the end of experiments the cells were incubated for 2 hrs in the presence of methionine-free medium and then treated with methionine^{-35}S (100 μCi/plate). After 5 hrs of incubation the medium was removed and cells were harvested and extensively washed in PBS containing $CaCl_2$ (1mM) and $MgCl_2$ (1mM). Labelled cells were lysed and the integrins immunoprecipitated with a polyclonal antiserum directed against the cytoplasmic domain of the β_1 integrin subunit and Protein A-Sepharose (Sigma), as described previously.[16,17] The antigen-antibody complexes were dissociated by boiling in sample buffer (200 mM Tris-HCl pH 6.8, containing 3% SDS, 10% glycerol and 0.001% bromophenol blue). Samples were analyzed by eletrophoresis in 7.5% SDS-polyacrylamide gels,[18] followed by autoradiography.

Immunocytochemical Assays

Indirect immunofluorescence was performed with ascitic fluids of monoclonals to high affinity 67 kDa LN-receptor (obtained by immunization of Balb/c mouse with N592 small cell lung cancer cells[19]) and the following integrin subunits: α_2 (Cl P1E6, Telios), α_3 (Cl P1B5, Telios), α_5 (Cl P1D6, Telios), β_1 (4B4, Coulter), β_4 (Cl 3E1, Telios) and a purified monoclonal antibody to α_6 integrin subunit (G_0H3, Immunotech) (1: 100 in the first case and 10 μg/ml in the second case) on live cells at 37°C for 30 min. Cells were then washed three times with Hank's solution and treated for 30 min at 37°C with biotinylated anti-mouse IgG or anti-rat IgG antibodies (Amersham Int., Little Chalfont, UK). After three washings with Hank's solution, the cells were treated with fluoresceinated streptavidin (Amersham Int., Little Chalfont, UK) at 0–4°C for 30 min. The fluorescence was evaluated by a FACScan (Becton Dickinson). Me1007 human melanoma cells were used as positive control.

Affinity Chromatography

Y-79 cell surfaces were radiolabelled with 250 μCi of $Na^{-125}I$ using lactoperoxidase, extracted and chromatographed on LN-Sepharose columns, as previously described.[12] Radiolabelled live cells were washed in PBS and then lysed in lysis buffer (25 mM n-octyl-β-D-glucopyranoside, 150 mM NaCl, 100 mM Tris pH 7.4, 2 mM phenylmethylsulfonyl fluoride, [PMSF] 1mM $MnCl_2$). Lysates were incubated with 1.5 ml of LN-Sepharose overnight at 4°C in lysis buffer and then transferred to a column and the unbound material collected. The column was washed twice with 4 column volumes of lysis buffer and once with 2 column volumes of lysis buffer until no detectable radioactivity was eluted. Bound material was then eluted from the column with 4 volumes of extraction buffer containing 20 mM EDTA and no $MnCl_2$, followed by extraction buffer with EDTA and 1 M NaCl. The various fractions were collected, dialyzed against 3 mM Tris pH 7.4, 1 mM EDTA and 0.5 mM PMSF, then distilled water, lyophilized and electrophoresed on 7% SDS-polyacrylamide gel.

RESULTS

Morphology and Attachment Studies

Y-79 cells adhere very poorly to uncoated plastic (Fig. 5.1, a) but up to 500-fold more cells attach to a plastic substratum coated with PN.[12] When

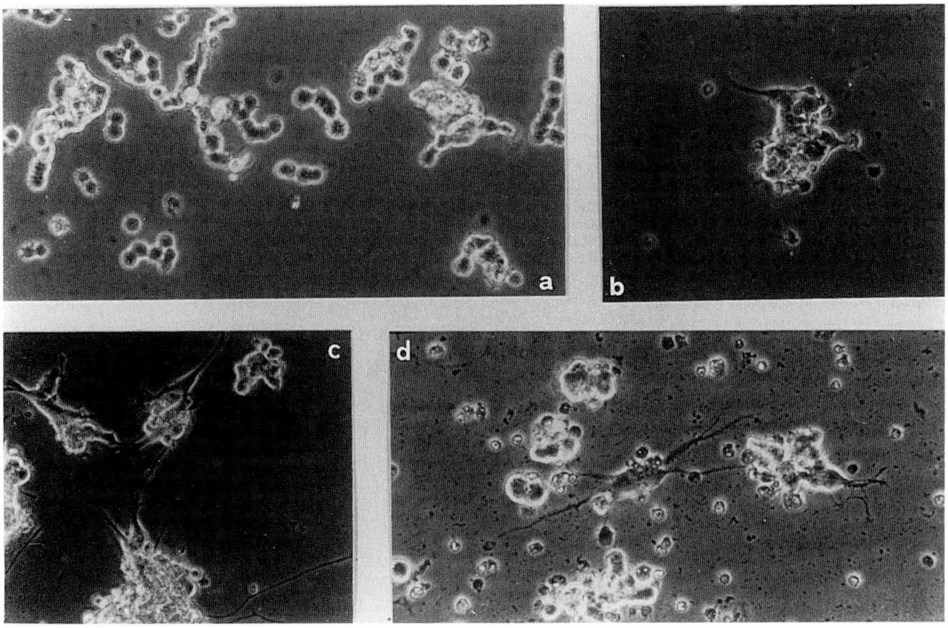

Figure 5.1 Y-79 cells grown on plastic (a) and on poly-D-lysine (2.5 $\mu g/cm^2$) substrate, in absence (b) or in presence (c, d) of laminin (12.5 $\mu g/ml$) in the medium. Morphological changes were documented 7–14 days after plating using phase contrast optics.

plated on a PN substrate, the cells flatten out, occasionally extending short, unbranched processes, reminiscent of photoreceptor cell morphology in culture (Fig. 5.1, b).[7,8] LN coating alone is not an effective substrate for Y-79 cell attachment, and overcoating of PN substrates with laminin inhibits attachment to PN (Fig. 5.2 and[12]). When LN is added to cultures on PN, after 5–7 days the cells show a more typical neuronal morphology, with numerous cells extending dendritic-like processes (Fig. 5.1, c and d).[8,20] However, LN, when added to the medium of Y-79 cells on a PN substrate, markedly increased Y-79 cell attachment: this effect is substantial after 24 hrs and it is dose dependent (Fig. 5.2). Addition of approximately 50 μg of LN/ml of medium appears to allow for a maximal rate of attachment, with 60% of plateau value reached at 12.5 $\mu g/ml$. The increased number of attached cells is not due to cell growth since the number of total cells (suspended plus attached) is invariant in all dishes after 24 hrs (not shown). The laminin effect appears to be specific since the increase in attachment to PN with LN can be fully inhibited by polyclonal anti-LN antibodies (Fig. 5.3). Control anti-FN antibodies are ineffective in inhibiting the LN-induced increase in attachment to PN. Y-79 cell adhesion to PN in the presence of LN in the medium is Ca^{++} dependent: when Ca^{++} free medium was used,

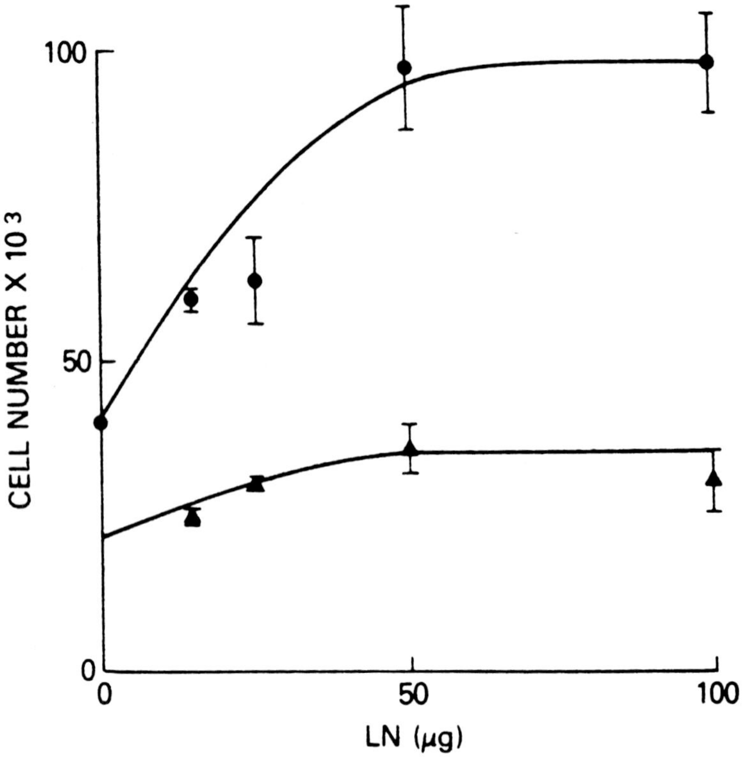

Figure 5.2 Effect of increasing concentrations of laminin (LN) on Y-79 cell attachment to poly-D-lysine (2.5 $\mu g/cm^2$). Attached cells were counted after 3 (△) or 24 (●) hrs. Values given are average values + S.E.M. from two separate experiments.

the Y-79 cell adhesion was about half (47%) of that in the presence of Ca^{++} (Fig. 5.4).

Migration Studies

Figure 5.5 illustrates the effects of various substances on Y-79 cell migration on PN-coated filters. Under these conditions, only about 0.2–1% of the total number of cells initially added to the upper chamber migrate to the lower chamber during a 7 hr experimental period. These studies indicate that Y-79 cells are poorly migratory ($1.9 \pm 0.1 \times 10^3$cells/filter) even on filters coated with PN. However, if one takes into consideration that only about 24% of the seeded cells attach to the filter, the percentage of effectively migrating cells is 0.8–4%. Addition of laminin (LN) or matrigel (MG) to the medium significantly increases retinoblastoma cell migration, about 2 fold. However,

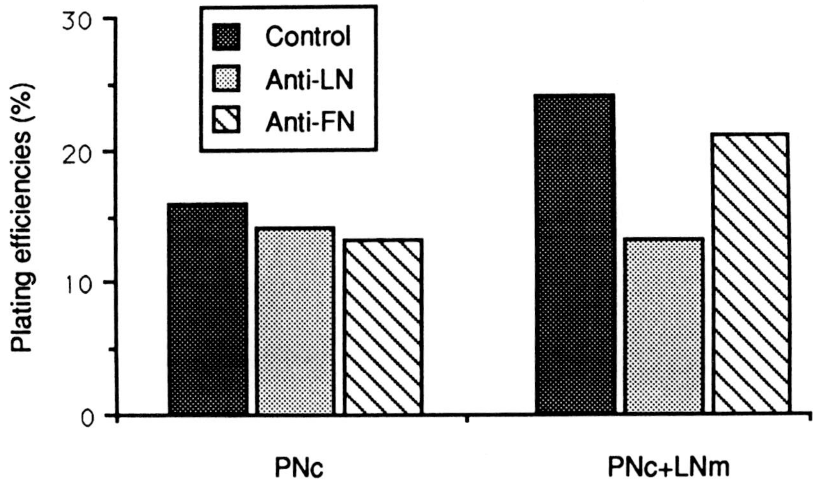

Figure 5.3 Effect of mouse polyclonal antibodies to laminin (LN) or fibronectin (FN) on laminin-induced increase in Y-79 cell attachment. Plates were coated with poly-D-lysine (PN_c) or with poly-D-lysine in the presence of laminin in the medium (PN_c + LN_m). No toxicity of the antibodies was detected under the present conditions. Data are expressed as the percentage (%) of plated cells attached after 3 hrs. Data are averages from duplicate samples in three separate experiments. Standard deviation (S. D.) was below 10%.

the effect is chemokinetic rather than chemotactic, as LN increases Y-79 cell migration independently if added in the lower chamber, upper chamber or both (not shown). Addition of fibroblast-conditioned medium (FCM) or purified fibronectin (FN) does not affect Y-79 cell migration (Fig. 5.5).

Immunoprecipitation

We have analyzed the presence of VLA integrin subunits in Y-79 cells, using anti-β_1 subunit antiserum. Immunoprecipitation of control A2058 melanoma cells (Fig. 5.6 panel B, lane 5) demonstrated a substantial amount of β_1 subunit and co-precipitated at least three α subunits, of m.r. approximately 180, 140 and 80 kDa. Immunoprecipitates from both Y-79 cells growing in suspension (lane 1) and Y-79 cells growing on PN coated dishes (lane 2), showed very little β_1 subunit (Fig. 5.6 panel A) and traces of an α subunit of 140 kDa, visible only on longer exposure (Fig. 5.6 panel B). Immunoprecipitates from Y-79 cells growing on PN coated dishes in the presence of LN in the medium (lanes 3 and 4), showed a 2-fold more intense band corresponding to the β_1 subunit and a faint 140 kDa band (Fig. 5.6 panel B).

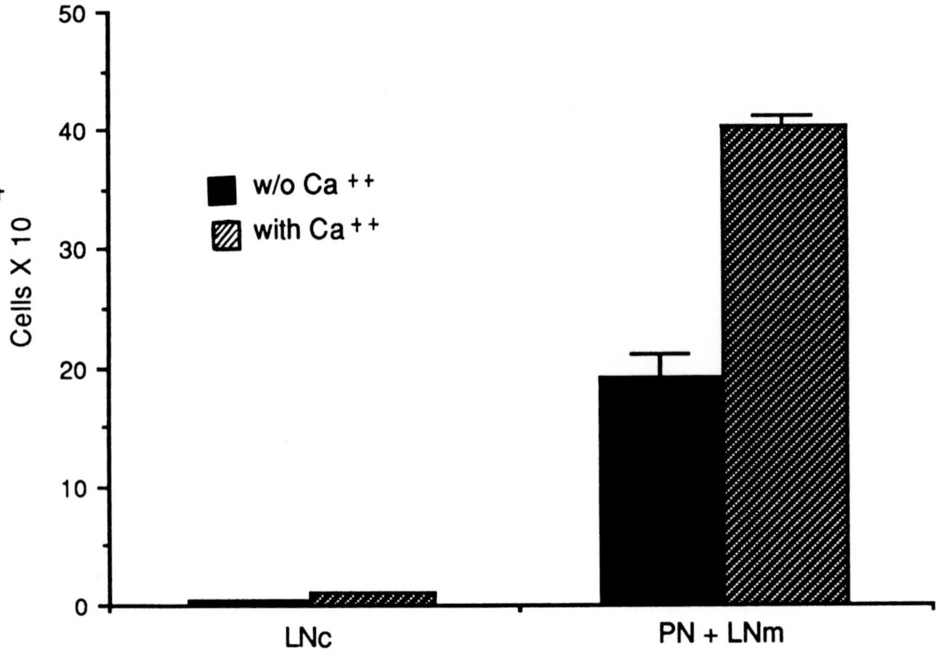

Figure 5.4 Effect of Ca^{++} (0.3 mM $CaCl_2$ in F_{12} medium supplemented with ITSS) on Y-79 cell attachment to laminin (50 μg/35 mm dish) (LN_c) or to poly-D-lysine (2.5 $\mu g/cm^2$) in presence of laminin (12.5 μg/ml) in the medium (PN + LN_M). Experiments were performed twice in triplicate. Bars represent S. D. of a typical experiment.

In all cases, Y-79 cells have extremely low quantities of β_1 and β_1 associated α integrin subunits as compared to the human melanoma cell controls.

Immunocytochemical Assays

By FACS analysis (Fig. 5.7), we have found that Y-79 cells growing in suspension do not exhibit LN-binding α_2 (Fig. 5.7, panel 1), α_3 (Fig. 5.7, panel 2) or β_4 (Fig. 5.7, panel 6) integrin subunits, whereas the peaks relative to β_1 (Fig, 5.7, panel 5) and α_6 (Fig. 5.7, panel 4) subunits, constituents of a specific LN-receptor, show a slight shift. We also tested α_1 even though no subunits of that m.r. (180 kDa) was co-precipitated with the β_1 subunit. No α_1 fluorescence could be detected (not shown). The quantity of β_1 and α_6 subunits is much lower with respect to control Me1007 human melanoma cells, which show a very strong shift (Fig. 5.7, panel 8 and 9). Also α_5, the α subunit of the "classic" FN-receptor, is not expressed (Fig. 5.7, panel 3). FACS analysis for the high affinity 67 kDa LN-receptor showed that this protein is also absent on Y-79 cells (Fig. 5.7, panel 7).

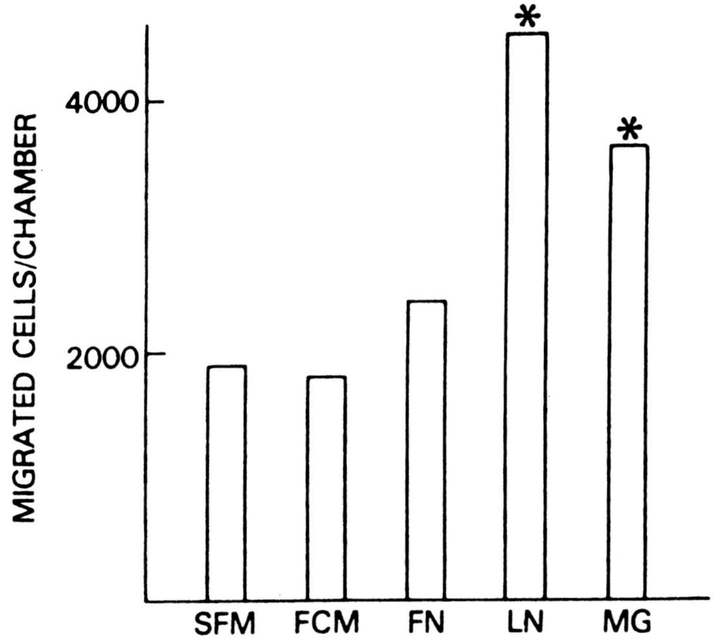

Figure 5.5 Effect of various matrix proteins on Y-79 cell migration. Y-79 cells were seeded in the upper well of a Boyden chamber in serum-free medium (SFM) containing BSA. The lower well contained the potential stimulatory substances. Theses are: fibroblast-conditionated medium (FCM), fibronectin (FN), laminin (LN) or matrigel (MG), at 50 nM. Four independent experiments were performed, each with duplicate samples. Asterisks indicate values significantly different from SFM control values, as determined by Student's t-test analysis.

Affinity Chromatography

By cell-surface iodination and LN-Sepharose affinity chromatography, we have investigated the nature of LN-binding sites on Y-79 cells. Extraction buffer containing 20 mM EDTA eluted one major radiolabelled band (Fig. 5.8). This single major cell-surface LN-binding protein migrated at approximately 100 kDa (LBM-100) and was not affected by reduction (data not shown). No other major bands were eluted under more stringent conditions using 1 M NaCl (data not shown), suggesting a possibly important role of LBM-100.

DISCUSSION

LN is able to alter the morphological differentiation of Y-79 retinoblastoma cells grown on poly-lysine[8,12,20,21] These effects are mediated by LN in suspension but not by a LN substrate[12,21] However, the induction of neuronal differentiation by LN appears to be attachment dependent, as LN and cAMP treatment of Y-79 cells in suspension increases photoreceptor marker synthesis.[21] LN as a substrate is ineffective for Y-79 cell attachment and reduces adhesion to PN.[12] In contrast, we have shown that LN in the culture medium increases Y-79 cell adhesion to PN in a dose and time dependent fashion. This LN effect on Y-79 cells is strongly affected by calcium. LN also enhances Y-79 cell migration. The presence of LN-receptors was analyzed and we found that neither the high affinity 67 kDa

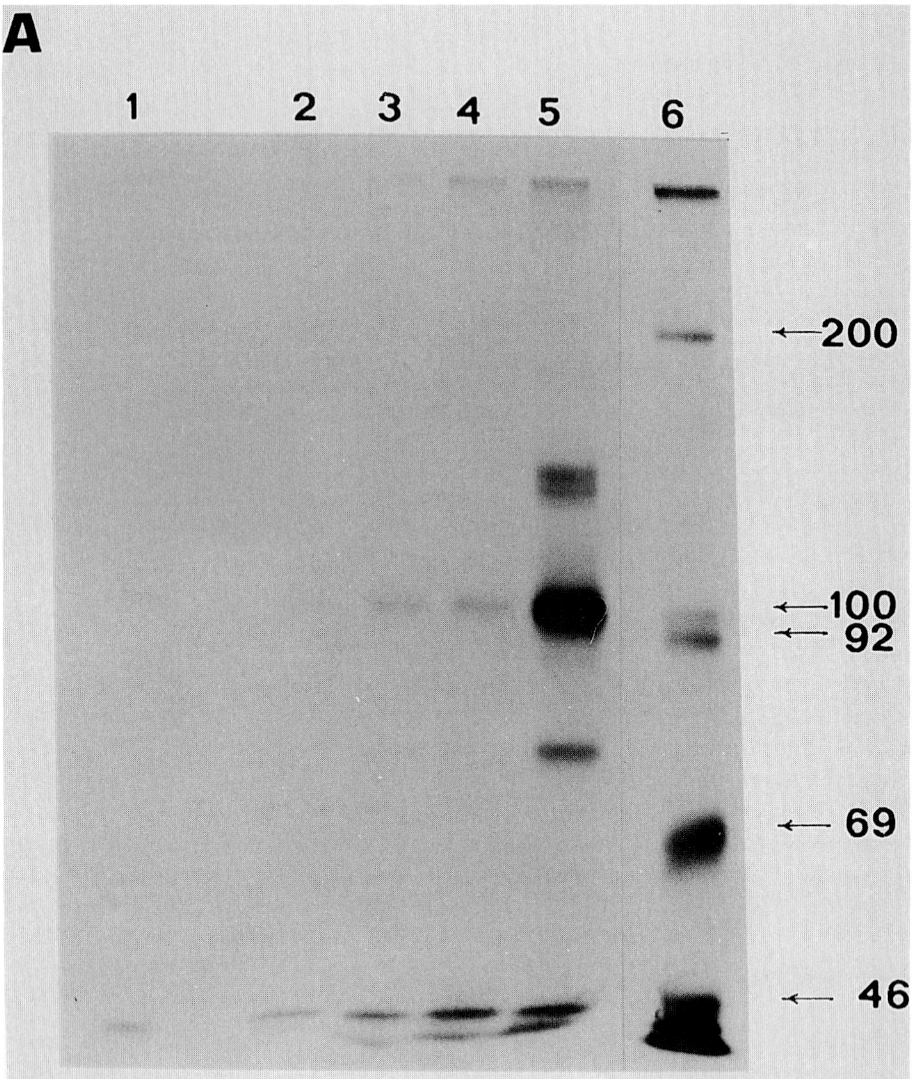

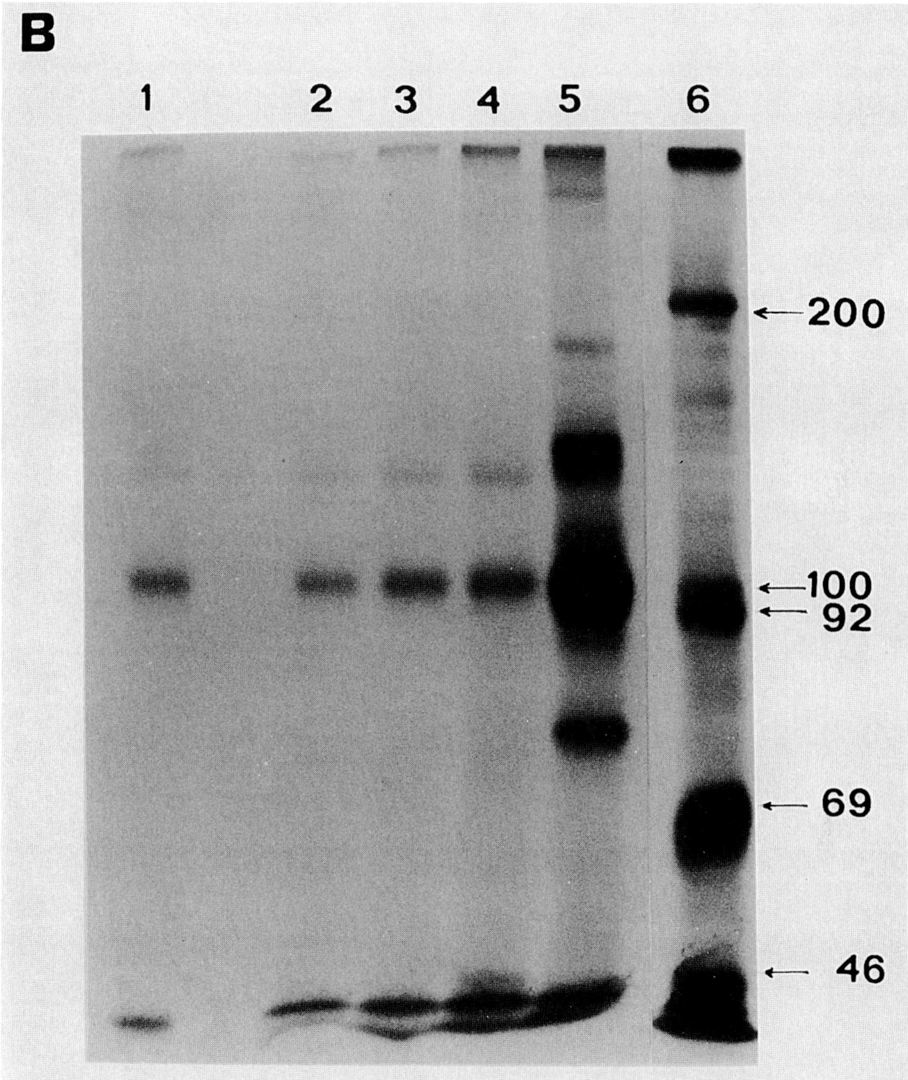

Figure 5.6 ^{35}S-labelled Y-79 cell lysates immunoprecipitated with the antiserum anti-β_1. Equal amounts of cells were labelled.
panel A: shorter exposure (2 weeks); *panel B*: overexposure (1 month):
lane 1: Y-79 cells grown in suspension;
lane 2: Y-79 cells grown on poly-D-lysine (5 μg/cm^2) coated dishes;
lane 3: Y-79 cells grown on poly-D-lysine (5 μg/cm^2) coated dishes; 2 days after seeding 12.5 μg/ml of laminin was added to the medium for 3 days;
lane 4: Y-79 cells grown on poly-D-lysine (5 μg/cm^2) coated dishes; 12.5 μg/ml of laminin was added to the medium for 5 days;
lane 5: A2058 melanoma cells, treated as lane 4;
lane 6: m.w. standards.

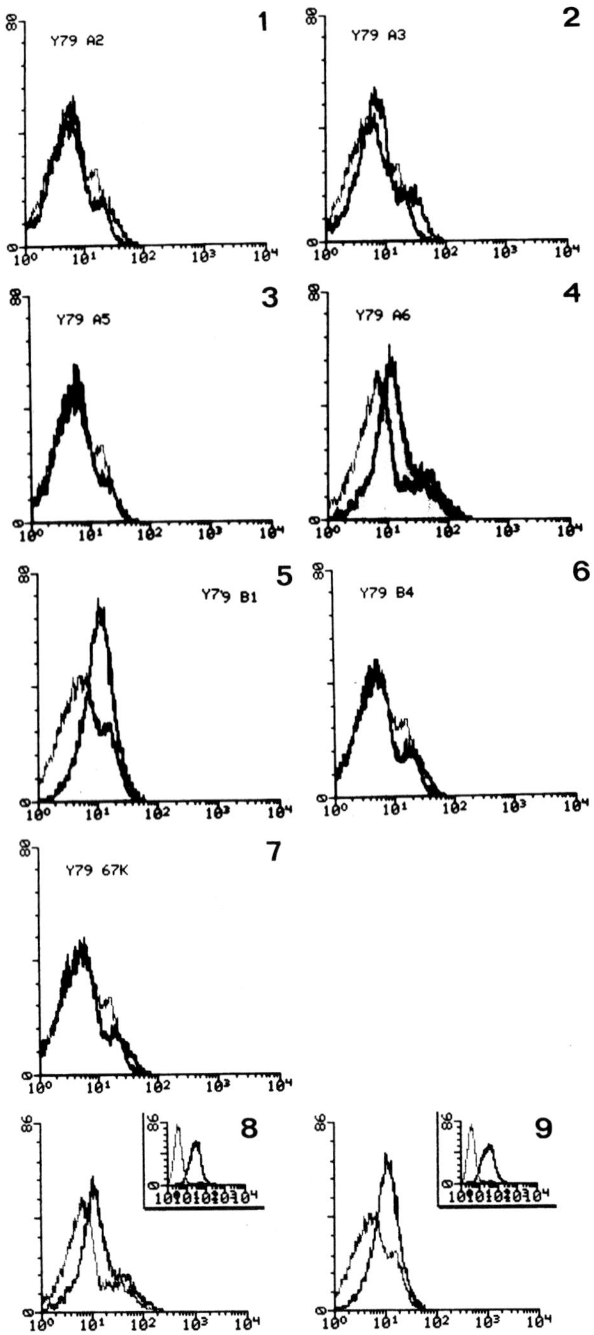

Figure 5.7 FACS analysis of Y-79 cells, performed with monoclonal antibodies to: 1) α_2 integrin subunit, 2) α_3 integrin subunit, 3) α_5 integrin subunit, 4) α_6 integrin subunit, 5) β_1 integrin subunit, 6) β_4 integrin subunit, 7) high affinity 67 kDa LN-receptor. In panels 8 and 9, the reaction of anti-α_6 (8) and anti-β_1 (9) integrin subunits with Y-79 cells is compared to the positive control Me1007 human melanoma cells (inset).

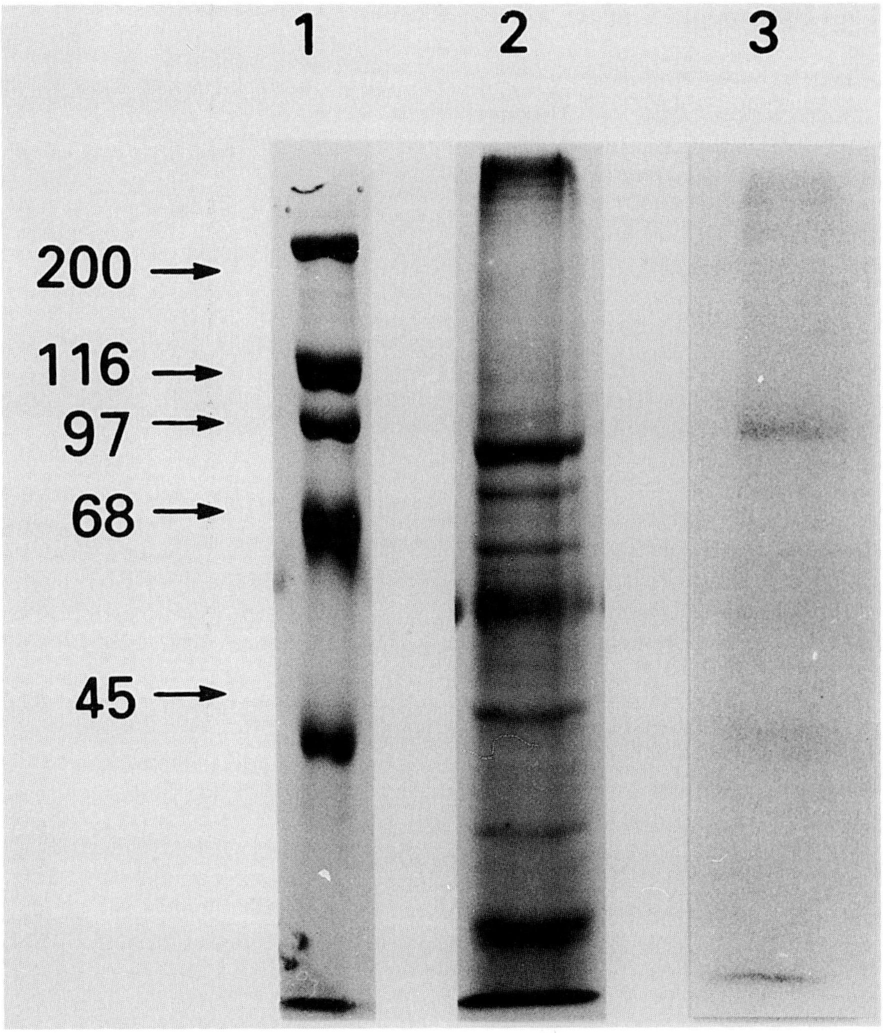

Figure 5.8 7% SDS-polyacrilamyde gel electrophoresis of fraction eluted from LN-Sepharose affinity columns, with extraction buffer containing 20 mM EDTA. Cell lysates of lactopexidase^{-125}I labelled Y-79 cells grown on poly-D-lysine were used. lane 1: m.w. standards; lane 2: Comassie-blue staining of 20 mM EDTA eluant fraction; lane 3: autoradiograph of 20 mM EDTA eluted fraction.

LN-receptor or the best known integrin receptors for LN are represented on these cells. Only trace amounts of β_1 and α_6 subunits could be detected, which probably can not account for the LN effects on the cells.We have shown that a single major LN-binding protein of 100 kDa was eluted by 20 mM EDTA from LN-Sepharose columns.[12] This molecule is a principal candidate as a mediator of Y-79-LN interactions. The nature and structure of this Ca^{++} dependent protein is under study.

Acknowledgements

The research was financed by a grant of AIRC "Geni Oncosoppressori" and grants from the CNR and Ministero della Sanit.

After submission of this study, similar data on low integrin levels on Y-79 cells has been reported by another group.

REFERENCES

1. Martin, G. R. and Timpl, R. (1987) Laminin and other basement membrane components. *Annual Review of Cell Biology,* **3**, 57–85.
2. Gundersen, R. W. (1987) Response of sensory neurites and growth cones to patterned substrata of laminin and fibronectin *in vitro. Developmental Biology,* **121**, 423–431.
3. Kleinman, H. K., Graf, J., lwamoto, Y., Kitten, G. T., Ogle, R. C., Sasaki, M., Yamada, Y., Martin, G. R. and Luckenbill-Edds, L. (1987) Role of basement membranes in cell differentiation, in *Cell Biology of the Testis and Epididymis,* Annuals of New York Academy of Sciences, **513**, 134–145.
4. Terranova, V. P., Liotta, L. A., Russo, R. G. and Martin, G. R. (1982) Role of laminin in the attachment and metastasis of murine tumor cells. *Cancer Research,* **42**, 2265–2269.
5. Mecham, R. P. (1991) Receptors for laminin on mammalian cells. *FASEB Journal,* **5**, 2538–2546.
6. Gallie, B., Albert, D. M., Wong, J. J., Buyukmichi, N. and Pullafito, C. A. (1977) Heterotransplantation of retinoblastoma into athymic "nude" mouse. *Investigative Ophthalmology and Visual Science,* **16**, 256–259.
7. Kyritsis, A. P., Tsokos, M., Triche, T. J. and Chader, G. J. (1984) Retinoblastoma: origin from a primitive neuroectodermal cell? *Nature,* **307**, 471–473.
8. Kyritsis, A. P., Tsokos, M. and Chader, G. J. (1987) Behaviour of human retinoblastoma cells in tissue culture, in *Progress in Retinal Research,* Vol. 6 (edited by Osborne, N. N. and Chader, G. J.) Pergamon Press, N.Y., 245–274.
9. Lee, E., Bookstein, R., Young, F., Lin, C., Rosenfeld, M. G. and Lee, W. H. (1988) Molecular mechanisms of retinoblastoma gene inactivation in retinoblastoma cell line Y-79. *Proceedings of the National Academy of Sciences of the USA,* **85**, 6017–6021.
10. Kyritsis, A. P., Tsokos, M. and Chader, G. J. (1984) Attachment culture of human retinoblastoma cells: long term culture conditions and effects of dibutyrylcyclic AMP. *Experimental Eye Research,* **38**, 411–421.
11. San Antonio, J. D. and Tuan, R. S. (1986) Chondrogenesis of limb bud mesenchyme in vitro: stimulation by cations. *Developmental Biology,* **115**, 313–324.
12. Albini, A., Noonan, D. M., Melchiori, A., Fassina, G. F., Percario, M., Gentlemen, S., Toffenetti, J. and Chader, G. J. (1992) Laminin induced retinoblastoma cell differentiation: possible involvement of a 100 kDa cell-surface laminin-binding protein. *Proceedings of the National Academy of Sciences of the USA,* **89**, 2257–2261.

13. Van Veen, T., Katial, A., Shinohara, T., Barrett, D. J., Wiggert, B., Chader, G. J. and Nickerson, J. M. (1986) Retinal photoreceptor neurons and pinealocytes accumulate mRNA for interphotoreceptor retinoid-binding protein (IRBP). *FEBS Letters*, **208**, 133–137.
14. Postlethwaite, A. E., Snyderman, R. and Kang, A. H. (1976) The chemotactic attraction of human fibroblasts to a lymphocyte-derived factor. *Journal of Experimental Medicine*, **144**, 1188–1203.
15. Kleinman, H. K., McGarvey, M. L., Hassell, J. R., Star, V. L., Cannon, F., Laurie, G. W. and Martin, G. R. (1986) Basement membrane complexes with biological activity. *Biochemistry*, **25**, 312–318.
16. Defilippi, P., van Hinsbergh, V., Bertolotto, A., Rossino, P., Silengo, L. and Tarone, G. (1991) Differential distribution and modulation of expression of alpha1/beta1 integrin on human endothelial cells. *Journal of Cell Biology*, **114**, 855–863.
17. Defilippi, P., Truffa, G., Stefanuto, G., Altruda, F., Silengo, L. and Tarone, G. (1991) Tumor necrosis factor α and interferon γ modulate the expression of the vitronectin receptor (integrin β_3) in human endothelial cells. *Journal of Biological Chemistry*, **226**, 7638–7645.
18. Laemmli, V. (1970) Cleavage of structure proteins during the assembly of the head of bacteriophage T_4. *Nature*, **227**, 680–685.
19. Tagliabue, E., Martignone, S., Mastroianni, A., Ménard, S., Pellegrini, R. and Colnaghi, M. I. (1991) Laminin receptors on SCLC cells. *British Journal of Cancer*, **63**-suppl 14, 83–85.
20. Campbell, M. A. and Chader, G. J. (1988) Effects of laminin on attachment, growth and differentiation of cultured Y-79 retinoblastoma cells. *Investigative Ophtalmology and Visual Science*, **29**, 1517–1522.
21. Fassina, G., Paglialunga, G., Noonan, D. M., Chader, G. J. and Albini, A. (1993) Modulation of Y-79 retinoblastoma cell differentiation and IRBP expression by dibutyryl cyclic AMP and laminin. *International Journal of Oncology*, **2**, 745–751.
22. Skubitz, A. P. N., Grossman, M. D., McCarthy, J. B., Wayner, E. A. and Cameron, J. D. (1994) The decreased adhesion of Y-79 retinoblastoma cells to extracellular matrix proteins is due to a deficit of integrin receptors. *Invest. Ophthalmol. Vis. Sci.* **35**, 2820–2833.

6. CRUCIAL INVOLVEMENT OF αv INTEGRINS IN MALIGNANT PHENOTYPE EXPRESSION OF HUMAN MELANOMA CELLS

B. FELDING-HABERMANN,[1] B. M. MUELLER,[2] L. C. SANDERS[2] and D. A. CHERESH[2]

The Scripps Research Institute, 10666 N. Torrey Pines Rd., La Jolla, CA 92037, USA

[1]Department of Molecular and Experimental Medicine, SBR 8

[2]Department of Immunology, IMM 13

Melanoma tumours originate in the skin and develop from neoplastically transformed melanocytes. Normal melanocytes are terminally differentiated cells and express a panel of adhesion receptors which are utilized to establish and maintain cell-cell contacts and cell matrix interactions. Amongst these adhesion receptors are specific integrins composed of α and β subunits that form non covalently linked heterodimers and mediate melanocyte attachment to extracellular matrix components as well as cell-cell interactions within the epidermal tissue.

Melanocytes characteristically express integrins of the $\beta 1$ family that enable them to interact with fibronectin, laminin and collagen.[1] In contrast to most tumours which show a general down regulation of integrins metastatic melanoma cells express new integrins that modify their ligand binding properties.[2] Altered laminin and collagen recognition is mediated by $\alpha 2\beta 1$ and $\alpha 6\beta 1$ expression[1,3–6] and $\alpha 4\beta 1$ expression results in melanoma cell interaction with an alternative cell attachment site in fibronectin as well as binding to activated endothelial cells.[7,8] One of the most striking differences between benign and malignant lesions derived from human melanocytes is the expression of integrin $\alpha v\beta 3$ since it is not detectable in nevi or horizontally spreading melanoma but occurs exclusively in cells that have progressed to the vertical growth phase and may cause metastatic melanomas.[9]

$\alpha v\beta 3$, the vitronectin receptor, is the most promiscuous of the integrins in terms of its ligand binding capacities. It recognizes Arg-Gly-Asp containing sequences in a variety of proteins, such as vitronectin, fibrinogen, von Willebrand factor,[10,11] thrombospondin,[12] osteopontin,[13] fibronectin[14] and laminin.[15] The expression of integrin $\alpha v\beta 3$ thus enables melanoma cells to interact with and potentially migrate in almost any tissue that the cells encounter during metastasis. Therefore, integrin $\alpha v\beta 3$ may play a crucial role in the development of a malignant phenotype in melanoma cells. In order to test this hypothesis, we generated an experimental model system that analyzes directly the requirement of αv integrin expression on human melanoma cells for their tumorigenic capacities.

M21 human melanoma cells express the αv integrins $\alpha v\beta 3$ and $\alpha v\beta 5$.[10,16] From this cell line a stable spontaneous mutant was selected based on the lack of αv expression using a monoclonal antibody that specifically recognizes the human αv subunit. These M21-L cells do not form either of the αv containing integrins due to a defect in αv gene expression at the transcriptional level.[10] M21-L cells, however, display indistinguishable ratios of $\beta 1$ integrins when compared to the parental M21 cell line. Since M21-L cells may differ from M21 cells not only in αv integrin expression but potentially in further unknown characteristics it was necessary to generate a valid control that allows conclusive statements on the role of the αv integrins when using the M21/M21-L cell lines. We therefore transfected M21-L cells with a vector containing a full length αv cDNA and a gene encoding neomycine resistance.[17] From drug resistant subpopulations M21-L4 cells were selected for αv integrin expression at levels comparable to the parental M21 cell line. M21-L12 cells were selected as non-αv-expressing control transfectants. In M21-L4 cells αv transfection restored cell surface expression of both, $\alpha v\beta 3$ and $\alpha v\beta 5$, as was demonstrated by FACS analysis and immunoprecipitation of surface labelled cells using αv specific as well as complex specific anti $\alpha v\beta 3$ or anti $\alpha v\beta 5$ monoclonal antibodies.[17] Interestingly, the ratio of $\alpha v\beta 3$ to $\alpha v\beta 5$ expression in M21-L4 cells was indistinguishable from the ratio found in the parental M21 cells with $\alpha v\beta 3$:$\alpha v\beta 5$ being approximately 20:1. This indicates that the expression levels of these integrins are controlled by β subunit synthesis when αv is present. In all four cell lines M21 (αv +), M21-L (αv-) and the transfectants, M21-L4 (αv +) and M21-L12 (αv-), the levels of $\beta 1$ integrin expression were unchanged as was demonstrated by using a $\beta 1$ specific monoclonal antibody.[17]

In order to test the requirement of αv integrins, $\alpha v\beta 3$ and/or $\alpha v\beta 5$, for malignant phenotype expression, the human melanoma cell variants M21, M21-L, M21-L4 or M21-L12 cells were injected subcutaneously into the flanks of athymic nude mice and were analyzed for their tumorigenicity.[17] Inoculation of 1×10^6 M21 cells resulted in rapidly growing tumours in 7/8 animals while animals injected with this number of M21-L cells developed relatively small tumours in only 2/8 animals and these were considerably delayed in their appearance. M21-L cells failed to produce tumours when smaller cell numbers were injected (1×10^5). The impaired tumorigenicity of αv lacking M21-L cells was restored upon αv transfection in M21-L4 cells. 1×10^6 M21-L4 cells produced fast growing tumours in 6/8 animals whereas the same number of αv-negative control transfectants, M21-L12, caused only a small tumour in 1/8 animals. Similar results were obtained in four independent experiments, thus showing that both tumour take rate as well as the median volume of the developing tumours depended to a large extent on the expression of integrin(s) $\alpha v\beta 3$ and/or $\alpha v\beta 5$ on the injected melanoma cells. It is important to point out that cells cultured from tumours induced by αv-negative M21-L or M21-L12 cells did not express αv integrins indicating that these tumours did not result from an *in vivo* selection of reverted αv-bearing cells. To rule out a general growth defect in the αv-lacking M21-L or M21-L12 melanoma cells their *in vitro* proliferation rates were analyzed and compared to those of the αv-positive M21 and M21-L4 cells. In the presence of fetal bovine serum all four cell lines displayed indistinguishable growth behaviours.[17] Therefore, the impaired tumorigenicity of αv-negative M21-L and M21-L12 cells must be due to a lack in their *in vivo* growth capacities.

Tumour cells are generally characterized by their altered dependence on growth factors.[18] An increase in malignancy correlates with a loss of growth factor requirement from the surrounding cells or tissue and may ultimately lead to complete autonomy of the tumour cells.[19] Since we found that the expression of αvβ3 and/or αvβ5 was crucial for M21 melanoma cell growth *in vivo*, this system provides an experimental basis to study the consequences of integrin occupancy on metabolic reactions of the cell, such as responsiveness to or production of growth factors and possibly the loss of exogenous growth factor dependence. Indeed, when αv-positive and αv-negative M21 cells were compared in *in vitro* proliferation tests in the absence of serum or other exogenous growth factors, but provided with immobilized extracellular matrix proteins, we could demonstrate that αv-expressing M21 and M21-L4 cells but not αv-negative M21-L or M21-L12 cells proliferated in response to immobilized vitronectin.[20] All four cell types showed increased proliferation when seeded onto fibronectin- or collagen type I- matrices and this is likely due to their equivalent expression of β1 integrins.[17] These results suggest that vitronectin can promote a proliferative signal to human melanoma cells by interacting with their αv integrin(s). Further evidence supports a role of αv integrin ligation not only for *in vitro* but also for *in vivo* tumour cell proliferation. When vitronectin was coinjected with M21 cell variants subcutaneously into nude mice, αv expressing M21 and M21-L4 cells induced tumours with a fourfold increase in proliferation over cells injected without vitronectin.[20] This increased proliferation was not observed when αv-negative M21-L or M21-L12 cells were coinjected with vitronectin. As a control, all four cell types showed increased tumour growth upon coinjection with collagen. Together, these results indicate that the presence of an αv integrin ligand, specifically vitronectin, in an *in vivo* microenvironment favours proliferation of cells that express αv integrins. Indeed, vitronectin deposition in the extracellular matrix of the skin has been shown to increase with age and with exposure to the sun.[21] Both, aging and sun exposure enhance the likelihood of contracting melanoma. Therefore, the presence of vitronectin and the expression of αv-containing vitronectin receptors on melanoma cells may stimulate their proliferation and thus favour melanoma tumorigenicity, which in turn may potentiate metastatic disease.

Other metabolic consequences of αv integrin ligation that impact upon tumorigenicity as well as metastatic dissemination could be the production of proteases that are necessary to invade surrounding tissue and underlaying basement membranes. Correlation of integrin expression, their ligand occupancy and protease secretion have been described at the molecular level for tumour cell lines[22] and using an experimental *in vivo* model for invasion mechanisms.[23] Moreover, the ability of tumour cells to form measurable tumours in an organism depends on their capacity to induce angiogenesis and thus ascertain sufficient supplies necessary for tumour growth beyond a critical size.[24] We demonstrated that αv integrin expression is crucially involved in human melanoma tumorigenicity. Therefore αv integrin expression and potentially αv integrin ligand occupancy may have direct effects on the production of proteases or angiogenic factors by the tumour cells. Our M21 melanoma cell model provides an experimental basis to test these possibilities.

An alternate explanation for the impaired tumorigenicity of αv-lacking melanoma cells may be the requirement of αv integrin-mediated *in vivo* adhesion at the site of injection. We therefore tested the function of the newly expressed αvβ3 and αvβ5 integrins on the αv transfected M21-L4 cells that showed restored tumorigenicity. Corresponding to the parental M21 cell line these cells readily attached to vitronectin in *in vitro* assays.[17] Inhibition studies using function blocking anti αvβ3 and/or αvβ5 monoclonal antibodies showed that both αv integrins, αvβ3 and αvβ5, were participating in M21-L4 cell attachment to vitronectin since both antibodies were required to block fully this adhesion. The *in vivo* ligand for the melanoma cell αv integrins in the skin is likely to be vitronectin, but this awaits confirmation. It is further unclear whether both integrins, αvβ3 and αvβ5, are crucially involved in melanoma cell tumorigenicity or which of both is responsible.

An experimental approach to identify the critical integrin, αvβ3 or αvβ5, or potentially β3 integrins in general, could be based on the specificity of subunit heterodimer formation within the β3 integrin family. αvβ3 and αIIbβ3 are the known receptors that exclusively share the β3 subunit.[25–27] αIIbβ3, the platelet fibrinogen receptor, does not occur on M21 melanoma cells. What has formerly been addressed as IIbβ3-like integrin on other human melanoma cell lines is likely αvβ3, as has later been established for most cases. Using the M21-L cell model it was shown that the platelet receptor α subunit GPIIb associates with β3 when transfected into M21-L cells which lack expression of a β3 complexing α subunit.[28] Moreover, we were able to demonstrate that upon IIb transfection into M21-L cells, IIb associated with β3 only, and not with β5.[29] Therefore, differences between the α subunits, αv and αIIb, account for specific recognition of the matching β subunit. The αIIb transfected cells, M21-LIIb, adhere well to IIbβ3 ligands, such as fibrinogen, von Willebrand factor and thrombospondin, but do not attach to vitronectin.[28,29] This finding is currently taken as an advantage to obtain information on the ligand involved in tumour formation after inoculation of M21 cell variants *in vivo*. If the IIb transfected M21-LIIb cells were able to proliferate *in vivo* to an extent comparable to M21 cells that express αvβ3 and αvβ5, vitronectin would then have to be excluded as the only critical ligand recognized by the injected cells *in vivo*. Furthermore, if the M21-LIIb cells were tumorigenic this would demonstrate that β5 expression is not crucial for melanoma cell growth *in vivo* and thus would focus on $\beta 3$ as the critical integrin β subunit. Experimental evidences so far indicate that expression of the $\beta 3$ subunit is linked to the progression of human melanoma. $\beta 3$ was found to be restricted to invasive melanoma.[9] An antibody to the $\beta 3$ subunit was reported to inhibit growth of human melanoma cells in nude mice.[30] Furthermore, Arg-Gly-Asp containing peptides which are known to block the function of $\alpha v\beta 3$ were shown to inhibit both melanoma tumour invasion *in vitro*[31] and the development of experimental metastasis in a murine melanoma model system *in vivo*.[32–34] Correspondingly, naturally occurring Arg-Gly-Asp encompassing peptides from snake venoms were demonstrated to block most efficiently melanoma cell adhesion[35] and experimental metastasis.[36]

Our data clearly demonstrate, that an αv containing integrin on human melanoma cells is critically involved in their tumorigenicity. The M21 melanoma cell system provides an experimental model to test further whether $\beta 3$ is the responsible β subunit and therefore help to identify the integrin which functions on human melanoma cells as a crucial receptor in establishing the malignant phenotype in these tumours. This knowledge will be of significant importance for the development of potential therapeutic measures to fight malignant melanoma, one of the most aggressive tumours in humans.

REFERENCES

1. Klein, C. E., Steinmayer, T., Kaufmann, D., Weber, L. and Broecker, E. B. (1991) Identification of a melanoma progression antigen as integrin VLA-2. *J. Invest. Dermatol.*, **96**, 281–284.
2. Buck, C., Albelda, S., Damjanovich, L., Edelman, J., Shih, D. T. and Solowska, J. (1990) Immunohistochemical and molecular analysis of $\beta 1$ and $\beta 3$ integrins. *Cell Differ. Dev.*, **32**, 189–202.
3. Kramer, R. H. and Marks, N. (1989) Identification of integrin collagen receptors on human melanoma cells. *J. Biol. Chem.*, **264**, 4684–4688.
4. Kramer, R. H., McDonald, K. A. and Vu, M. P. (1989) Human melanoma cells express a novel integrin receptor for laminin. *J. Biol. Chem.*, **264**, 15642–15649.
5. Ramos, D. M., Berston, E. D. and Kramer, R. H. (1990) Analysis of integrin receptors for laminin and type IV collagen on metastatic B16 melanoma cells. *Cancer Res.*, **50**, 728–734.
6. Mortarini, R., Anichini, A. and Parmiani, G. (1991) Heterogeneity for integrin expression and cytokine-mediated VLA modulation can influence the adhesion of human melanoma cells to extracellular matrix proteins. *Int. J. Cancer*, **47**, 551–559.
7. Mould, A. P., Wheldon, L. A., Komoriya, A., Wayner, E.A., Yamada, K. M. and Humphries, M. J. (1990) Affinity chromatographic isolation of the melanoma adhesion receptor for the IIICS region of fibronectin and its identification as the integrin $\alpha 4\beta 1$. *J. Biol. Chem.*, **265**, 4020–4024.

8. Martin-Padura, I., Mantovani, A., Anchini, A. and Dejana, E. (1991) Heterogeneity in human melanoma cell adhesion to cytokine activated endothelial cells correlates with VLA-4 expression. *Cancer Res.*, **51**, 2239–2241.
9. Albelda, S. M., Mette, S. A., Elder, D. E., Stewart, R., Damjanovich, L., Herlyn, M. and Buck, C. A. (1990) Integrin distribution in malignant melanoma: Association of the $\beta 3$ subunit with tumor progression. *Cancer Res.*, **50**, 6757–6764.
10. Cheresh, D. A. and Spiro, R. C. (1987) Biosynthetic and functional properties of an Arg-Gly-Asp-directed receptor involved in human melanoma cell attachment to vitronectin, fibrinogen and von Willebrand factor. *J. Biol. Chem.*, **262**, 17703–17711.
11. Cheresh, D. A. (1987) Human endothelial cells synthesize and express an Arg-Gly-Asp-directed receptor involved in attachment to fibrinogen and von Willebrand factor. *Proc. Natl. Acad. Sci. USA*, **84**, 6471–6475.
12. Lawler, J., Weinstein, R. and Hynes, R. O. (1988) Cell attachment to thrombospondin: the role of RGD and integrin receptors. *J. Cell. Biol.*, **107**, 2351–2361.
13. Reinholt, F. P., Hultenby, K., Oldberg, A. and Heinegard, D. (1990) Osteopontin, a possible anchor of osteoclasts to bone. *Proc. Natl. Acad. Sci. USA*, **87**, 4473–4475.
14. Charo, I. F., Nannizzi, L., Smith, J. W. and Cheresh, D. A. (1990) The vitronectin receptor $\alpha v\beta 3$ binds fibronectin and acts in concert with $\alpha 5\beta 1$ in promoting cellular attachment and spreading on fibronectin. *J. Cell Biol.*, **111**, 2795–2800.
15. Kramer, R. H., Cheng, Y. F. and Clyman, R. (1990) Human microvascular endothelial cells use $\beta 1$ and $\beta 3$ integrin receptor complexes to attach to laminin. *J. Cell. Biol.*, **111**, 1233–1243.
16. Wayner, E. A., Orlando, R. A. and Cheresh, D. A. (1991) Integrins $\alpha v\beta 3$ and $\alpha v\beta 5$ contribute to cell attachment to vitronectin but differentially distribute on the cell surface. *J. Cell Biol.*, **113**, 919–929.
17. Felding-Habermann, B., Mueller, B. M., Romerdahl, C. A. and Cheresh, D. A. (1992) Involvement of integrin αv gene expression in human melanoma tumorigenicity. *J. Clin. Invest.*, **89**, 2018–2022.
18. Dickson, C. and Peters, G. (1987) Potential oncogene product related to growth factors. *Nature*, **326**, 833.
19. Cross, M. and Dexter, T. M. (1991) Growth factors in development, transformation and tumorigenesis. *Cell*, **64**, 271–280.
20. Sanders, L. C., Felding-Habermann B., Mueller, B. M. and Cheresh, D. A. (1992) Role of αv integrins and vitronectin in human melanoma cell growth. In *Cold Spring Harbor Symposia on Quantitative Biology*, Vol. **LVII**. Cold Spring Harbor Laboratory Press, 233–240.
21. Dahlback, K., Lofberg, H., Alumets, J. and Dahlback, B. (1989) Immunohistochemical demonstration of age-related deposition of vitronectin (S-protein of complement) and terminal complement complex on dermal elastic fibers. *J. Invest. Dermatol.*, **92**, 727–733.
22. Werb, Z., Tremble, P. M., Behrendtsen, O. and Crowley, E. (1989) Signal transduction through the fibronectin receptor induces collagenase and stromelysin gene expression, *J. Cell Biol.*, **109**, 877–889.
23. Fisher, S. J., Cui, T. Y., Zhang, L., Hartman, L., Grahl, K., Zhang, G. T.,

Tarpey, J. and Damsky, C. H. (1989) Adhesive and degradative properties of human placental cytotrophoblast cells *in vitro. J. Cell Biol.*, **109**, 891–902.

24. Folkman, J. (1984) Angiogenesis: initiation and modulation. In *Cancer Invasion and Metastasis: Biological and Therapeutic Aspects* (edited by Nicolson, G. L. and Milas, L.), Raven Press, New York, pp. 201–209.
25. Fitzgerald, L. A., Steiner, B., Rall, S. C., Lo, S. and Phillips, D. R. (1987) Protein sequence of endothelial glycoprotein IIIa derived from a cDNA clone. Identity with platelet glycoprotein IIIa and similarity to 'integrin'. *J. Biol. Chem.*, **262**, 3936–3939.
26. Fitzgerald, L. A., Poncz, M., Steiner, B., Rall, S. C., Bennett, J. S. and Phillips, D. R. (1987) Comparison of CDNA-derived protein sequences of the human fibronectin and vitronectin receptor α-subunit and platelet glycoprotein IIb. *Biochemistry*, **26**, 8158–8165.
27. Suzuki, S., Argraves, W. S., Arai, H., Languino, L. R., Pierschbacher, M. D. and Ruoslahti, E. (1987) Amino acid sequence of the vitronectin receptor α subunit and comparative expression of adhesion receptor mRNAs. *J. Biol. Chem.*, **262**, 14080–14085.
28. Kieffer, N., Fitzgerald, L., Wolf, D., Cheresh, D. A. and Phillips, D. R. (1991) Adhesive properties of the β3 integrins: comparison of GPIIb-IIIa and the vitronectin receptor individually expressed in human melanoma cells. *J. Cell Biol.*, **113**, 451–461.
29. Felding-Habermann, B. and Cheresh, D. A. Unpublished results.
30. Boukerche, H., Berthier-Vergnes, O., Bailly, M., Dole, J. F., Leung, L. K. and McGregor, J. L. (1989) A monoclonal antibody (LYP18) directed against the blood platelet glycoprotein Iib/IIIa complex inhibits human melanoma growth *in vivo. Blood*, **74**, 909–912.
31. Gehlson, K. R., Argraves, W. S., Pierschbacher, M. D. and Ruoshlahti, E. (1988) Inhibition of *in vitro* tumor invasion by Arg-Gly-Asp-containing peptides. *J. Cell Biol.*, **106**, 925–930.
32. Humphries, M. J., Olden, K. and Yamada, K. M. (1986) A synthetic peptide from fibronectin inhibits experimental metastasis of murine cells. *Science*, **233**, 467–470.
33. Saiki, I., Murata, J., Lida, N., Sugimura, K. and Azuma, I. (1988) The inhibition of murine lung metastasis by synthetic polypeptides [poly (RGD) and poly (YIGSR)] with a core sequence of cell adhesion molecules. *Br. J. Cancer*, **59**, 194–197.
34. Humphries, M. J., Yamada, K. M. and Olden, K. (1988) Investigation of the biological effects of the anti-cell adhesion synthetic peptides that inhibit experimental metastasis of B16-F10 murine melanoma cells. *J. Clin. Invest.*, **81**, 782–790.
35. Knudsen, K. A., Tuszynski, G. P., Huang, T. F. and Nieviarowski, S. (1988) Trigramin, an RGD-containing peptide from snake venom, inhibits cell-substratum adhesion of human melanoma cells. *Exp. Cell Res.*, **179**, 42–49.
36. Soszka, T., Knudsen, K. A., Beviglia, L., Rossi, C., Poggi, A. and Niewiarowski, S. (1991) Inhibition of murine melanoma cell-matrix adhesion and experimental metastasis by albolabrin, an RGD-containing peptide isolated from the venom of *Trimeresurus albolabris. Exp. Cell Res.*, **196**, 6–12.

7. INTEGRINS IN RENAL CELL CARCINOMA: CORRELATIONS WITH HISTOLOGICAL GRADE

MATTI KORHONEN[1,2], VICTOR E. GOULD[3] and ISMO VIRTANEN[2]

[2]Department of Anatomy, University of Helsinki, Finland
[3]Department of Pathology, Rush Medical College, Chicago, IL, USA

Following our earlier studies on integrin expression in the developing and adult human kidney, we studied 41 renal cell carcinoma samples (RCC) for the expression of the $\alpha_{1-6,v}$ and $\beta_{1,3,4}$ integrin subunits, and the basement membrane components laminin and collagen type IV. We found that: 1) all RCCs expressed the $\alpha_3\beta_1$ complex; 2) Decreased expression of the $\alpha_6\beta_1$ complex and increased expression of the α_v subunit combined with β_5, correlate with increasing histologic grade of the tumours; 3) The tumour stroma, containing myofibroblasts and blood vessels, typically expresses the $\alpha_1\beta_1$ and $\alpha_3\beta_1$ complexes. Studies with RCC cells *in vitro* suggest that the α_v subunit may be complexed with the β_1, β_3 or β_5 subunits in RCC cells, one cell often expressing more than one heterodimer combination.

KEYWORDS: integrin, renal cell carcinoma, kidney, basement membrane, immunohistochemistry

INTRODUCTION

Integrins are a less than 10-year-old family of cell surface receptors, pertinent to the interactions of cells with their surroundings, viz. the extracellular matrix (ECM) and other cells (Hynes, 1992). Each integrin complex is a heterodimer of an α and a β subunit. The family is usually divided into three subgroups (β_1, β_2 and β_3), named after the principal β subunit that forms heterodimeric complexes with several α subunits. However, recent studies indicate that a single α subunit may also complex with various β subunits, for instance in the $\alpha_v\beta_1$, $\alpha_v\beta_3$ and $\alpha_v\beta_5$ complexes which may also differ in their affinities for various ligands (Felding-Habermann and Cheresh, 1993). Here, we intend to focus on the β_1 group, as well as on α_v subunit-containing integrins.

INTEGRINS *IN VITRO*

When cultured cells are seeded on a culture substratum coated with the appropriate ligand, many of the β_1 integrins are seen to organize into large supramolecular complexes termed focal adhesions (Burridge and Fath, 1989). Here, the actin cytoskeleton appears to be physically integrated with

[1] Correspondence. Matti Korhonen, MD, Department of Anatomy, University of Helsinki, P.O. Box 9 (Siltavuorenpenger 20 A), FIN-00014 University of Helsinki, Finland
Tel: +358-0-191 8468 Telefax: +358-0-191 8499

the ECM (hence the name 'integrin'). Such studies have revealed that integrins are required for cell adhesion, migration and, for instance, neurite outgrowth. Furthermore, studies of cell differentiation models have shown that the adhesive properties of cells, and integrin expression, may be modulated during differentiation. Moreover, integrin function may influence gene expression and play a part in controlling cell differentiation itself (Adams and Watt, 1993).

INTEGRINS *IN VIVO*

Several *in vivo* studies have been concerned with integrin distributions in developing and/or adult tissues, such as the skin (Peltonen *et al.* 1989), intestine (Choy *et al.* 1990), pancreas (Hall *et al.* 1991; Weinel *et al.* 1992), mammary gland (Koukoulis *et al.* 1991; Pignatelli *et al.* 1992), the liver (Volpes *et al.* 1991), cornea (Lauweryns *et al.* 1991; Tervo *et al.* 1991), placenta (Korhonen *et al.* 1991), and kidney (Cosio *et al.* 1990; Korhonen *et al.* 1990 a and b; Rahilly and Fleming, 1992). In a number of these studies, a temporo-spatially regulated expression of certain integrin complexes was reported.

These studies suggested that β_1 integrins may mediate cell adhesion, and possibly positional or differentiating signals during histogenetic processes. In experiments with embryos or tissue culture systems, a number of processes requiring cell adhesion have proved perturbable by either blocking antibodies directed against integrins, or by synthetic peptides derived from ECM ligands of integrins. Such processes include gastrulation (Naidet *et al.* 1987), neural crest cell migration (Perris *et al.* 1989), formation of capillary tubules by endothelial cells grown on a basement membrane substratum (Grant *et al.* 1989), and formation of polarized epithelium in an *in vitro* histogenesis model of the kidney (Sorokin *et al.* 1990).

INTEGRINS IN RCC

RCCs comprise about 2% of human cancers. They can be graded 1–3 according to histological and cytological criteria (Thoenes *et al.* 1986). The majority of cases are well-differentiated adenocarcinomas (G1), but more anaplastic forms (grades 2 and 3) are also found. The most often used prognostic criteria are tumour staging and size, but grading also correlates with the prognosis (Medeiros *et al.* 1988). Both proximal (Oosterwijk *et al.* 1990) and distal (Blouin *et al.* 1989) tubular origins have been suggested for RCCs.

In the developing and adult human kidney, specific nephron segments express a characteristic repertoire of integrins, the expression emerging at the time of differentiation of each particular structure (Korhonen *et al.* 1990 a and b). Integrin expression in various nephron segments is summarized in table 7.1. All kidney tubules, the presumed origin of RCC, express the $\alpha_6\beta_1$ integrin as their basement membrane receptor. Distal tubules, in contrast to their proximal counterparts, display additional immunoreactivity for the $\alpha_2\beta_1$ and $\alpha_3\beta_1$ complexes.

Following our studies on the developing and adult kidney, we became interested in the expression of integrins in RCCs. We studied 41 carcinoma samples for the expression of α_1-α_6, α_v, β_1, β_3 and β_4 subunits (Korhonen *et al.* 1992), and the results are outlined in the following.

Integrins in Tumour Cell-BM Interactions: $\alpha_3\beta_1$ and $\alpha_6\beta_1$

In general, a correlation between tumour architecture and integrin distribution was seen. In the well differentiated (G1) tumours, immunoreactivity for the BM components laminin and collagen IV gave the appearance of BM-like structures at the basal aspect of the tumour cell clusters (Figure 7.1b). In these tumours, the $\alpha_3\beta_1$ (Figure 7.1c) complex was distinctly, and α_6 β_1 (Figure 1a) exclusively concentrated at the tumour cell-BM interface. This suggests that in G1 RCCs the tumour cells have largely retained the epithelial cell-BM interaction profile characteristic of normal renal tubular cells.

In the anaplastic (G3) tumours, immunoreactivity for the BM components was mostly unevenly scattered (Figure 7.2b). Parallelling this finding, both the $\alpha_3\beta_1$ and $\alpha_6\beta_1$ (Figure 7.2a) complexes showed considerable or total disorganization.

Furthermore, with increasing histologic grade, a considerable proportion of the tumours totally lacked $\alpha_6\beta_1$ expression as detected by immunohistochemistry (Figure 7.2a, table 7.2). It would thus appear that in G3 carcinomas there is a decreased tendency for organized cell-BM interaction.

Other Integrins: the α_v Subunit

In the normal kidney, immunoreactivity for the α_v and β_5 subunits are seen in the glomerulus and some segments of the distal tubules (Athanasou *et al.*

Table 7.1 Distribution of integrin complexes in the adult nephron *in vivo*.

Cell type	*Expressed integrins*
Mesangium	$\alpha_1\beta_1$
podocytes	$\alpha_3\beta_1$,α_v
glomerular endothelium	several α's
Bowman's capsule	$\alpha_3\beta_1$,α_v
proximal tubule	$\alpha_6\beta_1$
distal tubule	$\alpha_6\beta_1$,$\alpha_3\beta_1$,$\alpha_2\beta_1$

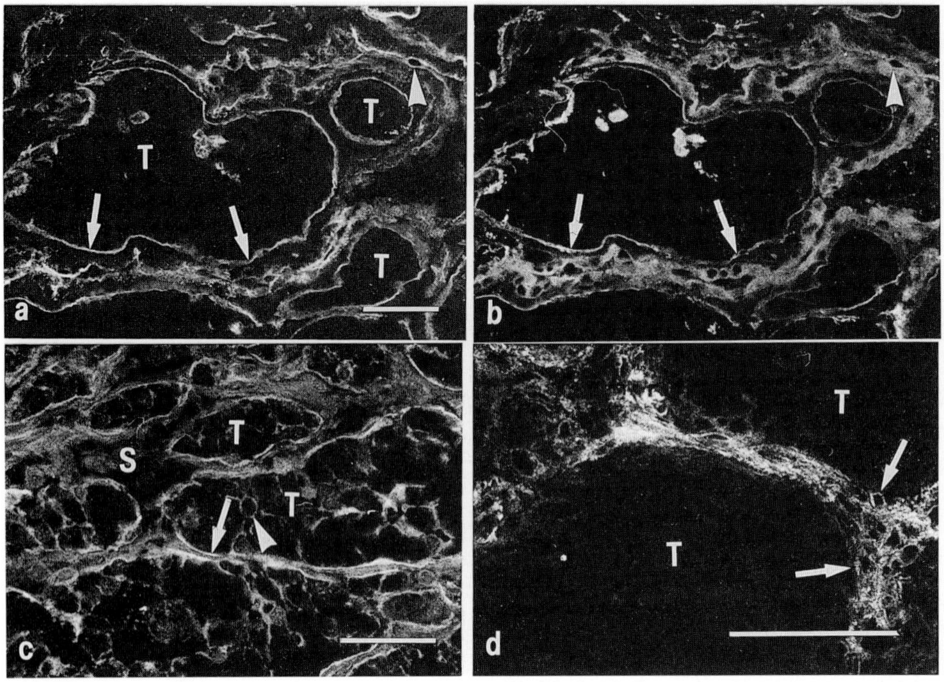

Figure 7.1 Indirect immunofluorescence figures of well-differentiated (G1) RCCs. Tumour cells (T) and stroma (S) are indicated. a and b: The tumour cell islands show a basally confined immunoreactivity (arrows) for the α_6 integrin subunit (a), coaligning with laminin (b). An α_6-reactive capillary is also indicated (arrowhead). c: Tumour cells display an overall cell surface immunoreactivity (arrowhead) for α_3, with a distinct concentration towards the basal aspect of the tumour cell islands (arrow). Stromal cells also react with anti-α_3. d: No immunoreactivity for the α_v subunit is seen on the tumour cells, but stromal cells (arrows) react. bars, 20 μm.

1990; Pasqualini *et al.* 1993). However, in RCCs increasing tumour grade correlated with increasing frequency for α_v expression. Thus, in 8% of G1 tumours and 80% of G3 tumours the α_v integrin subunit was detected on the tumour cells (Figures 7.1d and 7.2d, table 7.2). According to our preliminary results, a corresponding expression of the β_5 subunit was seen in several cases. As for the other integrin subunits (α_1, α_2, α_4, α_5, β_3, and β_4), only

Table 7.2 Integrin expression in RCCs of various grades.

tumour grade	α_3	α_6	α_V
1	100%	100%	8%
2	100%	79%	54%
3	100%	60%	80%

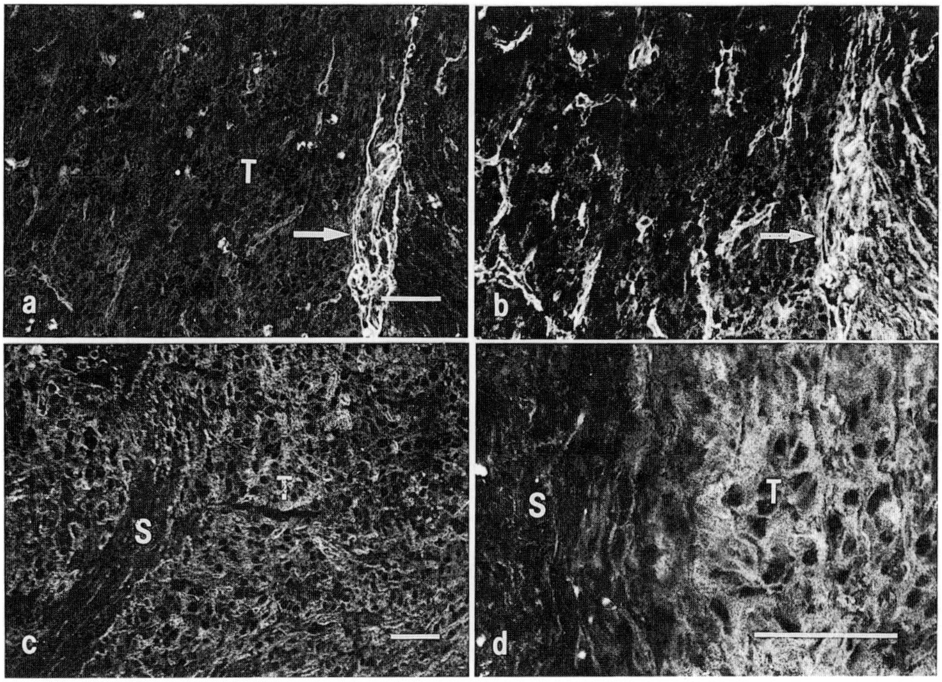

Figure 7.2 Poorly differentiated (G3) RCCs. Tumour cells (T) and stroma (S) are indicated. a: Anti-α_6 (a) reveals no immunoreactivity on the RCC cells, but a connective tissue septum (arrows) is positive. b: double immunostaining for laminin. c: This G3 tumour shows immunoreactivity for α_6 on the tumour cells, while the stromal elements react more weakly. Note the disorganization of the α_6 distribution compared to fig. 7.1a. Anti-α_v reveals immunoreactive tumour cells, while the stromal reactivity is weaker. bars, 20 μm.

sporadic expression, which did not correlate with tumour grade, was detected.

Tumour Stroma

The stromal elements of the tumours frequently expressed the $\alpha_1\beta_1$ and $\alpha_3\beta_1$ (Figure 7.1c) complexes, and sometimes immunoreactivity for $\alpha_5\beta_1$ as well as the α_v subunit (Figure 7.1d) was also detected. In addition to capillary endothelial cells, the stromal cells of tumours include reactive fibroblasts termed myofibroblasts (Sappino *et al.* 1990) responsible for the synthesis of stromal ECM components, including cellular fibronectins and tenascins. At present, the integrin repertoire of these cells is unknown, but our results suggest that they express at least the $\alpha_1\beta_1$ and $\alpha_3\beta_1$, and possibly the $\alpha_5\beta_1$ integrin complexes, as well as the α_v subunit.

DISCUSSION

Recently, several changes in the expression of certain integrins, correlating with malignancy, has been reported (reviewed in Juliano and Varner, 1993). In RCCs we found that an increased expression of the α_v subunit, and a decreased expression of the $\alpha_6\beta_1$ complex, correlated with malignant histologic grade.

One interesting note springs from the foregoing studies: either increased or decreased expression of certain integrins may accompany a more malignant tumour cell phenotype. It is tempting to hypothesize that a depletion of an integrin complex that plays a role in the histodifferentiation of the tissue in question could lead to a more anaplastic phenotype, while the aberrant emergence or upregulation of another type of integrin, that perhaps is associated with a more migratory phenotype, could lead to the same effect. The decreased expression of the $\alpha_6\beta_1$ complex in RCCs of high grade is also interesting with reference to the fact that it is the basement membrane receptor in normal renal tubules (Korhonen *et al.* 1990a) and is required for the formation of polarized epithelium during nephrogenesis (Sorokin *et al.* 1990). On the other hand, it should be noted that in other tumour systems, such as prostatic carcinoma (Bonkhoff *et al.* 1993) and pancreatic carcinoma (Weinel *et al.* 1992), the α_6 integrin subunit is expressed with increased intensity in metastatic lesions. Our results also suggest that in infiltrating cells of lobular breast carcinomas an increase in α_6 integrin expression is noted (Koukoulis *et al.* 1991).

In view of the reports linking the expression of the α_v and β_3 subunits with increased malignancy in melanomas (Albelda *et al.* 1990; Boukerche *et al.* 1989; Nip *et al.* 1992; Felding-Habermann *et al.* 1992), it is interesting that we found a correlation of α_v subunit expression and tumour grade. The α_v subunit may complex with several different β's, and at least the $\alpha_v\beta_3$ heterodimer is active in cell migration (Leavesley *et al.* 1992). Our preliminary results on 5 RCC cell lines that we have studied *in vitro* suggest that all five lines coexpress the α_v, β_3 and β_5 subunits. In our tissue series, the results indicate that the β_5 subunit is often seen in tumours expressing α_v but not β_3. Thus, it is possible that α_v is complexed not with β_3, but with one of the other β's in most RCCs *in vivo*. All the 5 cell lines expressed abundant $\alpha_3\beta_1$, while four were found to express $\alpha_6\beta_1$ *in vitro*.

Does integrin expression in RCCs take sides in the distal versus proximal tubule origin — controversy concerning RCCs? At present, the tentative answer would be no. The ubiquitous expression of $\alpha_3\beta_1$ would speak for, while the almost complete lack of $\alpha_2\beta_1$ speaks against, a distal tubular origin. The $\alpha_6\beta_1$ complex, on the other hand, is expressed in all normal renal tubules. It deserves mention, that while immunoreactivity for the α_2 subunit was detected in only two of the 41 RCCs studied, when this subunit was present the tumour cells displayed a very intense immunoreactivity indeed. Whether this subset of α_2-expressing tumours would also express other distal tubule markers, perhaps defining a subset of RCCs with a more pronounced distal tubule phenotype, remains to be elucidated.

In a recent study, Terpe *et al.* (1993) suggested that decrease of expression of α_2 and α_5 integrins would correlate with malignancy. We have not been able to confirm their results in our tumour samples, although 3 independently raised α_5 and 2 α_2 MABS were used. Of the five RCC lines, none expressed the α_2 subunit, and two expressed the α_5 subunit *in vitro*.

With this array of observations correlating integrin expression with tumour progression or grade etc., it is fair to ask: what is the biological significance of these observations? Are they mere epiphenomena, or does up- or down-regulated integrin expression actually influence such critical phenomena of cancer cell behaviour as invasion, accentuated growth, and metastasis *in vivo*? Also, it should be noted that our studies on 3 different carcinomas, breast, colon and kidney (Korhonen *et al.* 1992; Koukoulis *et al.* 1991; Koukoulis *et al.* 1993), have not revealed a singular and invariable correlation between carcinoma grade and integrin expression or metastasis and integrin expression.

Acknowledgements

Supported by a research contract with the Academy of Finland, and grants from the Finnish Cancer Research Fund, the Sigrid Juselius Foundation, and the Emil Aaltonen Foundation, the Finnish Medical Research Council, Finnish Medical Society "Duodecim" and the Ida Montin Foundation.

REFERENCES

1. Adams, J. C. and Watt, F. M. (1993) Regulation of development and differentiation by the extracellular matrix. *Development,* **117**, 1183–1198.
2. Albelda, S. M., Mette, S. A. and Elder, D. E. (1990) Integrin distribution in malignant melanoma: Association of the β_3 subunit with tumor progression. *Cancer Research,* **50**, 6757–6764.
3. Athanasou, N. A., Quinn, J., Horton, M. A. and McGee, J. O. (1990) New sites of cellular vitronectin receptor immunoreactivity detected with osteoclast-reacting monoclonal antibodies 13C2 and 23C6. *Bone and Mineral,* **8**, 7–22.
4. Blouin, P., Guiot, M.-C. and Jothy, S. (1989) Definition of the human renal cell carcinoma phenotype using monoclonal and polyclonal antibodies: a tumor marker study. *Experimental Pathology,* **36**, 147–163.
5. Bonkhoff, H., Stein, U. and Remberger, K. (1993) Differential expression of α_6 and α_2 very late integrins in the normal, hyperplastic, and neoplastic prostate: Simultaneous demonstration of cell surface receptors and their extracellular ligands. *Human Pathology,* **24**, 243–248.
6. Boukerche, H., Berthier-Vergnes, O., Bailly, M., Doré, J. F., Leung L. K. K. and McGregor, J. L. (1989) A monoclonal antibody (LYP18) directed against the blood platelet glycoprotein IIb/IIIa complex inhibits human melanoma growth *in vivo*. *Blood,* **74**, 909–912.

7. Burridge, K. and Fath, K. (1989) Focal contacts: transmembrane links between the extracellular matrix and the cytoskeleton. *Bio Essays*, **10**, 104–108.
8. Choy, M.-Y., Richman, P. I., Horton, M. A. and MacDonald, T. T. (1990) Expression of the VLA family of integrins in human intestine. *Journal of Pathology*, **160**, 35–40.
9. Cosio, F. G., Sedmak, D. D. and Nahman, S. N. (1990) Cellular receptors for matrix proteins in normal human kidney and human mesangial cells. *Kidney International*, **38**, 886–895.
10. Felding-Habermann, B., Mueller, B. M., Romerdahl, C. A. and Cheresh, D. A. (1992) Involvement of integrin α_v gene expression in human melanoma tumorigenicity. *J. Clin. Invest.*, **89**, 1–5.
11. Felding-Habermann, B. and Cheresh, D. A. (1993) Vitronectin and its receptors. *Current Opinion in Cell Biology*, **5**, 864–868.
12. Grant, D. S., Tashiro, K.-I., Segul-Real, B., Yamada, Y., Martin, G. R. and Kleinman, H. K. (1989) Two different laminin domains mediate the differentiation of human endothelial cells into capillary-like structures *in vitro*. *Cell*, **58**, 933–943.
13. Hall, P. A., Coates, P., Lemoine, N. R. and Horton, M. A. (1991) Characterization of integrin chains in normal and neoplastic human pancreas. *Journal of Pathology*, **165**, 33–41.
14. Hynes, R. O. (1992) Integrins: Versatility, modulation, and signaling in cell adhesion. *Cell*, **69**, 11–25.
15. Juliano, R. L. and Varner, J. A. (1993) Adhesion molecules in cancer: the role of integrins. *Current Opinion in Cell Biology*, **5**, 812–818.
16. Korhonen, M., Ylänne, J., Laitinen, L. and Virtanen, I. (1990a) Distribution of β_1 and β_3 integrins in human fetal and adult kidneys. *Laboratory Investigation*, **62**, 616–625.
17. Korhonen, M., Ylänne, J., Laitinen, L. and Virtanen, I. (1990b) The α_1-α_6 subunits of integrins are characteristically expressed in distinct segments of developing and adult human nephron. *Journal of Cell Biology*, **111**, 1245–1254.
18. Korhonen, M., Ylänne, J., Laitinen, L., Cooper, H. M., Quaranta, V. and Virtanen, I. (1991) Distribution of the α_1-α_6 integrin subunits in human developing and term placenta. *Laboratory Investigation*, **65**, 347–356.
19. Korhonen, M., Laitinen, L., Ylänne, J., Koukoulis, G. K., Quaranta, V., Juusela, H., Gould, V. E. and Virtanen, I. (1992) Integrin distributions in renal cell carcinomas of various grades of malignancy. *American Journal of Pathology*, **141**, 1161–1171.
20. Koukoulis, G. K., Virtanen, I., Korhonen, M., Laitinen, L., Quaranta, V. and Gould, V. E. (1991) Immunohistochemical localization of integrins in the normal, hyperplastic and neoplastic breast. *American Journal of Pathology*, **139**, 787–799.
21. Koukoulis, G. K., Virtanen, I., Moll, R., Quaranta V. and Gould, V. E. (1993) Immunolocalization of integrins in the normal and neoplastic colonic epithelium. *Virchows Archiv. B Cell Pathology*, **63**, 373–383.
22. Lauweryns, B., van den Oord, J. J., Volpes, R., Foets, B. and Missotten, L. (1991) Distribution of very late activation integrins in the human cornea. *Investigative Ophthalmology & Visual Science*, **32**, 2079–2085.

23. Leavesley, D. I., Ferguson, G. D., Wayner, E. A. and Cheresh, D. A. (1992) Requirement of the integrin $\beta3$ subunit for carcinoma cell spreading or migration on vitronectin and fibrinogen. *Journal of Cell Biology*, **117**, 1101–1107.
24. Medeiros, L. J., Gelb, A. B. and Weiss, L. M. (1988) Renal cell carcinoma. Prognostic significance of morphologic parameters in 121 cases. *Cancer*, **61**, 1639–1651.
25. Naidet, C., Semeriva, M., Yamada, K. M. and Thiery, J. P. (1987) Peptides containing the cell-attachment recognition signal Arg-Gly-Asp prevent gastrulation in Drodophia embryos. *Nature*, **325**, 348–350.
26. Nip, J., Shibata, H., Loskutoff, D. J., Cheresh, D. A. and Brodt, P. (1992) Human melanoma cells derived from lymphatic metastases use integrin $\alpha_v\beta_3$ to adhere lymph node vitronectin. *Journal of Clinical Investigation*, **90**, 1406–1413.
27. Oosterwijk, E., van Muijen, G. N. P., Oosterwijk-Wakka J. C. and Warnaar, S. O. (1990) Expression of intermediate-sized filaments in developing and adult human kidney and in renal cell carcinoma. *Journal of Histochemistry and Cytochemistry*, **38**, 385–392.
28. Pasqualini, R., Bodorova, J., Ye, S. and Hemler, M. E. (1993) A study of the structure, function and distribution of β_5 integrins using novel anti-β_5 monoclonal antibodies. *Journal of Cell Science*, **105**, 101–111.
29. Peltonen, J., Larjava, H., Jaakkola, S., Gralnik, H., Akiyama, S. K., Yamada, S. S., Yamada, K. M. and Uitto, J. (1989) Localization of integrin receptors for fibronectin, collagen, and laminin in human skin — variable expression in basal and squamous cell carcinomas. *Journal of Clinical Investigation*, **84**, 1916–1923.
30. Perris, R., Paulsson, M. and Bronner-Fraser, M. (1989) Molecular mechanisms of avian neural crest cell migration on fibronectin and laminin. *Developmental Biology*, **136**, 222–238.
31. Pignatelli, M., Cardillo, M. R., Hanby, A. and Stamp, G. W. H. (1992) Integrins and their accessory adhesion molecules in mammary carcinomas: Loss of polarization in poorly differentiated tumors. *Human Pathology*, **23**, 1159–1166.
32. Rahilly, M. A. and Fleming, S. (1992) Differential expression of integrin alpha chains by renal epithelial cells. *Journal of Pathology*, **167**, 327–334.
33. Sappino, A. P., Schurch, W. and Gabbiani, G. (1990) Differentiation repertoire of fibroblastic cells – expression of cytoskeletal proteins as marker of phenotypic modulations. *Laboratory Investigation*, **63**, 144–161.
34. Sorokin, L., Sonnenberg, A., Aumaillery, M., Timpl, R. and Ekblom, P. (1990) Recognition of the laminin E8 cell-binding site by an integrin possessing the α_6 subunit is essential for epithelial polarization in developing kidney tubules. *Journal of Cell Biology*, **111**, 1265–1270.
35. Terpe, H.-J., Tajrobchar, K., Günther, U. and Altmannsberger, M. (1993) Expression of cell adhesion molecules alpha-2, alpha-5 and alpha-6 integrin, E-cadherin, N-CAM and CD-44 in renal carcinomas: an immunohistochemical study. *Virchows Archiv. A Pathol. Anat.*, **422**, 219–224.
36. Tervo, K., Tervo, T., van Setten, G.-J. and Virtanen, I. (1991) Integrins in human corneal epithelium. *Cornea*, **10**, 461–465.
37. Thoenes, W., Störkel, St. and Rumpelt, H. J. (1986) Histopathology and classification of renal cell tumors (adenomas, oncocytomas and carcinomas). The basic cytological and histopathological elements and their use for diagnostics. *Path. Res. Pract.*, **181**, 125–143.

38. Volpes, R., van den Oord, J. J. and Desmet, V. J. (1991) Distribution of the VLA family of integrins in normal and pathological human liver tissue. *Gastroenterology*, **101**, 200–206.
39. Weinel, R. J., Rosendahl, A., Neumann, K., Chaloupka, B., Erb, D., Rothmund, M. and Santoso, S. (1992) Expression and function of VLA-α_2, -α_3, -α_5 and -α_6-integrin receptors in pancreatic carcinoma. *International Journal of Cancer*, **52**, 827–833.

8. ADHESION RECEPTORS IN THE DISSEMINATION OF NON-HODGKIN'S LYMPHOMAS

STEVEN T. PALS[1], EVELIENE MANTEN-HORST[1], PAUL DRILLENBURG[1], THADDAUS RADASZKIEWICZ[2]

[1]*Department of Pathology, Academic Medical Center, University of Amsterdam, Meibergdreef 9, 1105 AZ, Amsterdam, The Netherlands. Fax: 31-20-6960389*
[2]*Department of Pathology, University Hospital Vienna, Vienna, Austria*

INTRODUCTION

Cell adhesion receptors play a critical role in many cellular processes such as growth regulation, cell migration and organogenesis. In the immune system, these molecules enhance the efficacy of antigen-specific lymphocyte accessory-cell and target cell interactions which are required for the induction and regulation of an immune response and for effector cell function (Springer, 1990, Koopman and Pals, 1992). Moreover, they also mediate a variety of non-antigen specific interactions in the immune system including lymphocyte adhesion to endothelium and to the extracellular matrix during lymphocyte trafficking (Springer, 1990; Picker and Butcher, 1992).

Apart from their important role in the physiology of cellular interaction, cell adhesion receptors are also thought to play a key role in tumour metastasis and invasion. Activated lymphocytes and metastasizing tumour cells share several properties, including invasive behaviour, migration involving reversible adhesive contacts, accumulation and expansion in draining lymphoid tissue, release into the circulation and extravasation (Herrlich *et al.* 1993). The non-Hodgkin's lymphomas (NHLs) and lymphocytic leukemias represent transformed lymphocytes that have been arrested at a specific stage of maturation. Adhesion receptors that mediate physiologic lymphocyte homing may hence play an important role in their dissemination (Pals *et al.* 1989). In this paper the expression of adhesion receptors on lymphoid malignancies and their role in lymphoma dissemination is reviewed.

CELL ADHESION RECEPTORS

Four receptor families that are involved in cell-cell and cell-matrix interaction in the immune system have thus far been described. These are the immunoglobulin family (lg) (Williams and Barclay, 1988), the integrin family (Hemler *et al.* 1990, Springer, 1990; Hynes, 1992), the selectin family (Stoolman, 1989) and the CD44 family (Stamenkovic *et al.* 1989; Günthert *et al.* 1991; Pals *et al.* 1993). The role of the cell adhesion receptors in cellular interactions is dynamic; engagement does not only lead to cell adhesion persé, but transduces signals leading to cellular events including activation and proliferation. In turn, cytoplasmic signals regulate the cell surface expression and functional activity of adhesion receptors (Springer, 1990; Hynes, 1992). For detailed review of the structure and function of adhesion receptors, which is beyond the scope of this paper, the reader is referred to a number of excellent papers by Springer (1990), Hemler *et al.* (1990), and Hynes (1992). Table 8.1 summarizes the most important molecules discussed in the present paper.

LYMPHOCYTE HOMING

During lymphocyte homing, complex interactions between lymphocytes, other cells, and the extracellular matrix take place. Lymphocyte-endothelial recognition mechanisms, which display specificity with respect to lymphocyte

Table 8.1 Cell adhesion receptors*.

Receptor	*Mw (kD)*	*Family*	*Distribution*	*Ligand*
αLβ2 (LFA-1)	95/180	integrin	leucocytes	ICAM-1, ICAM-2, ICAM-3
α4β1	150/130	integrin	lymphocytes monocytes eosinofils	VCAM-1 fibronectin (CS-1)
α4β7	150/100	integrin	lymphocytes	MAdCAM-1 VCAM-1 fibronectin
ICAM-1	90-114	Ig	endothelium leucocytes FDC	αLβ2 αMβ2
VCAM-1	110	Ig	activated-endothelium FDC	α4β1
L-Selec tin	80–100	Selectin	leucocytes	peripheral lymph node addressin
CLA	200(?)	carbohydrate	T cell subset	E-Selectin (ELAM-1)
CD44 (multiple isoforms)	80–200	CD44	broad tissue distribution	hyaluronic acid, fibronectin

* only molecules discussed in the present paper are listed.

subpopulation, stage of the inflammatory response, and tissue site or organ, regulate this process. Engagement of lymphocytes and other leukocytes with endothelium requires at least three distinct steps (Butcher, 1991; Shimizu *et al.* 1992). First, recognition is initiated by binding of a constitutively active adhesion receptor to its endothelial counterstructure. This primary adhesion is mediated by lectin-carbohydrate interactions involving endothelial or leukocyte selectins and their cognate oligosaccharide ligands. Unless followed by a second event leading to activation of the lymphocyte, this initial adhesion is transient and reversible. Subsequent activation of the lymphocyte leads to triggering of activation dependent adhesion receptors of the integrin $\beta2$ and/or $\beta1$ families notably $\alpha L\beta2$ (LFA-1, CD11a/18) and $\alpha4\beta1$ (VLA-4, CD49d/29), which mediate stable adhesion and migration across the vessel wall. The exact triggers for integrin activation *in vivo* are presently unknown but presumably involve engagement of the primary adhesion receptor, crosslinking of other cell surface receptors e.g. CD44 (Koopman *et al.* 1990), or action of specific chemokines *e.g.* MIP-1β (Tanaka *et al.* 1993). In this model specificity is determined by unique combinations of primary and secondary receptor pairs on lymphocytes and endothelial cells, as well as on their activation by chemokines, the production of which may be site specific. Factors determining the adhesion receptor make-up of lymphocytes and endothelium are naive vs. memory phenotype (Mackay, 1991), and state of activation of lymphocyte and endothelium. In addition, endothelial cells may express tissue specific recognition elements, so called "addressins", that direct lymphocyte subpopulations to distinct anatomical sites like skin, gut, and peripheral lymph node (Picker and Butcher, 1992).

LYMPHOMA DISSEMINATION

Several clinical observations prompted us to suggest that conserved homing programmes play a pivotal role in determining the dissemination of NHLs (Pals *et al.* 1989). For example: 1) NHLs related to small resting lymphocytes generally are widely disseminated to multiple lymphoid sites at presentation. By contrast, tumours phenotypically related to large lymphocytes (*e.g.* activated germinal centre B cells, *i.e.* large centrocytes and centroblasts) are often localized initially. These differences presumably reflect the recirculating vs. sessile character of the normal counterparts. 2) Extranodal NHLs arising in the gastrointestinal tract or skin spread to or recur preferentially at mucosal sites and skin, respectively, a feature highly suggestive of tissue-specific homing. Below, we will summarize a number of recent studies which focus on the relation between adhesion receptor expression and lymphoma dissemination.

SELECTINS AND THEIR LIGANDS

L-Selectin

L-Selectin (LECAM-1, LAM-1, Mel-14) is a member of the selectin family of cellular adhesion molecules characterized by lectin, epidermal growth factor (EGF), and short consensus repeat (SCR) domains (Stoolman, 1989; Springer, 1990). L-selectin is broadly expressed by lymphocytes, monocytes and neutrophils but not by non-hematopoietic cells and is rapidly down-regulated upon cell activation. MAbs against L-selectin interfere with binding of lymphocytes to high endothelial venules (HEV) of peripheral lymph nodes (PLN) but not to mucosal-HEV, and, in the mouse, prefentially block lymphocyte migration to PLN. Hence, L-selectin on lymphocytes functions as a PLN homing receptor (Picker and Butcher, 1992). Analysis of NHL has shown that the vast majority of nodal NHL express L-selectin: Expression was found on most low-grade as well as aggressive B cell lymphomas and on virtually all nodal T-cell lymphomas. Low-grade B-cell lymphomas of the mucosa associated lymphoid tissues (MALT) also express L-selectin (Pals *et al.* 1991), however, expression is rare on aggressive (large-cell) mucosal B-cell lymphomas (Pals *et al.* 1991; Möller *et al.* 1991) and on mucosal T-cell lymphomas (Pals *et al.* 1991). Furthermore, most cutaneous T cell lymphomas are L-selectin low or negative (Abel *et al.* 1985). The absence of L-selectin from extranodal lymphomas might explain why these tumours disseminate preferentially to extranodal sites avoiding peripheral nodes.

Cutaneous lymphocyte antigen (HECA-452), a carbohydrate ligand of E-Selectin

Many T cells in the normal and inflamed skin express the cutaneous lymphocyte antigen (CLA), a carbohydrate epitope that is closely related to sialyl Lewis X antigen (sLEα) (Duijvestijn, 1988; Picker *et al.* 1990; Picker and Butcher, 1992). CLA is present on a minor subset of memory T lymphocytes in the peripheral blood, tonsils and lymph nodes (Picker *et al.* 1990) but on 40% of the T cells in the normal skin (Bos *et al.* 1993). CLA can bind E-Selectin (ELAM-1) on endothelium, and by interacting with this molecule functions as a skin homing receptor (Berg *et al.* 1991). Most cutaneous T cell lymphomas, mycosis fungoides as well as other types, express CLA. By contrast, expression of CLA on nodal T-NHL is rare (Meijer, 1989; Picker *et al.* 1990; Noorduyn *et al.* 1992). Gastrointestinal T-cell lymphomas lack CLA (Pals and Radaszkiewicz, unpublished). The selective expression of CLA on cutaneous T cell lymphomas suggests that CLA is a key factor in mediating the skin specific dissemination of cutaneous T cell lymphomas *in vivo*.

INTEGRINS

Integrins are heterodimeric proteins consisting of non-covalently associated α-and β-subunits. Most of the 8 β-chains can associate with several of the 14 different α-chains that have so far been discovered. Only a limited number of integrins are expressed on lymphocytes. Of these, the leucocyte integrins $\alpha L\beta 2$ (LFA-1), $\alpha 4\beta 1$ (VLA-4), and $\alpha 4\beta 7$ play an important role in lymphocyte interaction with endothelium during migration.

$\alpha L\beta 2$ (LFA-1)

$\alpha L\beta 2$ undoubtedly is one of the most important adhesion receptors of the immune system. It is involved in a wide variety of lymphocyte interactions with other cells including adhesion to endothelium and transendothelial migration (Haskard *et al.* 1986; Pals *et al.* 1989; Springer, 1990). Studies in the mouse have shown that $\alpha L\beta 2$ supports *in vitro* invasiveness of lymphoma cell lines and plays a role in lymphoma dissemination (Roossien *et al.* 1989). Expression of $\alpha L\beta 2$ in human non-Hodgkin's lymphomas is closely related to lineage derivation and stage of differentiation, but is not related to tumour grade (Horst *et al.* 1991). Clayburger *et al.* (1987) suggested that lack of expression of $\alpha L\beta 2$ on tumour cells might cause impaired immunosurveillance against those cells and hence be associated with an unfavourable prognosis. Subsequent studies have not born out this prediction (Horst *et al.* 1990b; Horst *et al.* 1991).

$\alpha 4\beta 1$ (VLA-4) and other $\beta 1$ integrins

The leucocyte integrin $\alpha 4\beta 1$ is unique among $\beta 1$ integrins since it is not only involved in cell-matrix but also in cell-cell interaction. It can bind to the CS-1 domain of fibronectin as well as to the vascular-cell adhesion molecule-1 (VCAM-1), a molecule expressed on several cell types including activated endothelium, dendritic cells, and follicular dendritic cells. Interactions between $\alpha 4\beta 1$ and VCAM-1 play an important role in lymphocyte migration to sites of inflammation (Van Dinther-Janssen *et al.* 1991; Yednock *et al.* 1992; Baron *et al.* 1993). Like $\alpha L\beta 2$, $\alpha 4\beta 1$ is expressed widely on NHL, expression being related to stage of differentiation (Möller *et al.* 1991 and 1992). It is not yet clear whether expression of $\alpha 4\beta 1$ or of other $\beta 1$ integrins is related to lymphoma dissemination or whether expression of these molecules has prognostic significance.

$\alpha 4\beta 7$, a novel integrin involved in mucosal homing

Recent studies have identified the integrin $\alpha 4\beta 7$ as a mucosal homing receptor that mediates lymphocyte migration to the intestinal mucosa by binding to MAdCAM-1, a vascular recognition molecule (addressin) selectively expressed on mucosal endothelium. MAdCAM-1 is an immunoglobulin family member with domains that display homologies to the vascular adhesion receptors for leukocytes ICAM-1 and VCAM-1, as well as to another mucosa associated Ig-family member, IgA1 (Briskin *et al.* 1993). Recent studies indicate that $\alpha 4\beta 7$ is the dominant lymphocyte receptor for MAdCAM-1 and for regulating lymphocyte homing to mucosal sites in the mouse (Holzmann *et al.* 1989). In man, $\alpha 4\beta 7$ appears to have a similar function. It is expressed on mucosal lymphocytes (Parker *et al.* 1992). Moreover, it is present on a subset of peripheral blood memory T-cells with putative gut homing properties (Schweighoffer *et al.* 1993).

We recently studied the expression of the $\alpha 4\beta 7$ mucosal homing receptor in 8 cases of malignant lymphomatous polyposis (MLP). This unusual presentation of non-Hodgkin's lymphoma of mantle cell type is characterized by multifocal involvement of the gastrointestinal tract. Unlike nodal mantle cell lymphomas, cases of MLP showed expression of $\alpha 4\beta 7$ (Pals *et al.* 1994). Also, low grade B-cell lymphomas of MALT and intestinal T-cell lymphomas express $\alpha 4\beta 7$ (Drillenburg *et al.* manuscript in preparation). These findings suggest that $\alpha 4\beta 7$ plays an important role in determining the characteristic mucosal dissemination that is often found in lymphomas of MALT.

THE CD44 FAMILY

CD44 is widely distributed on hematopoietic and non-hematopoietic tissues. Prior its uniform classification as CD44 the 80-90 kD form of these proteins which is dominant on many cells, were described under various names including PGP-1, ECMRIII, (In(Lu))-related p80, Hutch-1 and Hermes-1. A number of different functions, such as binding to collagen, fibronectin, and hyaluronate, and binding to post capillary venules as part of the process of lymphocyte recirculation were ascribed to CD44 (Haynes *et al.* 1989). In lymphoid differentiation, CD44 shows a trimodal distribution: the antigen is expressed at early stages of T and B cell differentiation, *i.e.* on prothymocytes and immature precursor B cells. Subsequently, at the stage of common thymocyte, and the pre- and early B cell, CD44 antigen(s) are lost from the cell surface to be reacquired during further maturation. At the activated germinal centre B-cell stage CD44 is heterogeneously expressed.

Interestingly, in diffuse large cell lymphomas, we found a strong correlation between CD44 expression and lymphoma dissemination (Pals *et al.* 1989; Horst *et al.* 1990b). Furthermore, high CD44 expression was identified as an unfavourable prognostic factor in these lymphomas (Horst *et al.* 1990b). Similar findings were also reported by Jalkanen *et al.* (1990, 1991). A number of CD44 variants have recently been described that are generated by alternative splicing (Günthert *et al.* 1991; Screaton *et al.* 1992). CD44 variant containing exon v6 encoded sequences have been shown to be causally involved in metastatic behaviour of rat carcinoma cell lines (Günthert *et al.* 1991) and are overexpressed in a variety of human tumours (Pals *et al.* 1993). We have shown that CD44 splice variants containing v6 are expressed on activated lymphocytes and on aggressive non-Hodgkin's lymphomas (Koopman *et al.* 1993). These variants may play a role in mediating their aggressive clinical behaviour.

CONCLUSIONS

Most non-Hodgkin's lymphomas express a variety of adhesion receptors. In general, expression of these receptors reflects the stage of maturation arrest of the tumour *i.e.* parallel the expression on normal lymphocytes at that particular stage. Extranodal lymphomas arising in the skin and the mucosal lymphoid tissues express the cutaneous and mucosal homing receptors CLA and $\alpha 4\beta 7$, respectively. This selective homing receptor expression implicates that NHL at different anatomical sites are derived from distinctive lymphocyte populations and presumably is a major factor in determining the tissue selective dissemination of those tumours. It should be noted, however, that other factors *viz.* local availability of specific cytokines or of specific antigen, might also contribute to the tissue-specific dissemination and outgrowth of NHL. Several independent studies have indicated that expression of CD44 on NHL is associated with an adverse prognosis. Further studies are needed to assess whether specific CD44 splice variants are involved and to unravel the mechanism by which CD44 influences lymphoma behaviour.

REFERENCES

1. Abel, E. A., Wood, G. S., Hoppe, R. T. and Warnke, R. A. (1985) Expression of Leu-8, a majority T-cell marker, is uncommon in mycosis fungoides. *J. Invest. Dermatol.*, **85**, 199.
2. Baron, J. L., Madri, J. A., Ruddle, N. H., Hashim, G. and Janeway, C. A., Jr. (1993) Surface expression of $\alpha 4$ integrin by CD4 T cells is required for their entry into brain parenchyma. *J. Exp. Med.*, **177**, 57.
3. Berg, E. L., Yoshino, T., Rott, L. S., Robinson, M. K., Warnock, R. A., Kishimoto, T. K., Picker, L. and Butcher E. C. (1991) The cutaneous lymphocyte antigen is a skin homing receptor for the vascular lectin endothelial cell-leukocyte adhesion molecule 1. *J. Exp. Med.*, **174**, 1461.
4. Bos, J. D., Boer, O. J. de, Tibosch, E., Das, P. K. and Pals, S. T. (1993) Skin homing T lymphocytes: Detection of cutaneous lymphocyte associated antigen (CLA) by HECA-452 in normal human skin. *Arch. Dermatol. Res.*, **285**, 179.
5. Briskin, M. J., McEvoy and Butcher, E. C. (1993) MAdCAM-1 has homology to immunoglobulin and mucin-like adhesion receptors and to lgA1. *Nature*, **363**, 461–464.

6. Butcher, E. C. (1991) Leukocyte-endothelial cell recognition: Three (or more) steps to specificity or diversity. *Cell*, **67**, 1033.
7. Clayberger, C., Meideros, L. J., Link, M. P., Warnke, R. A., Wright, A., Koller, T. D., Smith, M. D. and Krensky, A. M. (1987) Absence of Cell surface LFA-1 as a mechanism of escape from immunosurveillance. *Lancet.*, **ii**, 533.
8. Dinther-Janssen van, A. C. H. M, Horst, E., Koopman, G., Scheeper, R. J., Newman, W., Meijer, C. J. L. M. and Pals, S. T. (1991) The role of VLA-t/VCAM-1 in lymphocyte adhesion to endothelium in rheumatoid synovium. *J. Immunol.*, **147**, 4207.
9. Duijvestijin, A. M., Horst, E., Pals, S. T., Rouse, B. N., Steere, A. C., Picker, L. J., Meijer, C. J. L. M. and Butcher, E. C. (1988) High endothelial differentiation in human lymphoid and inflammatory tissues defined by monoclonal antibody HECA-452. *Am. J. Pathol.*, **130**, 147.
10. Elices, M. J., Osborn, L., Takada, Y., Crouse, C., Luhowskyj, S., Hemler, M. E. and Lobb, R. R. (1990) VCAM-1 on activated endothelium interacts with the leukocyte integrin VLA-4 at a site distinct from the VLA-4/fibronectin binding site. *Cell*, **60**, 577.
11. Günthert, U., Hofmann, M., Rudy, W., Reber, S., Zoller, M., Haussman, I., Matzku, S., Wenzel, A., Ponta, H. and Herrlich, P. (1991) A new variant of CD44 confers metastatic potential to rat carcinoma cells. *Cell*, **65**, 13.
12. Haskard, D., Cavender, D., Beatty, P., Springer, T. and Ziff, M. (1986) T-lymphocyte adhesion to endothelial cells: mechanisms demonstrated by anti-LFA-1 antibodies. *J. Immunol.*, **137**, 2901.
13. Haynes, B. F., Telen, M. J., Hale, P. L. and Denning, S. M. (1989) CD44-A molecule involved in leukocyte adherence and T cell activation. *Immunol. Today*, **10**, 423.
14. Hemler, M. E., Elices, M. J., Parker, C. and Takada, Y. (1990) Structure of the integrin VLA-4 and its cell-cell and cell-matrix adhesion functions. *Immunol. Rev.*, **114**, 45.
15. Herrlich, P., Zöller, M., Pals, S. T. and Ponta, H. (1993) CD44 splice variants: metastases meet lymphocytes. *Immunol. Today*, **14**, 395–399.
16. Holzmann, B., Mclntyre, B. W. and Weisman, I. L. (1989) Identification of a murine Peyer's patch-specific lymphocyte homing receptor as an integrin molecule with an α chain homologous to human VLA-4α. *Cell*, **56**, 37–46.
17. Horst, E., Meijer, C. J. L. M., Radaszkiewicz, T., van Dongen, J. J. M., Pieters, R., Figdor, C. G., Hooftman, A. and Pals, S. T. (1990a) Expression of a human homing receptor (CD44) in lymphoid malignancies and related stages of lymphoid development. *Leukemia*, **4**, 383.
18. Horst, E., Meijer, C. J. L. M., Radaszkiewicz, T., Ossekoppele, G. J., van Krieken, J. H. J. M. and Pals, S. T. (1990b) Adhesion molecules in the prognosis of diffuse large-cell lymphoma: expression of a lymphocyte homing receptor (CD44), LFA-1 (CD11a/CD18), and ICAM-1 (CD54). *Leukemia*, **4**, 595.
19. Horst, E., Radaszkiewicz, T., Hooftman-den Otter, A., Pieters, R., van Dongen, J. J. M., Meijer, C. J. L. M. and Pals, S. T. (1991) Expression of the leucocyte integrin LFA-1 (CD11a/CD18) and its ligand ICAM-1 (CD54) in lymphoid malignancies is related to lineage derivation and stage of differentiation but not to tumor grade. *Leukemia*, **5**, 848.
20. Hynes, R.O. (1992) Integrins: versatility, modulation, and signaling in cell adhesion. *Cell*, **69**, 11.

21. Jalkanen, S., Joensuu, H. and Klemi, P. (1990) Prognostic value of lymphocyte homing receptor and S value in non-Hodgkin's lymphoma. *Blood*, **75**, 1549.
22. Jalkanen, S., Joensuu, H., Soderstrom, K. O. and Klemi, P. (1991) Lymphocyte homing and the clinical behavior of non-Hodgkin's lymphomas. *J. Clin. Invest.*, **87**, 1835.
23. Koopman, G., van Kooyk, Y., de Graaff, M., Meijer, C. J. L. M., Figdor, C. G. and Pals, S. T. (1990) Triggering of the CD44 antigen on T lymphocytes promotes T cell adhesion through the LFA-1 pathway. *J. Immunol.*, **145**, 3589.
24. Koopman, G. and Pals, S. T. (1992) Cellular interactions in the germinal center: role of adhesion receptors and significance for the pathogenesis of AIDS and malignant lymphoma. *Immunol. Rev.*, **126**, 1–25.
25. Koopman, G., Heider, K. H., Horst, E., Adolf, G. R., Berg, F. van den, Ponta, H., Herrlich, P. and Pals, S. T. (1993) Activated human lymphocytes and non-Hodgkin's express a homoloque of the rat metastasis-associated variant of CD44. *J. Exp. Med.*, **177**, 897.
26. Mackay, C. R. (1991) T-cell memory: the connection between function, phenotype and migration pathways. *Immunol. Today*, **12**, 189.
27. Meijer, C. J. L. M., Beljaards, R., Horst, E., Willemze, R., van der Valk, P. and Pals, S. T. (1989) Differences in antigen expression between primary cutaneous and node based T-cell lymphomas. *J. Invest. Dermatol.*, **92**, 479.
28. Möller, P., Eichelmann, A., Mechtersheimer, G. and Koretz, K. (1991) Expression of β-1 integrins, H-CAM, and LECAM-1 in primary gastrointestinal B-cell lymphomas as compared to the adhesion receptor profile of the gut associated lymphoid system, tonsil and peripheral lymph node. *Int. J. Cancer*, **49**, 846.
29. Möller, P., Eichelmann, A. and Mechtersheimer, G. (1992) Adhesion molecules VLA-1 to define discrete stages of peripheral B lymphocyte development and characterize different types of B cell neoplasia. *Leukemia*, **6**, 256.
30. Noorduyn, L. A., Beljaards, R. C., Pals S. T., van Heerde, P., Radaszkiewicz, T., Willemze, R. and Meijer, C. J. L. M. (1992) Differential expression of the HECA-452 antigen (cutaneous lymphocyte associated antigen, CLA) in cutaneous and non-cutaneous T-cell lymphomas. *Histopathology*, **21**, 59.
31. Pals, S. T., Horst, E., Ossekoppele, G. J., Figdor, C. G., Scheper, R. J. and Meijer, C. J. L. M. (1989) Expression of lymphocyte homing receptor (CD44) as a mechanism of dissemination in non-Hodgkin's lymphoma. *Blood*, **73**, 885.
32. Pals, S. T., Horst, E., Scheper, R. J. and Meijer, C. J. L. M. (1989) Mechanisms of human lymphocyte migration and their role in the pathogenesis of disease. *Immunol. Rev.*, **108**, 111.
33. Pals, S. T., Meijer, C. J. L. M. and Radaszkiewicz, T. (1991) Expression of the human peripheral lymph node homing receptor (LECAM-1) in nodal and gastrointestinal non-Hodgkin's lymphomas. *Leukemia*, **5**, 628.
34. Pals, S. T., Koopman, G., Griffioen, A. J., Ponta H., Herrlich, P., Berg, F. van den and Horst, E. (1993) CD44: A variety of isoforms with functions in cell adhesion and metastasis. In *Lymphocyte adhesion molecules*, edited by Shimizu, Y. RG Landis Co. Austin USA, 135–151.
35. Pals, S. T., Drillenburg, P., Dragosics, B., Lazarovits, A. I. and Radaszkiewicz, T. (1994) Expression of the mucosal homing receptor $\alpha4\beta7$ in malignant lymphomatous polyposis of the intestine. *Submitted*.

36. Parker, C. M., Cepek, K. L., Russell, G. J., Shaw, S. K., Prosnett, D. N., Schwarting, R. L. and Brenner, M. B. (1992) A family of $\beta7$ integrins on human mucosal lymphocytes. *Proc. Natl. Acad. Sci. USA*, **89**, 1924–1928.
37. Picker, L. J., Mitchie, S. A., Rott, L. S. and Butcher, E. C. (1990) A unique phenotype of skin associated lymphocytes in humans. *Am. J. Pathol.*, **136**, 1053.
38. Picker, L. J. and Butcher, E. C. (1992) Physiological and molecular mechanisms of lymphocyte homing. *Annual Rev. Immunol.*, **10**, 561.
39. Roossien, F. F., Rijk, D., Bikker, A. and Roos, E. (1989) Involvement of LFA-1 in lymphoma invasion and metastasis demonstrated with LFA-1 deficient mutants. *J. Cell Biol.*, **108**, 1979.
40. Schweighoffer, T., Tanaka, Y., Tidswell, M., Erle, D. J., Horgan, K. J., Luce, G. E. G., Lazarovits., A. I., Buck, D. and Shaw, S. (1993) Selective expression of integrin $\alpha4\beta7$ on a subset of human CD4 + memory T cells with hallmarks of gut-tropism. *J. Immunol.*, **151**, 717–729.
41. Screaton, G. F., Bell, M. V., Jackson, D., Cornelius, F. B., Gerth, U. and Bell, J. (1992) Genomic structure of DNA encoding the lymphocyte homing receptor CD44 reveals at least 12 alternatively spliced exons. *Proc. Natl. Acad. Sci. USA*, **89**, 12160.
42. Shimizu, Y., Newman, W., Tanaka, Y. and Shaw, S. (1992) Lymphocyte interaction with endothelial cells. *Immunol. Today*, **13**, 106.
43. Springer, T. A. (1990) Adhesion receptors of the immune system. *Nature*, **346**, 425.
44. Stamenkovic, I., Amiot, M., Pesando, J. M. and Seed, B. (1989) A lymphocyte molecule implicated in lymphocyte homing is a member of the cartilage link protein family. *Cell*, **56**, 1057.
45. Stoolman, L. M. (1989) Adhesion molecules controlling lymphocyte migration. *Cell*, **56**, 907.
46. Tanaka, Y., Adams, D. H., Hubscher, S., Hirano, H., Siebenlist, U. and Shaw, S. (1993) T-cell adhesion induced by proteoglycan-immobilized cytokine MIP-1β. *Nature*, **361**, 79.
47. Williams, A. F. and Barclay, A. N. (1988) The immunoglobulin superfamily-domains for cell-surface recognition. *Annual Rev. Immunol.*, **6**, 381.
48. Yednock, T. A., Cannon, C., Fritz, L. C., Sanchez-Madrid, F., Steinman, L. and Karin, N. (1992) Prevention of experimental autoimmune encephalomyelitis by antibodies against $\alpha4\beta1$ integrin. *Nature*, **356**, 63.

9. DIFFERENTIAL EXPRESSION OF INTEGRINS BY ALVEOLAR EPITHELIA IN NORMAL AND FIBROTIC LUNG TISSUE

THOMAS PAPADOPOULOS, KARIN SIRTL, SUSANNE KERN, JOBST DÄMMRICH and HANS KONRAD MÜLLER-HERMELINK

Institute of Pathology, University of Würzburg, Federal Republic of Germany

The panel of integrins on human alveolar epithelium of normal and fibrotic lung tissue has been evaluated by using indirect immunofluorescence. Double staining techniques with a polyclonal rabbit anti-cytokeratin antibody and one of the monoclonal anti-integrin antibodies were applied to identify integrin-positive alveolar epithelia within the alveolar septa. Type I and type II alveolar epithelia in normal and fibrotic lung tissue consistently expressed $\alpha 1$, $\alpha 2$, $\alpha 3$, αv and $\beta 1$ integrin chains but they were negative for $\alpha 4$, $\alpha 5$, $\beta 3$ and $\beta 4$. The expression of the $\alpha 6$ chain on alveolar epithelium of normal lungs was restricted to single cuboidal cells. Immunofluorescence staining on fibrotic lung tissue revealed a strong de novo expression of the hemidesmosome associated integrins $\alpha 6$ and $\beta 4$ on alveolar epithelia. Our findings demonstrate an altered integrin expression on alveolar epithelium of fibrotic lungs which obviously reinforces the attachment of alveolar epithelia to the fibrotic alveolar tissue.

INTRODUCTION

Lung alveolar epithelium is a mosaic consisting of type I and type II pneumocytes. Both cell types serve distinct functions. The role of the extracellular matrix in differentiation[1,2] and various cellular functions[2–5] of the alveolar epithelia has been demonstrated in a large number of previous investigations. However only incomplete data on the expression of matrix receptors on these epithelia have been available up to now.[6,7] It was the aim of this study to characterize the matrix receptors on the cell surface of type I and type II pneumocytes in normal and fibrotic lung tissue.

MATERIALS AND METHODS

Tissue specimens and reagents

Immunofluorescence stainings were performed on cryostat sections of native lung specimens. Non fibrotic lung tissue was obtained from three lobectomy specimens of patients with primary or metastatic tumours in the lung. Fibrotic lung tissue was obtained from two open lung biopsies of patients with interstitial lung fibrosis.

The P1E6 monoclonal antibody (mAb)[8] directed against $\alpha 2$, the P1B5 mAb[8] directed against $\alpha 3$, the P4G9 mAb[9] directed aginst $\alpha 4$, the P1D6 mAb[8] directed against $\alpha 5$, the VNR147 mAb[10] directed against αv, and the 2E1 mAb[11] directed against $\beta 4$ were obtained from Biomol (Hamburg,

FRG). A polyclonal antibody directed against $\alpha 1$, the 102DF5 mAb[12] directed against $\beta 1$ and the BB10 mAb[13] directed against $\beta 3$ were obtained from Biermann GmbH (Bad Nauheim, FRG). The GoH3 mAb [14] directed against $\alpha 6$ was obtained from Dianova (Hamburg, FRG) and the polyclonal rabbit anti-cytokeratin antibody was obtained from Biogenex (Neufahrn, FRG).

Immunofluorescence

Immunofluorescence double stainings of lung tissue cryostat sections were performed in a two step procedure. The first step included: (i) incubation for 30 min with one of the anti-integrin antibodies mentioned above. All monoclonal antibodies were diluted in a 0.5% bovine serum albumin (BSA)-phosphate buffered saline (PBS) solution. The dilutions ranged between 1:50 and 1:100; (ii) incubation for 30 min with a biotin conjugated sheep anti mouse (Amersham Buchler, Braunschweig, FRG) or biotin conjugated donkey anti rat (Dianova, Hamburg, FRG) antiserum. Both antibodies were diluted 1: 200 in PBS; (iii) incubation for 30 min with a white egg-avidin-Cy3 complex, diluted 1: 600 in PBS. The second step included: (i) incubation for 30 min with a rabbit anti-cytokeratin antibody diluted 1:10 in a 0.5% BSA-PBS solution; (ii) incubation for 45 min with fluorescein-isothiocyanate (FITC)-conjugated swine anti-rabbit immunoglobulins (DAKO, Hamburg, FRG), diluted 1: 20 in PBS.

Immunofluorescence double stainings with a monoclonal 35βH11 mouse anti-cytokeratin antibody[15] and the rabbit anti-$\alpha 1$ antibody were performed in a slightly modified manner. In detail biotin conjugated goat anti rabbit antibodies and white egg-avidin-Cy3 were used for the detection of the rabbit antibodies, while FITC-conjugated goat anti mouse immunoglobulins were applied for the detection of the 35βH11 anti-cytokeratin antibody.

RESULTS

Integrins on alveolar epithelia of non fibrotic lung tissue

The epithelial layer in the pulmonary alveolus of non fibrotic and fibrotic lung tissue was identified by staining with an anticytokeratin antibody (Fig. 9.1a, 9.1c, 9.1e). Antibodies against the $\alpha 1$, $\alpha 2$, $\alpha 3$, αv and $\beta 1$ chains consistently stained type I and type II pneumocytes of adult lung tissue. The staining was intense and quite uniform.

$\alpha 6$ in the alveolar septa was strongly expressed by endothelial cells. The majority of the alveolar epithelia in non fibrotic lung tissue did not stain with the anti-$\alpha 6$ antibody at all. Only very few cuboidal alveolar epithelia showed a weak detectable immunostaining with the antibody to $\alpha 6$. Pneumocytes of

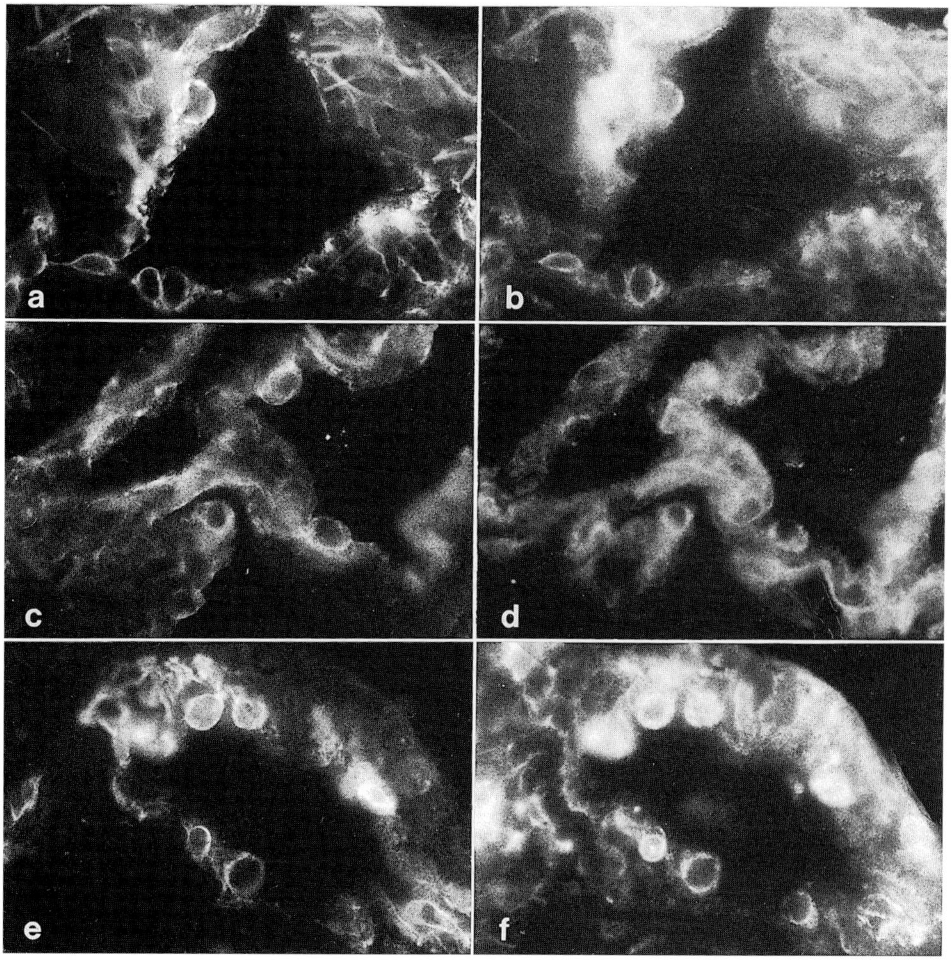

Figure 9.1 Cytokeratin/integrin immunofluorescent double staining of lung tissue. Cytokeratin positive alveolar epithelia of fibrotic lung tissue[a,c,e,] revealed a strong reaction with antibodies directed against the $\alpha2^{b}$, $\alpha3^{d}$ and $\alpha6^{f}$ integrin subunits.

adult lungs always remained negative when they were stained with antibodies directed against the $\alpha4$, $\alpha5$, $\beta3$ and $\beta4$ chains.

Integrins on alveolar epithelia of fibrotic lung tissue

Antibodies directed against the $\alpha1$, $\alpha2$ (Fig. 9.1b), $\alpha3$ (Fig. 9.1d), αv and $\beta1$ chains consistently stained type I and type II pneumocytes in fibrotic lung tissue as well. In contrast to alveolar epithelia of non fibrotic lung tissue pneumocytes of fibrotic lung tissue showed a strong expression of the $\alpha6$

(Fig. 9.1f) and $\beta4$ subunits. No immunoreactivity of alveolar epithelia was observed with antibodies directed against $\alpha4$, $\alpha5$ and $\beta3$.

DISCUSSION

It has been the aim of the present investigation to characterize the integrins which mediate the interaction between pneumocytes and the underlying extracellular matrix *in situ*. Among those integrins which are consistently expressed on alveolar epithelium of normal and fibrotic lung tissue, $\alpha1$, $\alpha2$ and $\alpha3$ belong to the molecules which in combination with the common $\beta1$ subunit preferentially bind on collagen.[16] The observed expression of different collagen receptors on alveolar epithelium is not surprising since type IV collagen is one of the major constituents of the alveolar basal lamina.[17] The immunohistochemical detection of different collagen receptors on alveolar epithelium is also consistent with previous data *in vitro* demonstrating high adhesion and spreading rates of isolated human type II pneumocytes on type I and type IV collagen as well.[4] The most interesting and rather surprising finding of the present study is related to the differential expression of $\alpha6$ and $\beta4$ in non fibrotic and fibrotic lung tissue. $\alpha6$ in combination with the $\beta1$ or $\beta4$ subunits is the classical receptor for laminin.[16] Expression of $\alpha6$ has been detected in the vast majority of epithelial tissues facing basement membrances,[7] including bronchial epithelium.[6,18] Since laminin is also a major component of the alveolar basal lamina[19] it was rather unexpected that we were unable to detect $\alpha6$ on most alveolar epithelia of non fibrotic lung tissue. The reason for the missing expression of $\alpha6$ in normal alveolar epithelium and its *de novo* expression in fibrotic lung tissue together with $\beta4$ is unclear. One possible explanation might be related to the continuously transmitted stretching and compressing forces at the interface between epithelium and the underlying basal lamina during inspiration and expiration. Previous investigations have demonstrated that the alveolar surface area may increase by as much as 80% during inflation to total lung capacity.[20] So far the continuous mechanical distension and relaxation of the alveolar epithelium represents a rather specific condition for an epithelial cell type since no comparable stretching forces are normally applied to other types of epithelia *in situ*. It is therefore tempting to speculate that the down regulation of $\alpha6$ and $\beta4$ expression in normal alveolar epithelium represents a specific adaptative mechanism of pneumocytes to the specific conditions in the pulmonary alveolus. The almost complete absence of $\alpha6$ on normal alveolar epithelium is also consistent with the previously reported low affinity of isolated human[4] or rat[5] type II pneumocytes for laminin.

The hypothesis that tension forces might represent one possible reason for the down regulation of $\alpha6$ in normal alveolar epithelium is also supported by the *de novo* expression of $\alpha6$ and $\beta4$ on alveolar epithelium of fibrotic lung tissue. Recent investigations have demonstrated that the $\alpha6/\beta4$ integrin complex is an integral part of the hemidesmosomal junctions between epithelia and the underlying basement membrane.[21] The observed *de novo* expression of $\alpha6$ and $\beta4$ on alveolar epithelium indicates that hemidesmosomal structures occur between alveolar epithelium and the underlying extracellular matrix in fibrotic lung tissue. Thus far our data correlate well and confirm at a molecular level previous ultrastructural observations of Kawanami *et al.*[22] He reported about the development of hemidesmosomes at the alveolar basal lamina of fibrotic lung tissue, which are normally absent in non fibrotic alveoli.

In conclusion, the present investigation has evaluated the integrin expression on human pulmonary alveolar epithelium showing obvious differences between normal and fibrotic lung tissue. Further studies and especially functional assays *in vitro* will provide valuable data concerning the molecular mechanisms that regulate the expression of these integrins *in situ* and *in vitro*.

Acknowledgments

The excellent technical assistance of Ms. Luzie Ionescu and the photographic expertise of E. Schmitt is gratefully acknowledged. This work was supported by the Deutsche Forschungsgemeinschaft (DFG), Grant Mu-579/1–4.

REFERENCES

1. Lwebuga-Mukasa, J. S. (1991) Matrix-driven pneumocyte differentiation. *American Review of Respiratory Diseases*, **144**, 452–457.
2. Rannels, S. R., Yarnell, J. A. and Fisher, C. S. *et al.* (1987) Role of laminin in maintainance of type II pneumocyte morphology and function. *American Journal of Physiology*, **253**, 835–845.
3. Rannels, S. R., Fisher, C. S. and Heuser, L. J. *et al.* (1987) Culture of type II pneumocytes on a type II cell-derived fibronectin-rich matrix. *American Journal of Physiology*, **253**, 759–765.
4. Papadopoulos, T., Ionescu, L. and Dämmrich, J. *et al.* (1990) Type I and type IV collagen promote adherence and spreading of human type II pneumocytes *in vitro*. *Laboratory Investigation*, **62**, 562–569.
5. Clark, R. A. F., Mason, R. J. and Folkvord J. M. *et al.* (1986) Fibronectin mediates adherence of rat alveolar type II epithelial cells via the fibroblastic cell-attachment domain. *Journal of Clinical Investigation*, **77**, 1831–1840.
6. Damjanovich, L., Albelda, S. M. and Mette, S. A. *et al.* (1992) Distribution of integrin cell adhesion receptors in normal and malignant lung tissue. *American Journal of Respiratory Cell and Molecular Biology*, **6**, 197–206.
7. Natali, P. G., Nicotra, M. R. and Bigotti, A. *et al.* (1992) Localization of the $\alpha6$ and $\beta4$ integrin subunits in normal human non-lymphoid tissues. *Journal of Cell Science*, **103**, 1243–1247.
8. Wayner, E., Carter, W. G. and Piotrowicz, R. S. *et al.* (1988) The function of multiple extracellular matrix receptors in mediating cell adhesion to extracellular

matrix: Preparation of monoclonal antibodies to the fibronectin receptors that specifically inhibit cell adhesion to fibronectin and react with platelet glycoproteins Ic-IIa. *The Journal of Cell Biology*, **107**, 1881–1891.
9. Wayner, E. A., Garcia-Pardo, A. and Humphries, M. J. *et al.* (1989) Identification and characterization of the T-lymphocyte adhesion receptor for an alternative attachment domain (CS-1) in plasma fibronectin. *The Journal of Cell Biology*, **109**, 1321–1330.
10. Freed, E., Gailit, J. and van der Geer, P. *et al.* (1989) A novel integrin β subunit is associated with the vitronectin receptor α subunit (αv) in a human osteosarcoma cell line and is a substrate for protein kinase C. *The EMBO Journal*, **8**, 2955–2965.
11. Hessle, H., Sakai, L. Y. and Hollister, D. W. *et al.* (1984) Basement membrane diversity detected by monoclonal antibodies. *Differentiation*, **26**, 49–54.
12. Ylänne, J., Hormia, M. and Järvinen, M. *et al.* (1988) Platelet glycoprotein IIb/IIIa complex in cultured cells. Localization in focal adhesion sites in spreading HEL cells. *Blood*, **72**, 1478–1486.
13. Ylänne, J. and Virtanen, I. (1989) The Mr 140,000 fibronectin receptor complex in normal and virus-transformed human fibroblasts and in fibrosarcoma cells: Identical localization and function. *International Journal of Cancer*, **43**, 1126–1136.
14. Sonnenberg, A., Janssen, H. and Hogervorst, F. *et al.* (1987) A complex of platelet glycoproteins Ic and IIa identified by a rat monoclonal antibody. *The Journal of Biological Chemistry*, **262**, 10376–10383.
15. Gown, A. M. and Vogel, A. M. (1982) Monoclonal antibodies to intermediate filament proteins of human cells: unique and cross-reacting antibodies. *The Journal of Cell Biology*, **95**, 414–424.
16. Hynes, R. O. (1992) Integrins: Versatility, modulation, and signaling in cell adhesion. *Cell*, **69**, 11–25.
17. Madri, J. A. and Furthmayr, H. (1980) Collagen polymorphism in the lung. An immunochemical study of pulmonary fibrosis. *Human Pathology*, **11**, 353-366.
18. Mette, S. A., Pilewski, J. and Buck, C. A. *et al.* (1993) Distribution of integrin cell adhesion receptors on normal bronchial epithelial cells and lung cancer cells *In vitro* and *In vivo*. *American Journal of Respiratory Cell and Molecular Biology*, **8**, 562–572.
19. Gil, J. and Martinez-Hernandez, A. (1984) The connective tissue of the rat lung: electron immunohistochemical studies. *The Journal of Histochemistry and Cytochemistry*, **32**, 230–238.
20. Gil, J., Bachofen, H. and Gehr, P. *et al.* (1979) Alveolar volume-surface area relation in air- and saline-filled lungs fixed by vascular perfusion. *The Journal of Applied Physiology*, **47**, 990–1001.
21. Stepp, M. A., Spurr-Michaud, S. and Tisdale, A. *et al.* (1990) $\alpha6\beta4$ integrin heterodimer is a component of hemidesmosomes. *Proceedings of the National Academy of Sciences USA*, **87**, 8970–8974.
22. Kawanami, O., Ferrans, V. J. and Roberts, W. C. *et al.* (1978) Anchoring fibrils. A new connective tissue structure in fibrotic lung disease. *American Journal of Pathology*, **92**, 389–403.

10. CELL ADHESION MOLECULES AS DIAGNOSTIC TOOLS IN CANCER

K. PATEL* and J. T. KEMSHEAD

The Imperial Cancer Research Fund, Paediatric and Neuro-Oncology Group, Frenchay Hospital, BRISTOL, BS16 1LE UK

The neural cell adhesion molecule (NCAM) is a major cell surface antigen in many human tumours. Differential exon usage by NCAM in normal tissues gives rise to tissue-specific isoforms. We have investigated the splicing of the tissue-specific mini-exons, VASE and MSD-1 in human tumours, to assess whether NCAM may be used as an aid in differential identification of human tumours. The sub-component MSD-1c was present only in rhabdomyosarcoma, but not SCLC or neuroblastoma, whereas the VASE mini-exon was present in all three tumour types. This suggests that an antibody to a synthetic peptide corresponding to MSD-1c sequence would be useful in differentiating rhabdomyosarcoma from neuroblastoma/SCLC.

NCAM is involved in cell-cell interactions during normal development via a homophillic mechanism. At present, it is not known whether the alternatively spliced forms of NCAMs have similar biological activities or not. Since their appearance during development is tightly controlled, one would expect that they do have different roles. Future studies will be directed at understanding the role of NCAMs in tumour biology using a combination of biochemical and molecular techniques.

INTRODUCTION

Pioneering work in neurobiology has identified a number of molecules involved in cell-cell and cell-substrate interactions. These molecules play key roles in development during cell migration and differentiation. The neural cell adhesion molecule (NCAM), is one such molecule.[1] During embryogenesis, NCAM is widely expressed in all three germ layers whereas its expression is probably restricted to the nervous system in the adult.[2]

Multiple forms of NCAM arise from both post-transcriptional and post-translational modifications. At the transcriptional level, NCAM pre-mRNA is subject to alternative splicing and differential usage of polyadenylation sites. As a result, in the brain, major NCAM proteins of 180, 140, 120 and 115 kDa are expressed whereas in skeletal muscle, isoforms of 155, 140, 125 and 115 kDa are present. Isolation of the corresponding cDNA clones for these isoforms has revealed that they differ in their carboxy termini.[3,4] In addition, in human brain, NCAMs of 170 and 95 kDa have been reported, although these species are poorly characterised at present.[5]

In addition, further heterogeneity in NCAM isoforms occurs through inclusion of mini-exons which result in differences in the extracellular segment. Inclusion of the VASE mini-exon (30 bp) between exon 7–8 junction leads to expression of neural-specific NCAMs.[6,7] Initially, this

* Correspondence.

mini-exon was thought to be neural-specific, but its presence in heart muscle has been reported recently.[8] The presence of VASE in the fourth Ig-like domain changes its resemblance from a C2-type domain to one that looks like the variable region of immunoglobulins. The functional significance of this is not known. However, the insertion of VASE in the fourth Ig-like domain occurs in the same position as a hypervariable loop in the variable domains of immunoglobulins. Since the hypervariable region in antibodies is involved in specific antigen binding, it is tempting to speculate that introduction of VASE in NCAM may be important for some unknown intermolecular interaction. The splicing of VASE is developmentally regulated and it is present in only a proportion of total NCAM.

Addition of the MSD-1 domain between exon 12–13 junction gives rise to skeletal muscle-specific NCAMs. MSD-1 consists of three mini-exons, MSD-1a (15 bp), -1b (48 bp) and -1c (42 bp).[9] Complete MSD-1 is present only in skeletal muscle, whereas MSD-1a can be found in neural NCAM. The inclusion of MSD-1 is developmentally regulated. Chick cardiac muscle contains a homologous region to MSD-1. In this instance, a 90 bp insert is present comprising of mini-exons 12A (15 bp), 12B (33 bp) and 12C (42 bp).[10] The functional consequences of the presence of MSD-1 are not clear, but a number of possibilities have been put forward. Inclusion of the MSD-1 domain introduces O-glycoslylation sites in NCAM. The presence of O-linked carbohydrates may affect NCAM-NCAM interactions in the plane of the membrane. Additionally, NCAM has been shown to co-associate with other adhesion molecules, e.g. L1, and influence their behaviour. The presence of MSD-1 may also affect this co-association. Finally, carbohydrates have been shown to influence protein turnover, so the presence of MSD-1 and its associated sugar residues could alter NCAM turnover. A further mini-exon, AAG (3 bp) is also found in cardiac and skeletal muscle, but it is also present in neural NCAM. It is, therefore, not a tissue-specific exon.

NCAM is also subject to post-translational modifications such as sulphation, phosphorylation and glycosylation, in particular, polysialylation.[11] The degree of polysialylation varies during development. In general, NCAM in embryonic tissues is highly polysialylated, whereas in adult tissue, less polysialylation occurs (however, there are some exceptions).[12]

The role of NCAM in development is clearly established. What is not clear at present is whether NCAM plays a role in disease, particularly in tumour biology. Two workshops on Neuroblastoma and Small Cell Lung Carcinoma (SCLC) have identified NCAM as the antigen for antibodies designated Cluster 1 in the SCLC workshop.[13] Subsequently, many reports have described the presence of NCAM in tumours. These include tumours of neuro-ectodermal,[5,14–16] mesodermal[17] and neuro-endocrine origin,[18] occasional haemopoietic malignancies[19,20] and tumours of unknown origin.[21] These studies have characterized the major forms of NCAM present in tumours.

Monoclonal antibodies in the Cluster 1 group were raised against a variety of immunogens with the idea of producing 'tumour-specific' reagents. However, such a reagent remains elusive. Many of the tumours that express NCAM can be difficult to diagnose diferentially by currently available antibodies. For example, neuroblastoma can be difficult to distinguish from rhabdomyosarcoma. In order to raise reagents of finer specificity, we have decided to investigate NCAM expression at the molecular level to identify the pattern of mini-exon usage. The rationale behind this approach is that if the mini-exon usage in tumours reflects that observed in normal tissues, it should be possible to raise reagents for differential identification of some tumour types.

Here, we report data on splicing of VASE and MSD-1 in representative rhabdomyosarcoma and SCLC cell lines, to illustrate that differences in the inclusion pattern of MSD-1c have been found suggesting that an antibody to a synthetic peptide corresponding to MSD-1c sequence would be useful in differential identification of rhabdomyosarcoma from SCLC/neuroblastoma.

METHODS

Cell Lines

The rhabdomyosarcoma cell line, JR1,[22] SCLC line, GLC-1,[23] and the haemopoietic line, GHI,[24] were grown in a 5% CO_2 incubator in RPMI 1640 medium supplemented with 10% foetal calf serum (fcs), 2 mM glutamine, 100 IU/ml penicillin and 100 μg/ml streptomycin. The JR1 line grows as a monolayer, whereas the others grow in suspension.

Polymerase Chain Reaction (PCR)

Total RNA was extracted by the method of Cathala *et al.* (1983),[25] 1.5 μg was reverse-transcribed in a total volume of 20 μl consisting of 50 mM Tris-HCl pH 8.5, 40 mM KCl, 10 mM $MgCl_2$, 0.4 mM β-mercaptoethanol, 1 μM appropriate 3′-primer, 20 U AMV-reverse transcriptase and 0.5 mM each dNTP at 42°C for 45 minutes.

The resulting cDNA was amplified using PCR in a total volume of 100 μl containing 10 mM Tris-HCl pH 8.3, 50 mM KCl, 1.5 mM $MgCl_2$, 0.02% gelatin, 200 μM each dNTP, 0.2 μM 5′-primer, 0.2 μM 3′-primer and 2.5 U Taq polymerase. Samples were overlaid with 50 μl mineral oil and subjected to denaturation at 94°C for four minutes. This was followed by 30 cycles of denaturation at 94°C for two minutes, annealing at 60°C for two minutes and extension at 72°C for 90 seconds. The last cycle had an extension period of seven minutes.

Analysis of PCR Products

Mineral oil was removed using 100 μl chloroform: isoamylalcohol (24:1). 10 μl of PCR product was run on a 2 or 2.5% agarose gel in X1 TAE buffer. The bands were blotted onto GeneScreen Plus membrane. Oligonucleotides were end-labelled by T4 polynucleotide kinase and used at 10^6 cpm/ml for hybridisation in a solution consisting of 10% dextran sulphate, 1% SDS, 1 M NaCl and 100 μg/ml salmon sperm denatured DNA. The blots were washed to high stringency and exposed to Fuji X-ray films for an appropriate period.

RESULTS

Splicing of MSD-1 Domain

RT-PCR was performed using primers flanking the exon 12–13 junction. A product of 255 bp would be expected if the MSD-1 was not included, whereas a 360 bp fragment would indicate complete MSD-1 inclusion. Blotting of the PCR products with NCAM cDNA clone λ9.5, a major band of 255 bp was detected in the JR1 and GLC-1 lines at short time exposure. This corresponds to the non-MSD-1 containing fragment. At longer exposures, a faint band of 360 bp was also observed (data not shown). This corresponds to an MSD-1-containing fragment. The blots were subsequently stripped and sequentially hybridised with oligonucleotides complementary to the whole MSD-1 or its sub-components a, b and c.

With oligonucleotides corresponding to MSD-1, 1a, 1b, and 1c, a band of 360 bp was identified in JR1 line. This corresponds to a fragment containing the MSD-1 domain. In addition, with oligonucleotides to MSD-1, 1a and 1b, smaller bands were also identified which presumably are fragments containing one or more of the sub-domains. Oilgonucleotide to MSD-1c did not identify any species smaller than 360 bp (Fig. 10.1, lane B).

In contrast, with the SCLC lines, no fragment of 360 bp was identified using any of the oligonucleotides. Only smaller fragments corresponding to species containing 1a or 1b were identified (Fig. 10.1, lane A) with oligonucleotides to MSD-1, MSD-1a or 1b. Furthermore, no signal was obtained with MSD-1c oligonucleotide, indicating that the SCLC line does not contain any PCR fragments with MSD-1c (Fig. 10.1, lane A).

No signal was obtained with any of the oligonucleotides using GHl cell line (Fig. 10.1, lane C). The date from neuroblastoma was similar to that obtained with SCLC (data not shown). Fragments containing MSD-1c were not detected.

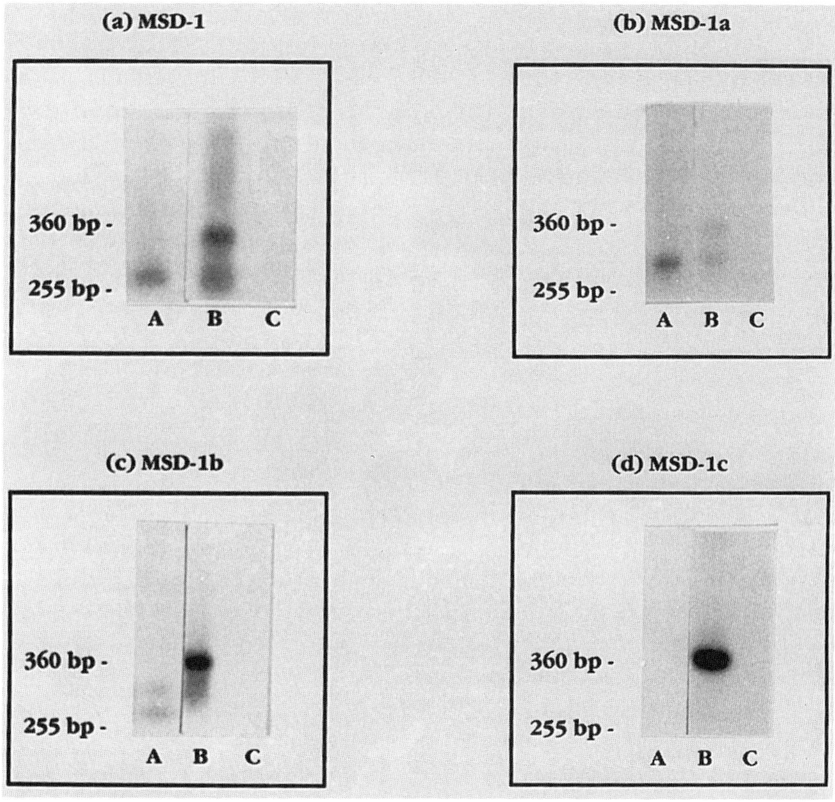

Figure 10.1 Splicing of the MSD-1 Domain in Human Rhabdomyosarcoma and SCLC Cell Lines.
RT-PCR was performed with primers flanking exon 12–13 junction. Aliquots of PCR products were separated by agarose gel electrophoresis, transferred to GeneScreen plus membranes and hybridised with radiolabelled oligonucleotides complementary to whole of the MSD-1 domain (panel A), MSD-1a only (panel B), MSD-1b only (panel C) or MSD-1c (panel D). Blots were washed to high stringency and exposed to autoradiography films.

Lane A = GLC-1 (SCLC) cell line.
B = JR-1 (rhabdomyosarcoma) cell line.
C = GHI (haemopoietic) cell line.

Splicing the VASE Domain

RT-PCR was performed using primers flanking exon 7–8 junction. A product of 194 bp would indicate the presence of NCAM-lacking VASE, whereas a 224 bp would indicate expression of VASE.

Using an oligonucleotide complementary to mouse VASE mini-exon, a fragment of 224 bp was identified, whereas with an oligonucleotide complementary to exon 7, bands of 194, 224 and 260 bp were detected in both JR1 and SCLC lines (Fig. 10.2). Therefore, the 194 bp fragment and 224 bp fragments correspond to VASE-lacking and VASE-containing species respectively. The presence of the 260 bp fragment was an unexpected finding. It could be due to the presence of a novel mini-exon at exon 7–8 junction or it could be artifactual. Separation of the PCR products under denaturing conditions did not reveal the presence of the 260 bp fragment showing that the presence of this band is due to aggregation (data not presented). Screening of neuroblastoma lines revealed that VASE-lacking, as well as VASE-containing, NCAM are present (data not shown). No bands were detected with the GHI line as expected (Fig. 10.2).

DISCUSSION

NCAM, a member of the immunoglobulin gene superfamily, is a major antigen in a variety of tumours. These tumours can be difficult to distinguish from each other by currently available techniques. In an attempt to raise reagents of finer specificity, we have investigated the presence of tissue-specific mini-exons of NCAMs in small round cell tumours occurring in children and in SCLC occurring in adults.

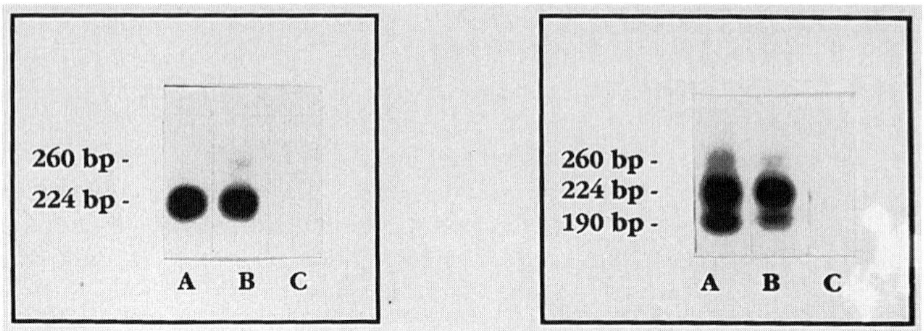

Figure 10.2 Splicing of the VASE Domain in Human Rhabdomyosarcoma and SCLC Cell Lines.
RT-PCR was performed using primers to exon 7–8 junction. PCR products were separated by agarose gel electrophoresis, transferred to GeneScreen plus membrane and hybridised with radiolabelled oligonucleotides complementary to mouse VASE (L. H. S.* panel) or exon 7 (R. H. S.** panel). Blots were washed to high stringency and exposed to autoradiographic films for an appropriate period.

Lane A = GLC-1 (SCLC) cell line.
B = JR-1 (rhabdomyosarcoma) cell line.
C = GHI (haemopoietic) cell line.

* = Left Hand Side ** = Right Hand Side

The splicing of MSD-1 was found to be complex, but in the cell lines investigated, MSD-1c was present only in rhabdomyosarcoma but not in neuroblastoma or SCLC lines. This suggestts that an antibody to a synthetic peptide corresponding to MSD-1c sequence would be useful in differential diagnosis.

The splicing of VASE may not be useful diagnostically since it is present in rhabdomyosarcoma, SCLC and neuroblastoma as well as Wilms' tumour (data shown). The wider expression of VASE in tumour lines probably reflects the wider distribution observed in normal tissues. We have described the presence of both VASE and MSD-1 in NCAMs from tumour cell lines. It is not known if the presence of these domains reflects the differentiation status of tumours or if there is aberrant splicing of NCAM. It remains to be determined what the functions of the alternatively spliced domains are and whether NCAM plays a role in tumour biology.

During normal development, NCAM participates in cell-cell interactions via homophillic binding. In addition, it can also affect the binding activity of other adhesion molecules, e.g. L1, by co-association. It is not known whether the major isoforms of NCAM have different biological activities or not and whether these activities are affected by the presence of VASE and MSD-1. The availability of full-length cDNA clones containing the VASE and/or MSD-1 domains and appropriate functional assay systems should, in the future, yield some clues as to the function of these mini-exons.

For our future studies, we will be extending our investigation on splicing of VASE and MSD-1 in other tumours such as melanoma, Wilms' tumour and Ewing's sarcoma. Futhermore, an antibody to a synthetic peptide corresponding to MSD-1c will be raised and assessed for its usefulness in differential diagnosis of tumours. Studies will also be conducted to investigate the role of NCAM in tumour biology.

Acknowledgements

We are grateful to the Imperial Cancer Research Fund for their financial support and Ms Murphy for typing this manuscript.

REFERENCES

1. Cunningham, B. A., Hemperly, J. J., Murray, B. A., Prediger, E. A., Brackenbury, R. and Edelman, G. M. (1987) Neural cell adhesion molecule: struture, immunoglobulin-like domains, cell surface modulation, and alternative RNA splicing. *Science*, **236**, 799–806.
2. Edelman, G. M. (1985) Cell-adhesion and the molecular processes of morphogenesis. *Ann. Rev. Biochem.*, **54**, 135–169.
3. Goridis, C. and Wille, W. (1988) The three size classes of mouse NCAM proteins arise from a single gene by a combination of alternative splicing and use of different polyadenylation sites. *Neurochem. Int.*, **12**, 269–272.

4. Walsh, F. S. (1989) Generation of multiple neural cell adhesion molecules proteins from a single gene. *Biochem. Soc. Trans.*, **17**, 975–976.
5. Frost, G., Patel, K., Bourne, S., Coakham, H. B. and Kemshead, J. T. (1991) Expression of alternative isoforms of the neural cell adhesion molecule (NCAM) on normal brain and a variety of brain tumours. *Neuropath. & Neurobiol.*, **17**, 207–217.
6. Small, S. J., Haines, S. L. and Akeson, R. A. (1988) Polypeptide variation in an N-CAM extracellular immunoglobulin-like fold is developmentally regulated through alternative splicing. *Neuron*, **1**, 1007–1017.
7. Santoni, M. J., Barthels, D., Vopper, G., Boned, A., Goridis, C. and Wille, W. (1989) Differential exon usage involving an unusual splicing mechanism generates at least eight types of NCAM cDNA in mouse brain. *EMBO J.*, **8**, 385–392.
8. Reyes, A. A., Small, S. J. and Akeson, R. (1991) At least 27 alternatively spliced forms of the neural cell adhesion molecule mRNA are expressed during rat heart development. *Mol. Cell Biol.*, **11**, 1654–1661.
9. Thompson, J., Dickson, G., Moore, S. E., Gower, H. J., Putt, W., Kenimer, J. G., Barton, C. H. and Walsh, F. S. (1989) Alternative splicing of the neural cell adhesion molecule gene generates variant extracellular domain structure in skeletal muscle and brain. *Genes & Development*, **3**, 348–357.
10. Prediger, E. A., Hoffman, S., Edelman, G. M. and Cunningham, B. A. (1988) Four exons encode a 93-base-pair insert in the three neural cell adhesion molecule mRNAs specific for chicken heart and skeletal muscel. *Proc. Natl. Acad. Sci. USA*, **85**, 9616–9620.
11. Sorkin, B. C., Hoffman, S., Edelman, G. M. and Cunningham, B. A. (1985) Sulfation and phosphorylation of the neural cell adhesion molecule, N-CAM. *Science*, **225**, 1476–1488.
12. Rothbard, J. B., Brackenbury, R., Cunningham, B. A. and Edelman, G. M. (1982) Differences in the carbohydrate structures of neural cell adhesion molecules from adult and embryonic chicken brains. *J. Biol. Chem.*, **257**, 11064–11069.
13. Patel, K., Moore, S. E., Dickson, G., Rossell, R. J., Beverley, P. C., Kemshead, J. T. and Walsh, F. S. (1989) Neural cell adhesion molecule (NCAM) is the antigen recognized by monoclonal antibodies of similar specificity in small cell lung carcinoma and neuroblastoma. *Int. J. Cancer*, **44**, 573–578.
14. Patel, K., Bourne, S., Phimister, E., Coakham, H. and Kemshead, J. T. (1990) Presence of neural cell adhesion molecule on human embryonic and brain tumours. *Biochem. Soc. Trans.*, **18**, 408–410.
15. Phimister, E., Kiely, F., Kemshead, J. T. and Patel, K. (1991) Expression of neural cell adhesion molecule (NCAM) isoforms in neuroblastoma. *J. Clin. Pathol.*, **44**, 580–585.
16. van Duijnhoven, H. L. P., Helfrich, W., de Leij, L., Roebroek, A. J. M., van de Ven W. J. M., Healey, K., Culverwell, A., Rossell, R. J., Kemshead, J. T. and Patel, K. (1992) Splicing of the VASE exon of neural cell adhesion molecule (NCAM) in human Small Cell Lung Carcinoma (SCLC). *Int. J. Cancer*, **50**, 118–123.
17. Zuber, C. and Roth, J. (1990) The relationship of polysialic acid and the neural cell adhesion molecule (NCAM) in Wilms' tumour and their subcellular distributions. *Eur. J. Cell Biol.*, **51**, 313–321.
18. Long, J., Hemperly, J. J. and Lloyd, R. V. (1991) Expression of neural cell

adhesion molecule in normal and neoplastic human neuroendocrine tissues. *Am. J. Pathol.*, **38**, 961–969.

19. Lanier, L. L., Testi, R., Bindl, J. and Phillips, J. H. (1989) Identity of Leu-19 (CD56) leucocyte differentiation antigen and neural cell adhesion molecule. *J. Exp. Med.*, **169**, 2233–2238.
20. Patel, K., Frost, G., Rossell, R. J., Pizer, B. L., Gee, A., Sugimoto, T., Phimister, E. and Kemshead, J. T. (1992) Expression of the neural cell adhesion molecule (NCAM) on the haemopoietic cell line Nalm-16. *Leukaemia Res.*, In Press.
21. Lipinski, M., Hirsch, M-R., Deagostini-Bazin, H., Yamada, O., Tursz, T. and Goridis, C. (1987) Characterization of neural cell adhesion molecules (NCAM) expressed by Ewing and neuroblastoma cell lines. *Int. J. Cancer*, **40**, 81–86.
22. Clayton, J., Pincott, J. R., van den Berghe, J. A. and Kemshead, J. T. (1986) Comparative studies between a new human rhabdomyosarcom cell line, JR1, and its tumour of origin, *Br. J. Cancer*, **54**, 83–90.
23. De Leij, L., Postmus, P. E., Buys, C. H., Elema, J. D., Ramaekers, F., Poppema, S., Bralwer, M., Van Der Veen, A., Mesander, G. and The, T. H. (1985) Characterisation of three new variant type cell lines derived from small cell carcinoma of the lung. *Cancer Research*, **45**, 6024–6033.
24. Minowada, J., Dnuma, T. and Moore, G. E. (1972) Brief communication: Rosette-forming human lymphoid cell lines. I. Establishment and evidence for origin of thymus-derived lymphocytes. *J. Nat. Cancer Inst.*, **49**, 891–895.
25. Cathala, G., Savouret, J. F., Mendez, B., West, B. L., Kauin, M., Martrial, J. A. and Baxter, J. D. (1983) A method for isolation of intact, transitionally active ribonucleic acid. *DNA*, **2**, 329–335.

11. EXCHANGE OF INHIBITORS AND PROTEASES AT TUMOUR CELL SURFACES

FRANK S. STEVEN[1], MARGARET M. GRIFFIN[1], BERNARD S. BROWN[1], IAN C. TALBOT[2] and CHRIS HANSKI[3]

[1]*Department of Biochemistry and Molecular Biology, University of Manchester, Manchester M13 9PT, UK*
[2]*ICRF Colorectal Unit, Department of Pathology, St. Mark's Hospital, London EC1V 2PS, UK*
[3]*Universitatsklinikum Stegliz, Freie Universitat Berlin, Hindenburgdamm 30, D-1000 Berlin 45, Germany*

INTRODUCTION

In this study we are concerned with a cell surface protease, referred to as guanidinobenzoatase (GB) on account of this enzyme's ability to cleave guanidinobenzoates as true substrates.[1,2] Trypsin-like proteases are normally inhibited by guanidinobenzoates,[3,4] these compounds being used in the standard assay of trypsin-like enzymes employing the kinetics of active site titration.[3,4] Guanidinobenzoates are used by these enzymes as suicide substrates and this leads to the rapid quantitation of the enzymes' active sites. GB has now been demonstrated to be functionally identical[5] with tissue type plasminogen activator (t-PA). The study of cell surface GB may therefore have relevance to tumour cell surface chemistry especially as raised PA levels have been associated with transformed cells in culture and tumour cells when compared to the levels of activity of PA in the corresponding normal cells.[6,7,8] Recently t-PA has been shown to cleave a guanidinobenzoate as a true substrate[9] again pointing out the similarity of GB and t-PA. There can be many isoforms of t-PA[10,11] which differ in their sites of glycosylation with the possible result that these isoenzymes will have different properties. It has recently been shown that the cell surface GB is present as different isoenzymic forms on normal colonic epithelial cells and carcinoma cells of the same colon.[12] These isoenzymic forms of GB were detected by their selective inhibition by cytoplasmic protein inhibitors of GB extracted from the cells under study. These inhibitors may play a role *in vivo* in regulating the activity of cell surface GB.[13] We were able to follow these inhibition reactions on frozen sections of tissue employing affinity techniques normally used to study protein-protein recognition in the solution or on affinity columns. These techniques involve the same type of affinity reactions taking place in antibody-antigen recognition although we are not using antibodies but naturally occurring protein inhibitors of GB. Having developed this reversible inhibition process we were then able to detach the enzyme-inhibitor complex (GB-I) from the cell surface and replace the GB by genetically engineered t-PA, which then behaved like cell surface bound GB.[14] We outline the methods used for demonstrating cell surface GB activity, reversible inhibition of cell surface GB by cytoplasmic proteins and human plasma leading to the evidence for isoenzymic forms of GB. This leads to the technology for the exchange of inhibitors with cell surface GB.[15,16] We then move to the exchange of enzymes on the receptors on cells with GB activity, by means of the affinity of fibrin for cell surface GB[17] and the subsequent recapture of inhibitor molecules. This study has been carried out with both tumours of the head and neck region as targets for these exchange reactions[14] and also employing colonic epithelial cells. What follows is mainly concerned with the normal and carcinomatous colorectal epithelial cells.

MATERIALS

Fresh frozen sections (200) of normal colon and carcinoma of the colon were dispatched in dry ice to Manchester by air from Berlin, Germany.

Cultured adhering human colonic tumour cells (LS174T) were kindly provided by Dr. David Blakey, ICI Pharmaceuticals Division, Mereside, Alderley Park, Macclesfield, Cheshire SK10 4TG, UK. Normal human serum was provided by FSS when required.

9-amino acridine (9AA) and Texas Red acid chloride, human fibrinogen and human thrombin were purchased from Sigma Chemical Co., St. Louis, MO, USA. PD-10 columns of G-25M Sephadex were supplied by Pharmacia LKB Biotechnology AB Uppsala, Sweden.

METHODS

Preparation of Texas Red labelled inhibitor (TR-I)

The culture fluid was drained from the bottles of cultured colonic tumour cells (LS174T). The cells were washed four times in excess isotonic saline to remove traces of culture fluid and non-cellular material. The cells were then ruptured in water and the soluble fraction collected, filtered and freeze-dried. This total cell extract was reacted with Texas Red acid chloride in 2% sodium bicarbonate solution pH7.8 for 2 h, and the Texas Red protein fraction was then collected by passage over a PD-10 column. The Texas Red-protein fraction contained the colonic carcinoma cell cytoplasmic inhibitor of GB and many other proteins. This crude preparation is referred to here as TR-I since all we are concerned with in this work is its ability to recognise the GB on the surface of colonic carcinoma cells. The further purification of this inhibitor will be described elsewhere.

Texas Red labelled t-PA (TR-t-PA)

Pure t-PA was treated with Texas Red acid chloride and the protein fraction isolated after passage over a PD-10 column.

Fluorescent labelling of cells with active GB

The sections were place in a tank containing 200ml isotonic saline and 10^{-3} M 9-amino acridine for 2 min. The excess reagents were removed by washing in a tank of isotonic saline for 10 s. Cells possessing active GB bound 9AA fluoresced yellow.

Preparation of sections

Three types of frozen sections obtained from normal and colorectal carcinoma tissue were used in this study. Direct 9AA staining was carried out on fresh frozen sections to ascertain the status of GB in the tissue *in vivo.* When we were involved in the transfer of inhibitors between slides, we needed sections in which the GB was active but which lacked cytoplasmic inhibitors of cell surface GB. Two methods were used to make such sections. The first process was described as protection.[16] *Protected* sections refer to frozen sections which were initially saturated with 9AA under conditions in which the inhibitor proteins are extracted from the section and removed prior to the reversible removal of all the 9AA by prolonged washing in fresh isotonic saline.[16] Such *protected* sections contain active cell surface GB and can be used as targets for the challenge of a suitable protein inhibitor of GB, the effectiveness of this challenge being monitored by 9AA staining. A second process was developed to provide sections with active GB loaded with 9AA as a target for potential protein inhibitors of GB. These *loaded* sections were prepared by placing the sections in 200 ml of fresh isotonic saline containing 10^{-3} M 9AA for 30 min, prior to rinsing off the excess reagent and drying the slide before applying the inhibitor. Again, the effectiveness of the inhibitor was assayed with subsequent 9AA staining.

(i) Direct inhibition of GB

Direct inhibition with the cytoplasmic inhibitors present in the cells within the section was carried out as follows. The fresh frozen sections were covered with 10 μl isotonic saline and left in a wet box for 10 min. The surface fluid was rinsed from this slide and the sections stained with 9AA.

(ii) Transfer of inhibitors

The target sections for these inhibition studies were either fresh frozen sections presented as *protected* or as *loaded* sections. When demonstrating the reversible exchange of inhibitors, we used formaldehyde treated frozen sections in the later stages of these cyclic operations. The donor sections which provided the potential inhibitors were fresh frozen sections, the surfaces of which were covered with 10 μl isotonic saline to extract soluble inhibitors. After 20 min the surface fluid from 3 sections of donor tissue was transferred by microsyringe to the surface of a target section (with active GB) ready to receive the potential inhibitor.

(iii) Serum inhibitors of GB

The target sections were covered with 10 μl fresh human serum and left in a wet box for 10 min. The excess serum was rinsed from the slide and the section stained with 9AA.

(iv) Cultured cell inhibitor of GB

Unlabelled cultured cell extract and crude TR-I were used as potential inhibitors, like the serum inhibitors.

(v) Formaldehyde displacement of inhibitor from GB

Sections were placed in 4% formaldehyde[13] in isotonic saline overnight, rinsed in isotonic saline and stained with 9AA. These sections could then be used for further inhibition studies, once the 9AA had been removed from the active centre of GB by exhaustive washing in isotonic saline.

Transfer of GB from cell surface to fibrin fibrils

This technique has already been described in detail.[17] Essentially, this employed *protected* sections with active GB. These were placed under a fibrin overlay for 1h, the fibrin removed from the surface of the section prior to 9AA staining and subsequent use of the sections to bind TR-t-PA on the receptors which originally bound GB. We also used the fibrin overlay technique to provide soluble inhibitor, obtained from sections in which the GB had been previously complexed with inhibitor. The GB-I was dissociated by the fibrin, the GB remaining on the fibrin whilst the soluble inhibitor protein could then be used for transfer studies (see above).

Microscopy

The yellow fluorescence of 9AA bound to active GB was observed using the Leitz Diaplan fluorescence microscope, employing filter cube [G] (Catalogue No. 513602) and barrier filter K490. Colour photography was carried out with an Olympus Om-2_N camera fitted with an automatic exposure meter. We used Kodak Ektachrome ASA 400 colour film throughout. The red fluorescence of TR-labelled proteins was observed in the Leitz fluorescence microscope using filter cube [N_2] (Catalogue No. 513609), see Fig. 11.8, and the Zeiss fluorescence microscope fitted with a rhodamine cube, see Fig. 11.9.

RESULTS AND DISCUSSION

The active centre of the cell surface enzyme GB binds 9-amino acridine (9AA) as a competitive inhibitor.[2] The binding of 9AA to cells possessing active GB can be visualised by fluorescence microscopy;[2] these cells exhibit yellow fluorescence. 9AA can thus be used as a simple fluorescent probe to investigate the enzymic status of GB on cells in frozen sections; if the enzyme is inhibited prior to exposure to 9AA, the cells no longer fluoresce.[13]

In this study we illustrate our observations on epithelial cells of normal mucosa and of carcinomas in the human colon. These observations have been made in other tumour locations and reported elsewhere, for example, in tumours of the head and neck region.[14] The normal colonic epithelial cell possesses an active form of cell surface GB *in vivo*, as shown by the binding of 9AA to the surface of these cells in fresh frozen sections. (Fig. 11.1). The cell surfaces fluorescence yellow when the enzyme is active but after exposure to the protein inhibitor extracted in isotonic saline from fresh frozen sections of normal colon, these normal epithelial cells no longer bind the yellow fluorescent probe (Fig. 11.2). The inhibitor can be displaced from the GB-I complex by treatment with formaldehyde[13] with consequent regain in the ability of the cell surface GB to bind 9AA, and also to bind a fresh

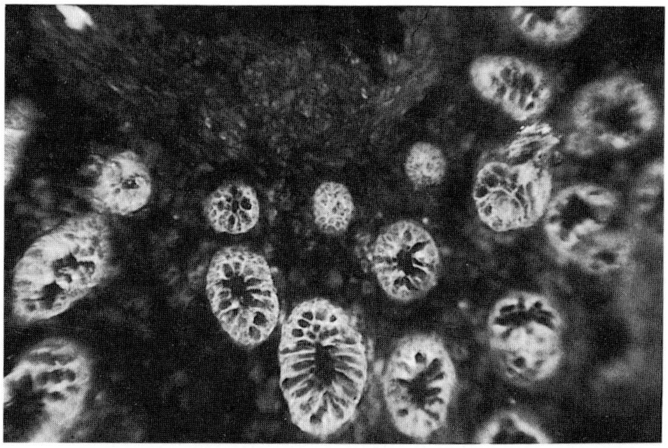

Figure 11.1 *Frozen section of normal colon stained with 9AA.* The surfaces of the colonic epithelial cells possess active GB, bind 9AA and fluoresce yellow. The surrounding lymphocytes lack active GB and appear as non-fluorescent green dots. Magnification × 165. (See Colour Plate IX at the back.)

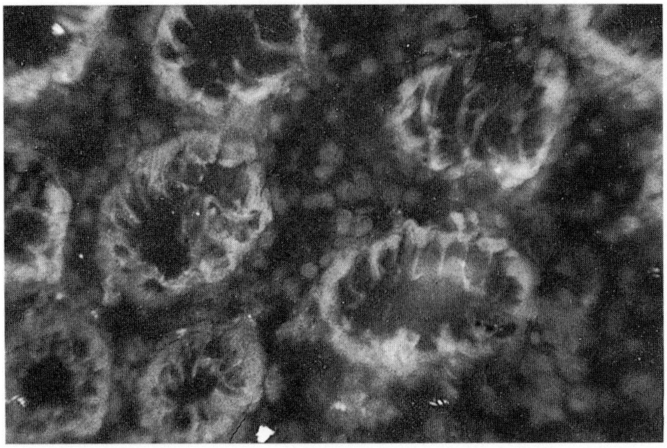

Figure 11.2 *Frozen section of normal colon, placed under 10 μl isotonic saline, prior to 9AA staining.* The epithelial cells fail to bind 9AA and appear green. The GB on these cells has been inhibited by the cytoplasmic inhibitor of GB extracted in isotonic saline. The GB-I cannot bind 9AA and the cells fail to fluoresce yellow. This inhibitor can be displaced with formaldehyde with a regain in ability to bind 9AA. Magnification × 250. (See Colour Plate X at the back.)

preparation of inhibitor. This whole process can be repeated many times on the same slide and may be considered to be reversible and cyclic. The advantage of this displacement of the protein inhibitor is that the enzyme on a particular cell surface can be challenged repeatedly by different inhibitors, the successful binding of inhibitor being registered by the failure of 9AA to bind to the cell surface, a process reversed by formaldehyde.[13]

The colonic carcinoma cells in fresh frozen sections also bind 9AA and fluorescence yellow (Fig. 11.3). Clearly these cells possess a form of active GB and behave similarly to the normal epithelial cells, since the fluorescent staining is also inhibited by a protein extracted from fresh colonic carcinoma tissue (Fig. 11.4). Again, this is reversed in a cyclic manner by formaldehyde treatment.[13]

It could well be asked what is the significance of cell surface GB in terms of the tumour cells when the normal colonic epithelial cell also possesses an obvious cell surface GB (Fig. 11.1). This question led to the study of the specificity of molecular recognition which may be demonstrated when an inhibitor obtained from normal colon is used to challenge GB on both normal and carcinoma cells in the colon using frozen sections. Both normal colonic epithelial cells and carcinoma cells possess effective inhibitors of their cell surface GB (Figs 11.2, 11.4). These inhibitors have to be removed from the slide before meaningful experiment can be carried out with transferred inhibitors from other cells or sections. Two approaches have been used to provide frozen sections, lacking cytoplasmic inhibitors but possessing active

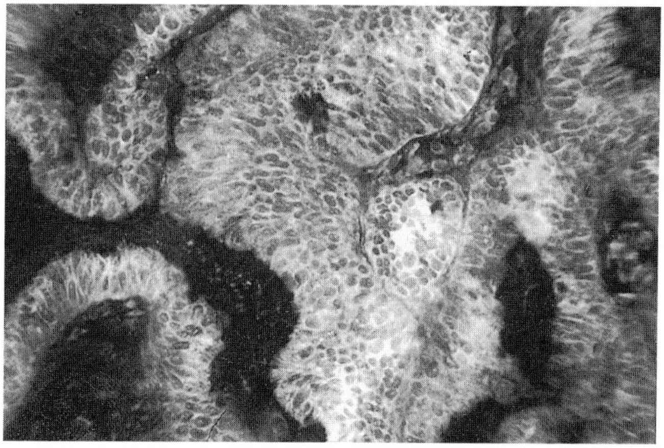

Figure 11.3 *Frozen section of colonic carcinoma stained with 9AA.* The surfaces of the carcinoma cells possess activate GB which binds 9AA and the cells fluoresce yellow. Magnification × 250. (See Colour Plate XI at the back.)

native GB on the cell surfaces. The first procedure developed was the *protected section*[16] in which the cell surface GB is first filled with 9AA and the inhibitors extracted in a large volume of 9AA-containing isotonic saline; the 9AA was then removed in excess saline to leave active GB. In a second

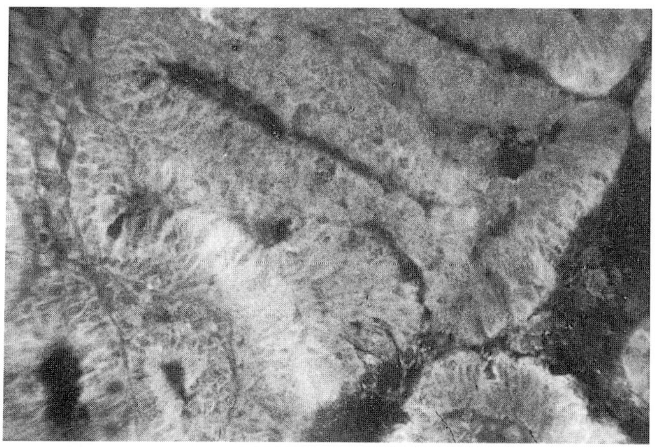

Figure 11.4 *Frozen section of colonic carcinoma, placed under 10μl isotonic saline, prior to 9AA staining.* The GB on the carcinoma cells now lacks activity since it is in the form of the GB-I complex. GB in GB-I complex does not bind 9AA and these cells do not exhibit yellow fluorescence. The inhibitor can be displaced with formaldehyde with regain in GB-activity as judged by the regain in ability to bind 9AA. Magnification × 250. (See Colour Plate XII at the back.)

procedure the sections were treated with 9AA, as above, but the 9AA was not washed out after removal of the soluble protein inhibitors — these preparations are referred to as *loaded sections* and have 9AA filling the active centre of all GB molecules. When we used *protected sections* and challenged these with the extracted inhibitor, recognition of GB by the inhibitor was direct. When we used *loaded sections*, the GB-9AA required to be dissociated by the higher affinity of the protein inhibitor for GB before GB-I complex formation took place. Both techniques provided identical results when the cell surface GB was challenged by an appropriate protein inhibitor, followed by location of cells bearing active GB by 9AA staining. We observed that normal colonic epithelial GB was inhibited in these sections when exposed to normal colonic cytoplasmic protein inhibitors (similar to Fig. 11.2). The same observations were made with *protected* and *loaded* sections of colonic carcinoma exposed to cytoplasmic protein inhibitors obtained from carcinoma sections (similar to Fig. 11.4). Both systems were cyclically reversible after formaldehyde treatment. The most interesting observations were obtained when *protected* and *loaded* sections of carcinoma tissues were exposed to the inhibitors extracted from normal colonic tissue (Fig. 11.5). Much of the GB on the cell surface of the carcinoma cells were not recognised by the inhibitors obtained from normal colonic epithelial cells; these cells were totally non-fluorescent after subsequent exposure to the carcinoma cell extract followed by 9AA staining (result similar to Fig. 11.4).

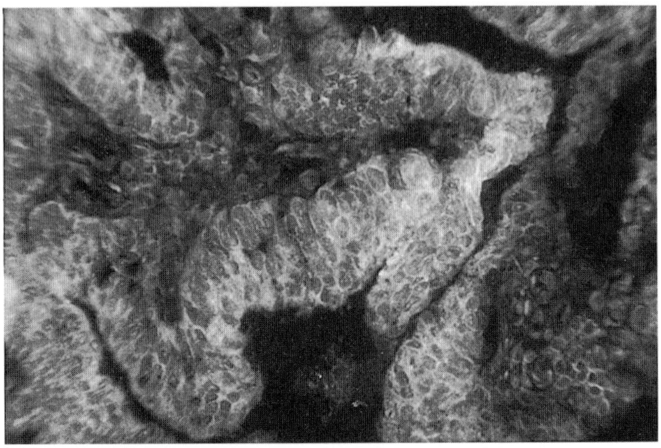

Figure 11.5 *Frozen section of colonic carcinoma, placed under 20μl isotonic saline extract of inhibitors obtained from normal colonic section, prior to 9AA staining.* Isotonic saline extracts inhibitors from normal colonic cells which recognise and inhibit GB on normal epithelial cells (Fig. 11.2). These inhibitors do not inhibit all the GB on colonic carcinoma cells surfaces. Thus carcinoma cell GB still binds 9AA and fluoresces yellow. Magnification × 250. (See Colour Plate XIII at the back.)

The evidence presented in Figs 11.1–5 indicates that the cytoplasmic inhibitors extracted from normal and carcinoma cells of the colon exhibit selective inhibition of cell surface GB on these cells. Since the active centre of the GB on both cell types bind 9AA equally well, this selective inhibition indicates differences in the conformations of these two-iso-enzymic forms of GB which could possibly be due to alterations in glycosylation.[10,11] Similar differences in isoenzymes have been demonstrated in head and neck tumours[14] and breast carcinoma. It is of particular interest that this evidence for isoenzymic forms of GB on the normal and carcinoma cells can be confirmed when normal human serum is used to challenge the cell surface GB on *protected* and *loaded* sections of colonic tissue. Inhibitors in normal serum rapidly recognise and inactivate GB on normal colonic cells (data similar to Fig. 11.2) but do not inhibit the GB on carcinoma cell surfaces (Fig. 11.6).

The data presented above have been concerned with the transfer of inhibitors obtained from extracts of frozen sections or tissue which contain many cell types. It could be argued that we do not know the cellular origin of our inhibitors. This criticism can be overcome by employing cultured colonic carcinoma cells as the source of the cytoplasmic inhibitor. The Texas Red-labelled inhibitor fraction (obtained post PD-10 column chromatography (TR-I)) was used to challenge 9AA *loaded* sections of colonic carcinoma tissue. Partial replacement of 9AA took place within 10 minutes exposure to the TR-1, in which dual fluorescence of the tumour cells can be observed (Fig. 11.7 yellow fluorescence of 9AA and Fig. 11.8 red fluorescence of GB-TR-I complex). After the 9AA had been exhaustively washed from these

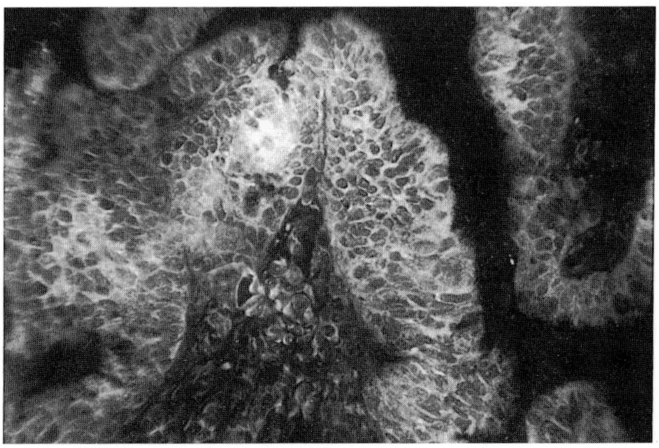

Figure 11.6 *Frozen section of colonic carcinoma, placed under 10μl normal serum, prior to 9AA staining.* The carcinoma cells still possess some active GB which binds 9AA and these cells exhibit yellow fluorescence. Magnification × 250. (See Colour Plate XIV at the back.)

sections the effect of TR-I labelling of the carcinoma cells is obvious (Fig. 11.9) when examined in the Zeiss microscope using a filter combination as for Rhodamine. The transfer of inhibitors to active GB on the surface of these cells is not confined to the cytoplasmic and serum inhibitors described above. Plasminogen-activator inhibitor type I (PA1-1) can also be demonstrated to bind to GB on *protected sections* of tumour tissue.[15]

It has long been known that fibrils of fibrin have a high affinity for t-PA and the complex of this enzyme and the fibrin greatly increases the rate of plasminogen activation.[17] GB and GB-I on *protected* and *9AA loaded* sections of carcinoma tissue are transferred to fibrin leaving *no* latent GB which can be reactivated by formaldehyde.[14] Indeed the latter fibrin clot can be centrifuged to leave a supernatant fraction containing active inhibitor protein which can subsequently be transferred to another target slide containing active GB on the surface of tumour cells. So in this case we are able to transfer native inhibitor from one cell to another by means of a fibrin overlay, by dissociating the original GB-I into GB-fibrin and I, prior to transfer of I. This fibrin transfer technique may present possibilities for the preparation of tumour GB specific inhibitors and corresponding tumour specific isoenzymes of GB. The technique may also provide tumour cells stripped of cell surface GB but still possessing the native receptor protein for binding GB to the cell surface. This possibility has been demonstrated,[14] employing squamous cell carcinoma of the head and neck regions. The latter studies demonstrated the binding of a fluorescent labelled pure t-PA to the cell surface after removal of GB by fibrin. The newly-bound t-PA also bound 9AA in a manner similar to the original GB on these cells. We suggest that this may be a useful method of studying the binding and exchange of GB and t-PA to cell surface receptors.[14] Similar studies have now been carried out with colonic carcinoma as the target cell for receptor binding of t-PA and GB. A flow diagram representing these reactions at the cell surface is presented as Scheme 11.1. We conclude that, although fluorescent affinity probes for the active centre of GB and the use of cytoplasmic protein inhibitors is not a conventional way of studying cell surface chemistry in frozen sections, these techniques can undoubtedly provide interesting information which conventional histology is unable to do.

Acknowledgements

FSS wishes to thank both the Wigan and District Cancer Research Fund/ J. T. Starkey Memorial Fund and the Association of International Cancer Research for their generous provision of funds enabling this work to be carried out in Manchester.

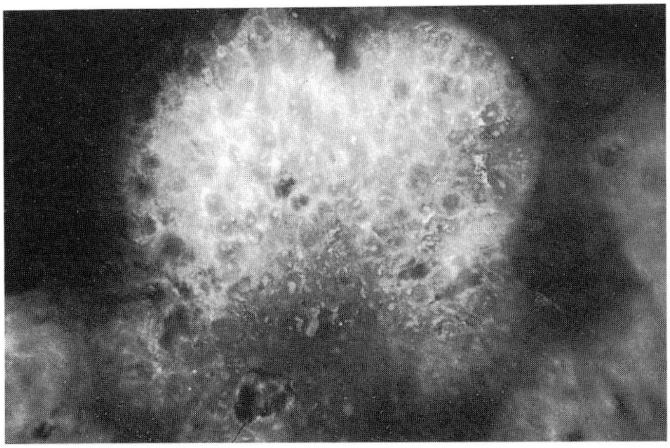

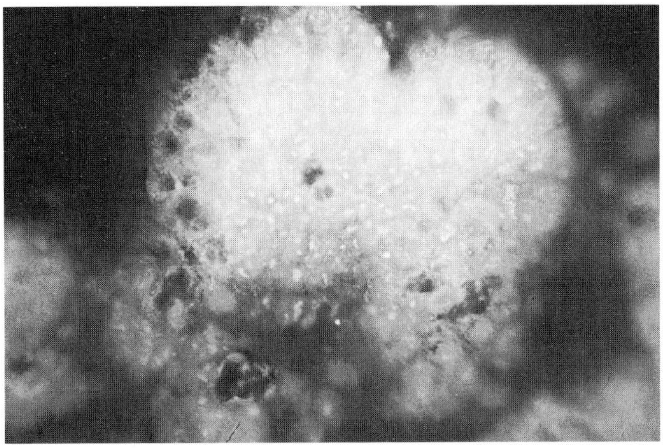

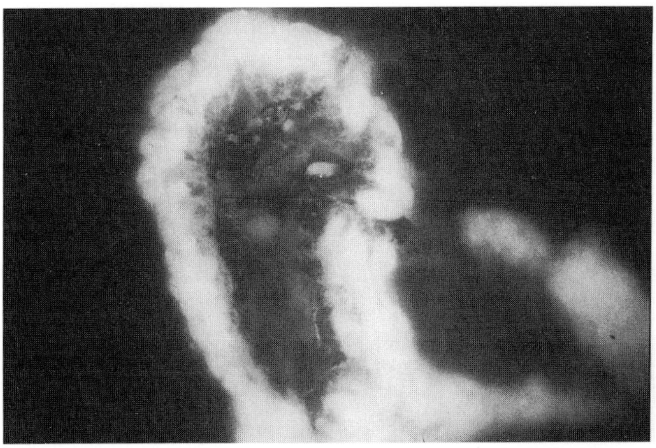

Figures 11.7,8,9 *Frozen section of colonic carcinoma co-stained with 9AA and TR-I.* In this study the section was first stained with 9AA and excess reagent rinsed off, so as to leave the GB loaded with 9AA. At this stage 10 μl TR-I was added to compete with the 9AA for the GB, i.e. to form GB-9AA or GB-I complex for 10 min. The section was photographed before the 9AA was washed from the GB in Figs 11.7,8, but after washing in Fig. 11.9.
In Fig. 11.7, the binding of 9AA to the carcinoma cells resulted in yellow surface fluorescence with filter cube [G] in place whilst the TR-I bound to those cells exhibited red fluorescence with filter cube [N_2] in place.
After the section was exhaustively washed in isotonic saline to remove 9AA, the TR-I remained attached to the carcinoma cells which exhibited red fluoresence on the Zeiss microscope fitted with a Rhodamine cube. Magnification × 500. (See Colour Plate XV at the back.)

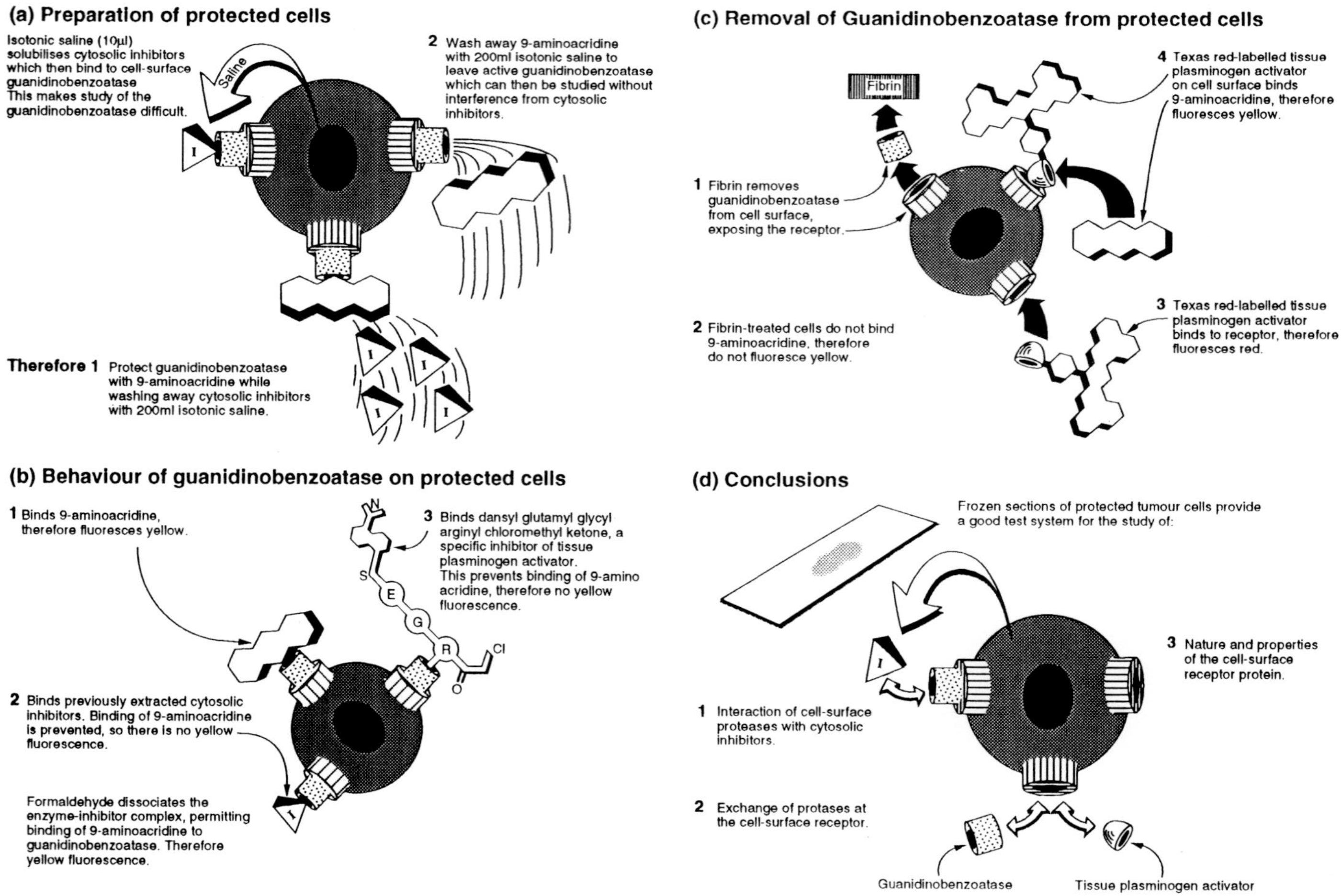
(a) Preparation of protected cells
Isotonic saline (10µl) solubilises cytosolic inhibitors which then bind to cell-surface guanidinobenzoatase
This makes study of the guanidinobenzoatase difficult.
Saline
I
2 Wash away 9-aminoacridine with 200ml isotonic saline to leave active guanidinobenzoatase which can then be studied without interference from cytosolic inhibitors.
Therefore 1 Protect guanidinobenzoatase with 9-aminoacridine while washing away cytosolic inhibitors with 200ml isotonic saline.
(b) Behaviour of guanidinobenzoatase on protected cells
1 Binds 9-aminoacridine, therefore fluoresces yellow.
3 Binds dansyl glutamyl glycyl arginyl chloromethyl ketone, a specific inhibitor of tissue plasminogen activator.
This prevents binding of 9-amino acridine, therefore no yellow fluorescence.
N
S
E
G
R
Cl
O
2 Binds previously extracted cytosolic inhibitors. Binding of 9-aminoacridine is prevented, so there is no yellow fluorescence.
Formaldehyde dissociates the enzyme-inhibitor complex, permitting binding of 9-aminoacridine to guanidinobenzoatase. Therefore yellow fluorescence.
(c) Removal of Guanidinobenzoatase from protected cells
Fibrin
4 Texas red-labelled tissue plasminogen activator on cell surface binds 9-aminoacridine, therefore fluoresces yellow.
1 Fibrin removes guanidinobenzoatase from cell surface, exposing the receptor.
2 Fibrin-treated cells do not bind 9-aminoacridine, therefore do not fluoresce yellow.
3 Texas red-labelled tissue plasminogen activator binds to receptor, therefore fluoresces red.
(d) Conclusions
Frozen sections of protected tumour cells provide a good test system for the study of:
1 Interaction of cell-surface proteases with cytosolic inhibitors.
2 Exchange of protases at the cell-surface receptor.
3 Nature and properties of the cell-surface receptor protein.
Guanidinobenzoatase
Tissue plasminogen activator

REFERENCES

1. Steven, F. S. and Al-Ahmad, R. K. (1983) Evidence for an enzyme which cleaves the guanidinobenzoate moiety from active site titrants specifically designed to inhibit and quantify trypsin. *Eur. J. Biochem.*, **130**, 335–339.
2. Steven, F. S., Griffin, M. M. and Al-Ahmad, R. K. (1985) The design of fluorescent probes which bind to the active centre of guanidinobenzoatase. *Eur. J. Biochem.*, **149**, 35–40.
3. Chase, T. and Shaw, E. N. (1967) p-Nitrophenyl-p-guanidinobenzoate HC1. A new active site titrant for trypsin. *Biochem. Biophys. Res. Commun.*, **29**, 508–514.
4. Coleman, P. L., Latham, H. G. and Shaw, E. N. (1976) Some sensitive methods for the assay of trypsin-like enzymes. *Methods Enzymol.*, **45**, 12–26.
5. Steven, F. S., Griffin, M. M., Cederholm-Williams, S. A., Mangel, W. F. and Maier, H. (1991) Evidence for the functional similarity between tumour cell surface guanidinobenzoatase and tissue plasminogen activator. *Anticancer Res.*, **11**, 641–648.
6. Unkeless, J. C., Dano, K., Kellerman, G. M. and Reich, E. (1974) Fibrinolysis associated with oncogenic transformation. *J. Biol. Chem.*, **249**, 4295–4300.
7. Quigley, J. P., Goldfarb, R. H., Scheiner, C., O'Donnell-Tormey, J. and Yeo, T. K. (1980) In *Tumour cell surfaces and Malignancy*, edited by Hynes, R. O. and Fox, C. F. Published by Alan R. Liss, New York, pp. 773–796.
8. Moscatelli, D. and Rifkin, D. B. (1988) Membrane and matrix localisation of proteinases: a common theme in tumour cell invasion and angiogenesis. *Biochem. Biophys. Acta.*, **948**, 67–85.
9. Geiger, M. and Binder, B. R. (1987) Tissue type plasminogen activator and urokinase: differences in the reaction pattern with the active site titrant 4-methylumbelliferyl-p-guanidinobenzoate hydrochloride. *Biochim. Biophys. Acta.*, **912**, 34–40.
10. Parekh, R. B., Dwek, R. A., Rudd, P. M., Thomas, J. R., Rademacher, T. W., Warren, T., Wun, T-C., Herbert, B., Reitz, B., Palmier, Ramabhadran, T. and Tiemeier, T. W. (1989) N-glycosylation and *in vitro* enzymatic activity of human recombinant tissue plasminogen activator expressed in chinese hamster ovary cells and a murine cell line. *Biochemistry*, **28**, 7670–7679.
11. Parekh, R. B., Dwek, R. A., Thomas, J. R., Opdenakker, G., Rademacher, T. W., Wittwer, A. J., Howard, S., Nelson, R., Siegel, N. R., Jennings, M. G., Harakas, N. K. and Feder, J. (1989) Cell-type specific and site specific N-glycosylation of Type I and Type II human tissue plasminogen activator. *Biochemistry*, **28**, 7644–7662.
12. Steven, F. S., Griffin, M. M. and Talbot, I. C. (1991) Evidence for distinct isoenzymic forms of a cell surface protease on normal colonic cells and on carcinoma cells from the same colon. *Anticancer Res.*, **11**, 143–150.
13. Steven, F. S., Maier, H. and Arndt, J. (1989) Inhibition of a protease associated with tumour cells and the probable mechanism of regulation of this interaction *in vivo*. *J. Enz. Inhibition*, **3**, 145–157.
14. Steven, F. S., Griffin, M. M., Bulleid, N. J., Maier, H. and Born, I. A. (1991) Studies on the interaction and exchange of inhibitors and proteases on the surface of tumour cells in frozen sections. *Anticancer Res.*, **11**, 2117–2122.
15. Steven, F. S. and Booth, N. A. (1991) The inhibitor reacting with a tumour cell surface protease can be exchanged with plasminogen activator inhibitor (PA1-I). *J. Enz. Inhib.*, **4**, 173–279.

16. Steven, F. S., Griffin, M. M. and Maier, H. (1991) Temporary competitive inhibition of a tumour cell surface protease as a protective mechanism in the preparation of membrane bound native enzyme in the presence of excess cytoplasmic inhibitors. *J. Enz. Inhib.*, **5**, 77–85.
17. Steven, F. S. and Blakey, D. C. (1991) The role of fibrin fibrils in the dissociation of a cell surface protease-inhibitor complex and evidence for the recapture of the inhibitor protein. *J. Enz. Inhibition*, **5**, 299–315.
18. Stryer, J. (1988) *Biochemistry* Third Edition p. 256, Freeman and Company, New York.

12. THE REQUIREMENT OF ADHESION MOLECULES FOR CELL MEDIATED TUMOUR CELL DESTRUCTION

ANDREW M. JACKSON*, ANTON ALEXANDROFF and KEITH JAMES

Department of Surgery, University of Edinburgh Medical School, Teviot Place, Edinburgh, EH8 9AG, UK

The expression of adhesion molecules on eight human bladder cancer cell lines was studied. Tumour lines of low grade did not constitutively express intercellular adhesion molecule-1 (ICAM-1), but were found to express intercellular adhesion molecule-2 (ICAM-2). High grade cells, however, only expressed ICAM-1. The conjugation and subsequent lysis of the cell lines by activated lymphocytes was found to be dependent upon the expression by the target cell of either ICAM-1 or ICAM-2. When effector cells were pre-treated with antibodies to leucocyte function associated antigen-1 (LFA-1, CD11a) the inhibition of lysis was greater and ranged from 72 to 96% with a mean of 87% inhibition. These results suggest that the sensitivity of bladder cancer cell lines to activated lymphocytes is partially attributable to the expression of ICAM-1 on the tumour cell surface. However, blocking of ICAM-1 with antibodies could only partially inhibit LFA-1 dependent lysis thus supporting recent evidence for the existence of a third ligand for LFA-1.

Expression of ICAM-1 was induced and augmented on the tumour cell lines following stimulation with interferon-gamma (IFNγ). A corresponding increase in susceptibility to cell mediated cytolysis was observed. These results demonstrate the requirement for intimate contact between tumour and effector cell and the role of adhesion molecules in this interaction. Furthermore, important therapeutic questions arise as the susceptibility of tumour cells to cell mediated destruction is dictated by the tumour cell as well as the lymphocyte.

ABBREVIATIONS: ICAM-1(Intercellular adhesion molecule-1)
ICAM-2(Intercellular adhesion molecule-2)
LFA-1(Leukocyte function associated antigen-1)
IL-2(Interleukin-2)
IFNγ (Interferon-gamma)
LAK(Lymphokine activated killer)
TCC(Transitional cell carcinoma)

INTRODUCTION

Lymphokine activated killer (LAK) cells are capable of lysing a wide variety of otherwise NK resistant tumour cells.[1] This cytolytic activity is distinct from T cell mediated killing in that it is non-MHC restricted and does not affect normal cells.[2] Previously we have demonstrated differential susceptibility to LAK killing of a panel of four bladder cancer cell lines.[3]

The structure(s) involved in non-specific target recognition by activated lymphocytes remain largely unknown. However, molecules have been identified on the surface of leucocytes which are involved in cell-cell adhesion.[4,5] Prominent among these structures are members of the

* Correspondence.

β2-integrin family, CD11a, b, and c/CD18.[6,7] LFA-1 (CD11a/CD18, leucocyte function-associated antigen-1) is the major leucocyte integrin on lymphocytes.[8] There are three characterized ligands for LFA-1, ICAM-1 (intercellular adhesion molecule-1) and ICAM-2 and ICAM-3.[9,10,11]

Conflicting evidence exists as to the role of adhesion molecules in non-antigen-specific cell-mediated killing of tumour cells. Quillet-Mary and colleagues demonstrated that target lysis by human LAK cells was dependent upon target binding properties, but LFA-1, LFA-3 and ICAM-1 were not the major adhesion ligands on targets as differences in adhesion molecule expression did not account for the marked differences in susceptibility.[12] In contrast, other workers have shown that the susceptibility of colon carcinoma cells to LAK mediated lysis is associated with elevated expression of ICAM-1 and LFA-3.[13]

Several pro-inflammatory cytokines induce and augment the expression of cell adhesion molecules. Interferon-gamma (IFNγ) increases the expression of MHC-class I, MHC-class II, and ICAM-1 on many cell types.[14–16] Studies on CTL antigen-specific lysis of keratinocytes, show that both ICAM-1/LFA-1 and Ag/CD3-TcR interactions are important when target cells are stimulated with IFNγ. In fact, enhanced cytolysis of keratinocytes after short term treatment with IFNγ could be explained solely on the basis of ICAM-1 induction.[17]

We have investigated the role of adhesion molecules found to be constitutively expressed by a panel of bladder carcinoma cell lines with particular emphasis on their susceptibility to cell mediated lysis. Furthermore, we have studied the ability of IFNγ to modulate adhesion molecule expression and examined the functional role of increased adhesion molecule expression. We demonstrate that ICAM-1 and ICAM-2 molecules expressed by target cells contribute towards the sensitivity to LAK cells and that LFA-1 expressed on effector cells is the major adhesion ligand involved in the killing of these tumour cells. Furthermore, the observation of LFA-1 dependent, ICAM-1 and ICAM-2 independent cell interactions adds weight to the existing evidence for a third ligand for LFA-1.

MATERIALS AND METHODS

Cell lines

Eight human transitional cell carcinoma lines RT4, RT112, MGH-U1 (J. Masters, Institute of Urology, London), SD, J82, 5637, UMUC3 (P. Perlmann, Stockholm, Sweden) and EJ18 (J. Boyd, University of Edinburgh) were used as a tissue culture model of bladder cancer. The cell lines have proved to be stable and their cytological appearances and growth characteristics have remained representative of the parent tumours. The cell lines were routinely cultured in RPMI 1640 (Gibco, Uxbridge, UK) supplemented with 5% FCS (Sera-Lab, Crawley Down, UK), sodium pyruvate (5mM) and L-glutamine (2mM).

Cytokines

Highly purified recombinant IFNγ was purchased from Boehringer Mannheim (Lewes, UK) and contained 2.5×10^7 Umg^{-1} of protein and less than 2 Endotoxin Umg^{-1}. Purified recombinant IL-2 was generously provided by Euro-Cetus BV (Amsterdam, The Netherlands) and contained 1.8×10^7 Umg^{-1} of protein.

Monoclonal Antibodies

The following murine monoclonal antibodies were generously provided by Dr. T. A. Springer; RR1/1 (anti-ICAM-1, IgG1), CBR-IC2 (anti-ICAM-2, IgG1). Anti-LFA-1 and anti-HLA-DR monoclonal antibodies were obtained from Dako, High Wycombe, UK. Anti-LFA-2 monoclonal antibody was provided by the Scottish Antibody Production Unit (Carluke, UK).

Effector Cells

Mononuclear cells were isolated from fresh, heparinized peripheral blood from normal, healthy laboratory volunteers. Briefly, peripheral blood was centrifuged on Ficoll-Isopaque (density = 1.086 g/cm^3) for 30 minutes at $400 \times g$. The interface was collected and subsequently washed thrice, adherent cells removed by culture on plastic for 1 hr at 37°C and non-adherent cells collected and washed twice in PBS.

Activation of lymphocytes

Mononuclear cells were cultured at 1×10^6 ml^{-1} in complete medium supplemented 1,000 Uml^{-1} recombinant IL-2. The cells were incubated at 37°C in a humidified atmosphere of 5% CO_2 in air for 6 days. After incubation, cells were washed and resuspended at 4×10^6 ml^{-1} and assessed for conjugation and cytolytic activity against labelled target cells.

Cytotoxicity Assay

Cytotoxicity was measured by means of a standard 4-hr chromium-release assay. All assays were performed in triplicate and the data expressed as the percentage specific cytotoxicity, calculated as below:

$$\%\text{Specific Cytotoxicity} = \frac{\text{cpm exp release} - \text{cpm spontaneous release}}{\text{cpm total release} - \text{cpm spontaneous release}}$$

Quantitation of surface antigen expression

Single colour immunofluorescence was performed on bladder cancer cell lines stained indirectly. Cells were incubated at 4°C for 30 minutes with optimal concentrations of primary monoclonal antibody. Following two washes, the binding of monoclonal antibodies was detected using a goat anti-mouse-FITC conjugate. The cells were washed and resuspended in 1% formaldehyde prior to analysis using a flow-cytometer.

Antibody blocking studies

The above cytotoxic assay protocol was modified as follows: ^{51}Cr-labelled cells or effector cells were incubated with or without monoclonal antibodies (10 μg/ml) in a final volume of 100 μl for 30 minutes at room temperature. After two washes, cells were added to each well.

Conjugation studies

Conjugates were formed between tumour and activated lymphocytes by spinning the cells together at a ratio of 1:10 gently onto a glass slide. Our previous studies have shown that conjugation is an extremely rapid event (personal observation). Following fixation in acetone the cells were stained with monoclonal antibodies to ICAM-1 and subsequently with a secondary HRP-conjugated antibody to murine IgG. The addition of an enzyme substrate gave a coloured reaction and the cells were lightly counter-stained with haematoxylin.

A total of 300 tumour cells were counted for each cell line and the number of lymphocytyes conjugated to each tumour cell was also determined with discrimination between ICAM-1 positive and negative tumour cells.

RESULTS

Constitutive expression of ICAM-1 and ICAM-2 on TCC cells

The cell surface phenotype of the eight TCC cell lines was studied using flow cytometry. The results are summarised in Table 12.1. None of the cell lines constitutively expressed both ICAM-1 and ICAM-2. Only two of the eight cell lines failed to express ICAM-1, both of which corresponded to tumour grade G1 however, these lines did express ICAM-2. A small percentage of RT112 (G2) cells, expressed of ICAM-1. The remaining five cell lines, all of which were G3, expressed higher levels of ICAM-1 (17–72% of cells positive). None of the cells expressed the other adhesion molecules LFA-1, LFA-2 or VCAM or NCAM (data not shown). The mean level of specific cytotoxicity against each cell line is shown also in table 12.1. A highly

Table 12.1 Summary of the expression of ICAM-1 and ICAM-2 and the sensitivity of eight bladder cancer cell lines to LAK mediated cytolysis. Results expressed as percent specific cytotoxicity at a standard effector to target ratio of 40:1. Each value represents the mean of at least six experiments, each performed in triplicate. The standard deviations were within 10%.

Cell line	*RT4*	*UMUC3*	*RT112*	*MGH-U1*	*EJ18*	*J82*	*SD*	*5637*
Grade	G1	G1	G2	G3	G3	G3	G3	G3
% cell ICAM – 1 +	0	0	5	17	72	63	30	50
% cells ICAM – 2 +	6	33	0	0	0	0	0	0
Specific Cytotoxicity	40	76	24	40	65	80	50	61

significant correlation exists between constitutive ICAM-1 or ICAM-2 expression and the sensitivity to killing by LAK cells ($r = 0.809$, $p < 0.01$).

Modulation of adhesion molecule expression by IFNγ

The effect of IFNγ on the expression of ICAM-1 and ICAM-2 molecules was studied. IFNγ induced or augmented the expression of ICAM-1 molecules on all cell lines except UMUC3. Sensitive cell lines, however, responded to as little as 1Uml^{-1} IFNγ and maximum induction was achieved with 100 Uml^{-1} (Fig. 12.1). Representative kinetics of induction are also shown in Fig. 12.2 indicating maximal ICAM-1 expression after 24 hours. As expected, expression of ICAM-2 was not augmented or induced by IFNγ. Furthermore, LFA-1, LFA-2, VCAM and NCAM expression were not induced on any of the cell lines following treatment with IFNγ (data not shown).

Immunohistochemical detection of adhesion molecules between effectors and target cells

Cytospins of effector and target cells (10:1) were stained for ICAM-1 and the number of conjugates per target cell was counted; the data is shown in Fig. 12.3. There was a significantly greater number of effector cells conjugated to a single ICAM-1 expressing target cell than to a ICAM-1 negative target cell for all four cell lines ($p < 0.02$). Differences were also evident between the individual cell lines in the case of ICAM-1 expressing targets, however, these differences failed to reach significance ($p < 0.2$). When the data was corrected for the percentage of cells known to express ICAM-1, differences were evident between the cell lines (Table 12.2). A greater number of effector cells was seen to conjugate to EJ18 and J82 than to the MGH-U1 and RT112 cell lines ($p < 0.01$). No difference was detected between RT112 and MGH-U1. The correlation between the susceptibility of targets to LAK activity and the number of effector cells conjugated to each target cell was $R = 0.971$, $p < 0.01$.

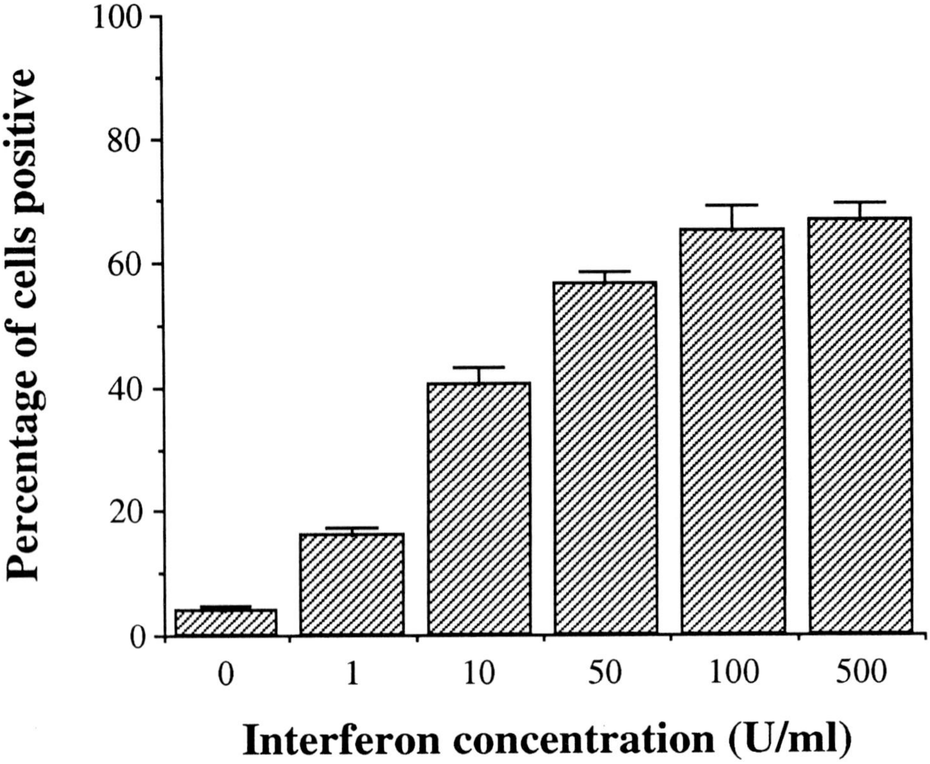

Figure 12.1 The effect of interferon-gamma stimulation on ICAM-1 expression by all cell lines. Cells were cultured with increasing concentrations of interferon-gamma for up to 24 hours and ICAM-1 expression determined by flow-cytometry. All standard deviations were within 5%.

Table 12.2 Summary of the conjugation data shown in figure 12.3. The results show the mean number of effector cells (± 1sd) for four target cell lines. This number is calculated from the number of effector cells per ICAM-1 positive and negative target cell and is adjusted for the mean percentage of cells expressing ICAM-1.

Cell line	*Number of effector cells per target* *ICAM – 1 +*	*ICAM – 1 –*	*Number of effector cells per target*
MGH-U1	1.9 ± 1.2	0.3 ± 0.5	0.40 ± 0.6
EJ18	2.6 ± 1.6	0.3 ± 0.5	1.96 ± 1.2
RT112	2.3 ± 1.3	0.2 ± 0.4	0.3 ± 0.4
J82	3.7 ± 2.0	0.3 ± 0.5	2.41 ± 1.4

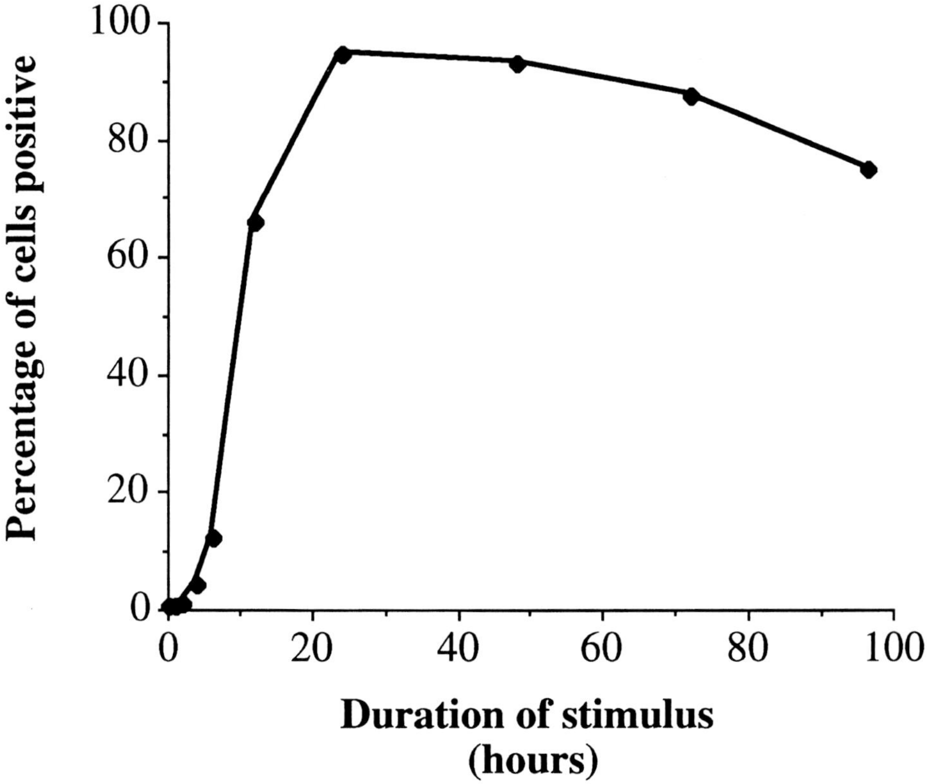

Figure 12.2 The time course of ICAM-1 expression by the RT4 cell line following stimulation with 100 U/ml interferon-gamma. Expression was measured by flow cytometry.

Susceptibility of TCC cells to LAK killing

We have previously reported the differential susceptibility of four TCC cell lines to cell mediated killing.[3] The specific cytotoxicity observed against all eight cell lines is summarised in Table 12.1. No correlation was evident between the susceptibility to LAK cell killing and grade of the parent tumour.

TCC cells were stimulated with different concentrations of IFNγ for 24 hours and their sensitivity to lysis by LAK cells was examined. After treatment with as little as $10 \mathrm{Uml}^{-1}$ IFNγ a significant ($p < 0.01$) increase in specific cytotoxicity was observed. Treatment with $100 \mathrm{~Uml}^{-1}$ resulted in a further increase in specific cytotoxicity ($p < 0.001$). Correlation between ICAM-1 expression and LAK killing was evident (Fig. 12.4). Previously we have reported the cytotoxic and cytostatic effects of IFNγ against bladder cancer cells, however, there was no significant difference in spontaneous release of ^{51}Cr by untreated or treated cells.[17]

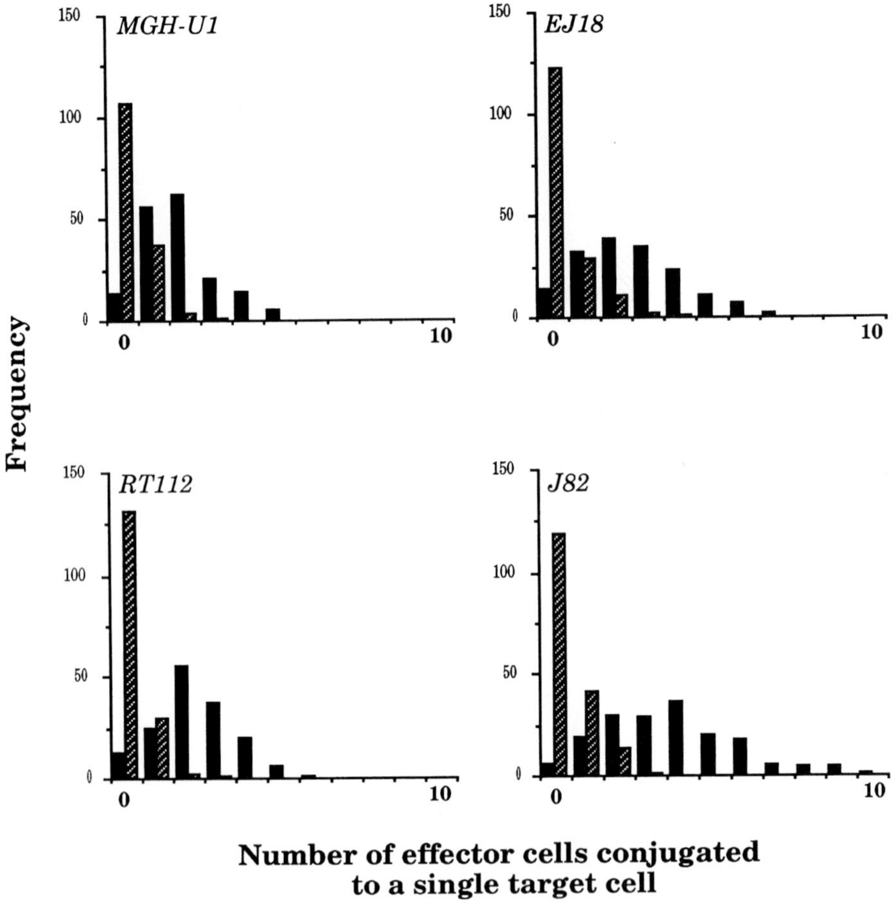

Figure 12.3 The frequency of the number of effector cells conjugated to a single target cell is shown. Targets and effector cells were mixed at a 1:10 ratio and cytospins were made. Following immunohistochemical staining for ICAM-1, the number of effector cells conjugated to either ICAM-1 positive (solid bars) or ICAM-1 negative (hatched bars) cells were counted.

Blocking of adhesion molecules on target cells with antibodies decreases sensitivity to LAK killing

The functional roles of constitutive ICAM-1 and ICAM-2 expression were analysed by monitoring the effects of antibodies to these structures on the lysis of TCC cell lines. Figure 12.5 shows representative blocking studies for

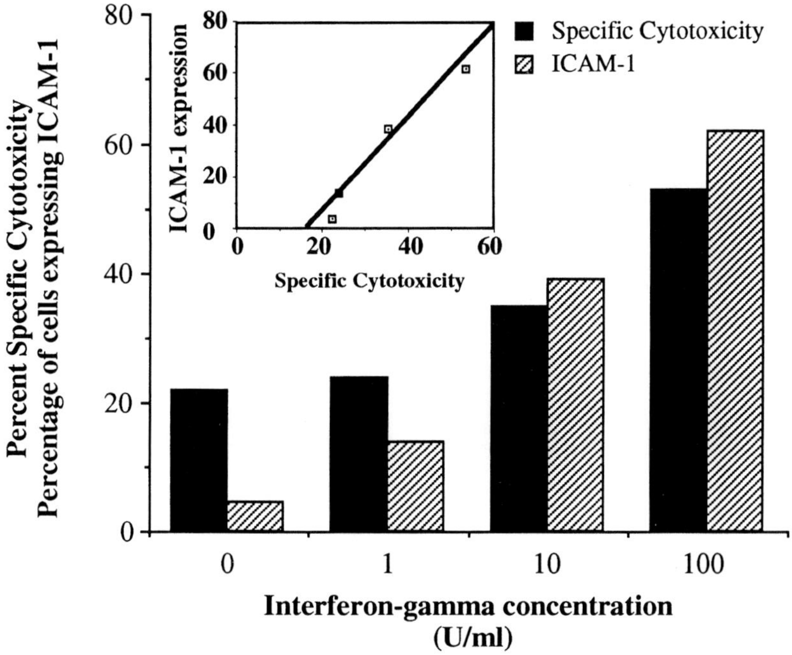

Figure 12.4 The relationship between ICAM-1 induction and LAK susceptibility in the RT112 cell line. Increased target cell lysis by LAK cells parallels interferon-gamma induced increases in cell surface ICAM-1 expression. Main figure shows the effects of increasing interferon-gamma concentration on both ICAM-1 expression and specific cytotoxicity of RT112 cells by LAK cells. Insert shows correlation between ICAM-1 expression and specific cytotoxicity.

all cell lines. Monoclonal antibody blocking was performed prior to the assay or during the assay. Specific cytotoxicity against J82, MGH-U1, EJ18, 5637 and SD was decreased by blocking with anti-ICAM-1 antibody. This effect was more marked if the antibody was present throughout the assay period. Anti-ICAM-2 antibodies had no effect on sensitivity of these five lines.

LAK killing of UMUC3 (one of the two ICAM-2 positive cell lines) was decreased only when anti-ICAM-2 antibodies were present during the assay. The sensitivity of RT4 and RT112 could not be altered by attempted blocking with antibody. Monoclonal antibodies to HLA-DR did not modify the sensitivity of any of the lines.

Although antibodies to adhesion molecules ICAM-1 and ICAM-2 were seen to decrease the sensitivity of TCC targets to LAK cells, the killing could not be completely negated, even in the presence of increased antibody concentration.

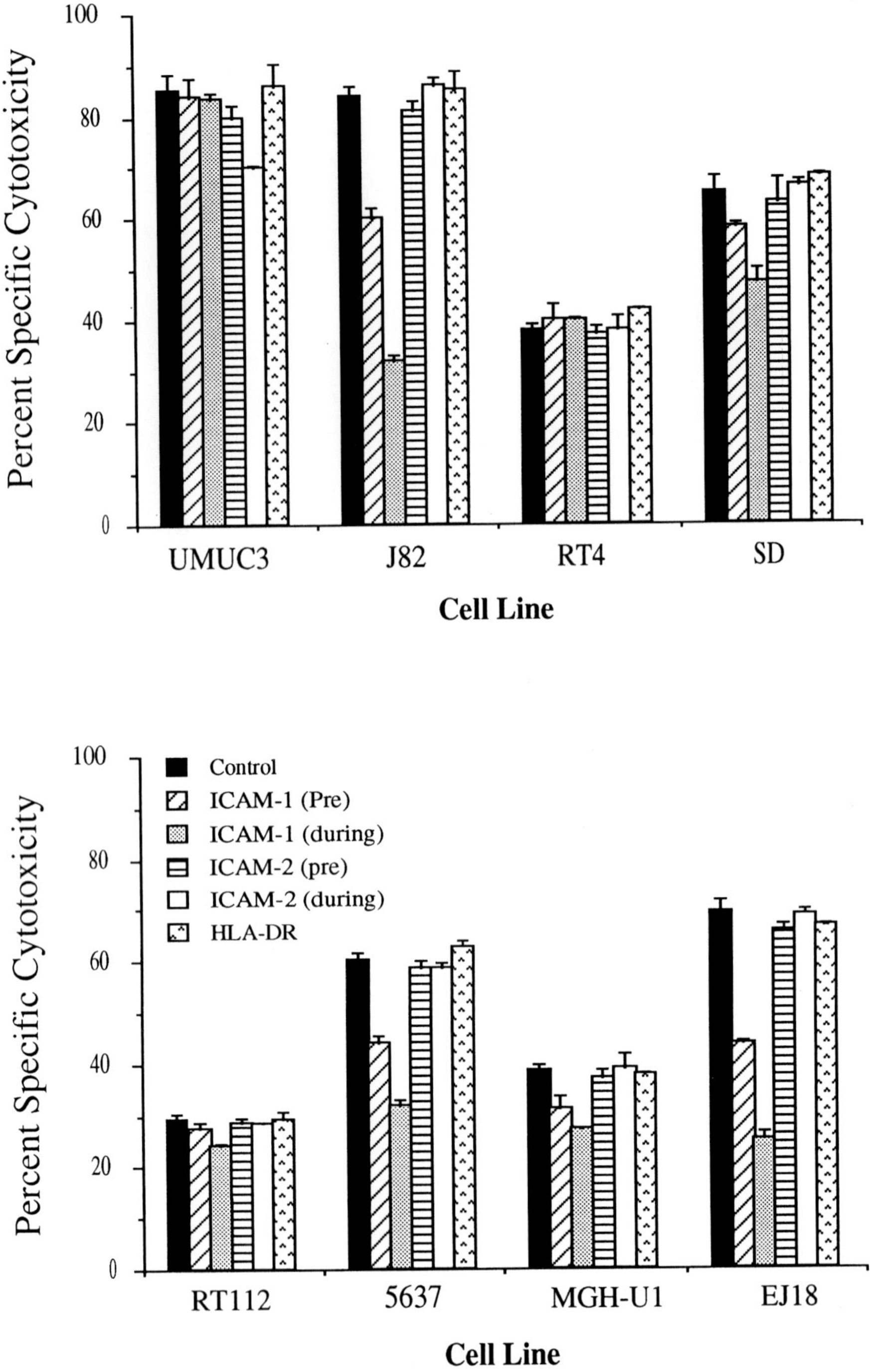

Figure 12.5 Effects of anti-ICAM-1, anti-ICAM-2 and anti-HLA-DR antibodies on the lysis of untreated TCC target cells by LAK cells. Target cells were incubated with antibodies (10 μgml^{-1}) either prior to (Pre) or during the assay period. Bars represent the mean specific lysis for triplicate assays. Error bars show standard deviation.

Blocking of LFA-1 on LAK cells with antibodies decreases their potential for target killing

To further examine the role of adhesion molecules in LAK killing of bladder cancer targets, we studied the effect of blocking adhesion structures present on the effector cells. Figure 12.6 details the effects of blocking LAK cells on their ability to lyse four TCC targets. When effector cells were treated with monoclonal antibodies to LFA-2 ICAM-1 or ICAM-2, no change was observed in LAK potential against any of the four targets. However, when LAK cells were coated with anti-LFA-1 antibodies, a marked and significant decrease ($p < 0.0001$) in specific cytotoxicity was seen. The percentage decrease ranged from 72% (RT112) to 96% (J82) with a mean of 84% for all eight cell lines.

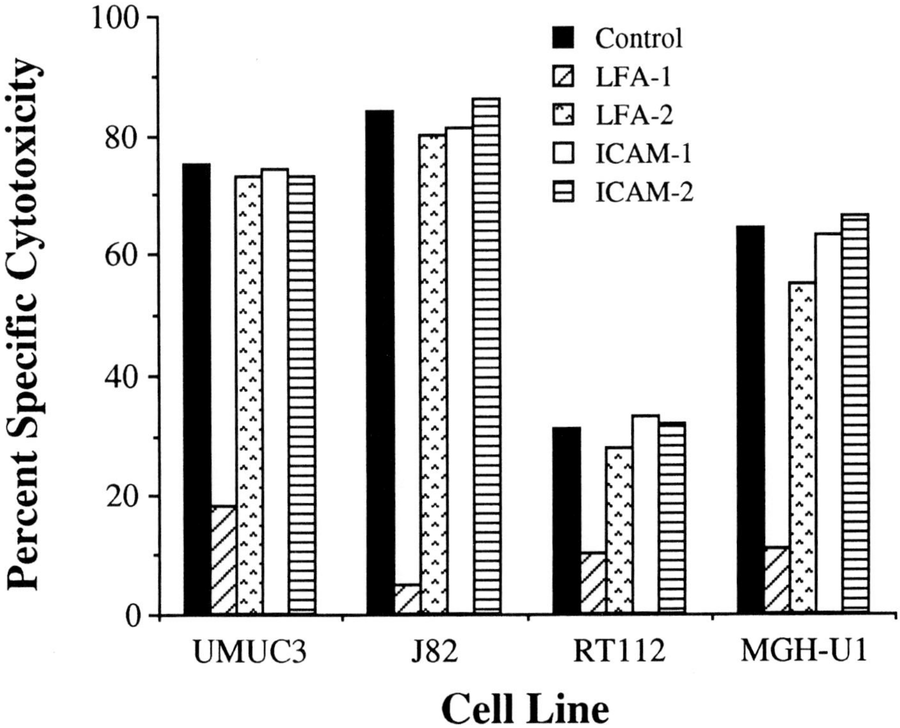

Figure 12.6 Effects of anti-LFA-1 antibodies on the lysis of untreated TCC target cells by LAK cells. Target cells were incubated with antibodies (10 $\mu g ml^{-1}$) prior to the assay period. Standard deviation was within 10%.

DISCUSSION

The constitutive expression of adhesion molecules on the surface of eight bladder cancer cell lines was studied in conjunction with their susceptibility to LAK mediated cytolysis. We have also studied the effect of interferon-gamma treatment on adhesion molecule expression and susceptibility to LAK killing. Our results show that all eight cell lines constitutively express either ICAM-1 or ICAM-2 but not both. Furthermore, the consititutive expression of ICAM-2 appears to be restricted to G1 tumour lines and ICAM-1 expression limited to G2 and G3 cells.

Treatment with IFNγ augmented ICAM-1 expression but not that of ICAM-2. This is important in view of the fact that several cytokines including IFNγ, TNFα, and IL-1 are produced by patients undergoing immunotherapy for superficial bladder cancer.[18] Treatment with IFNγ resulted in increased sensitivity to cell mediated lysis that was shown not due to the known cytotoxic effects of IFNγ.[19] Anti-ICAM-1 or ICAM-2 antibodies partially inhibited the sensitivity of the targets, however, treatment of effector cells with anti-LFA-1 antibody more-completely inhibited LAK activity.

The increased susceptibility of IFNγ treated target cells to LAK mediated killing correlated with increased expression of ICAM-1. Although IFNγ also induces the expression of HLA-class II molecules on the surface of the TCC cell lines[13] these molecules are not thought to be involved for two reasons: firstly, antibodies to HLA-class II molecules failed to alter the sensitivity of the targets to LAK killing. Secondly, the kinetics of HLA-class II induction are markedly different from those of ICAM-1 induction, being slower and less sensitive to IFNγ.[13]

In marked contrast to our present studies, previous reports have demonstrated that IFNγ treatment of tumour cells decreased the susceptibility to LAK mediated lysis.[20] It has been suggested that protection from LAK and NK mediated lysis of target cells occurred in the post-binding stage, preventing triggering and activation of the LAK cell[21] or programming for lysis or secretion of factors.[22] This is clearly not the case for IFNγ treated bladder cancer cell lines.

The blocking of LFA-1 on effector cells caused almost complete abrogation of basal TCC cell killing and IFNγ induced killing in most cell lines. However, blocking of ICAM-1 (ligand for LFA-1) on the target cells failed to result in a similar degree of inhibition. These LFA-1 dependent, ICAM-1 independent phenomena could not be accounted for by the presence of ICAM-2 on the target cell. Therefore, a portion of cytolytic activity may be due to conjugation via and LFA-1/ICAM-3 pathway.[11] It is possible that other post-conjugation events were modified by treatment with IFNγ and these events may play a role in increased killing. Future studies will attempt to further characterise the reported LFA-1 dependent, ICAM independent activity.

Acknowledgements

The authors are particularly grateful to Dr. Timothy Springer (Harvard Medical School, Boston, USA) for the generous gift of murine monoclonal antibodies to human ICAM-1 and ICAM-2. This work was funded by the Cancer Research Campaign of Great Britain.

REFERENCES

1. Grimm, E. A., Mazumder, A., Zhang, H. G. and Rosenberg, S. A. (1982) Lymphokine-activated killer cell phenomenon: Lysis of natural killer resistant fresh solid tumour cells by interleukin-2 activated autologous human peripheral-blood lymphocytes. *J. Exp. Med.*, **155**, 1823.
2. Rosenberg, S. A. and Loetz, M. T. (1986) Cancer immunotherapy using interleukin-2 and interleukin-2 activated lymphocytes. *Ann. Rev. Immunol.*, **4**, 681.
3. Jackson, A. M., Hawkyard, S., Prescott, S., Ritchie, A. W. S., James, K. and Chisholm, G. D. (1992) An investigation of factors influencing the induction of LAK activity aginst a variety of human bladder cancer cell lines. *J. Urol.*, **147**, 207.
4. Kavanaugh, A. F., Lightfoot, E., Lipsky, P. E. and Oppenheimer-Marks, N. (1991) Role of CD11/CD18 in adhesion and transendothelial migration of T cells. *J. Immunol.*, **146**, 4149.
5. Butcher, E. C. (1990) Cellular and molecular mechanisms that direct leukocyte traffic. *Am. J. Pathol.*, **136**, 3.
6. Hogg, N. (1989) The leukocyte integrins. *Immunol. Today*, **10**, 111.
7. Patarroyo, M. and Makgoba, M. W. (1989) Leukocyte adhesion to cells: Molecular basis, physiological relevance, and abnormalities. *Scand. J. Immunol.*, **30**, 129.
8. Larson, R. S. and Springer, T. A. (1990) Structure and function of the leukocyte integrins. *Immunol. Rev.*, **114**, 181.
9. Wang, P., Vanky, F., Li, S. L., Patarroyo, M. and Klein, E. (1990) Functional characteristics of the intercellular adhesion molecule-1 (CD54) expressed on cytotoxic human blood lymphocytes. *Cellular Immunol.*, **131**, 366.
10. de Fougerolles, A. R., Stacker, S. A., Schwarting, R. and Springer, T. A. (1991) Characterisation of ICAM-1 and evidence for a third counter-receptor for LFA-1 *J. Exp. Med.*, **174**, 253.
11. de Fougerolles, A. R. and Springer, T. A. (1992) Intercellular adhesion molecule-3, a third counter receptor for lymphocyte function associated molecule-1 on resting lymphocytes. *J. Exp. Med.*, **175**, 185.
12. Quillet-Mary, A., Cavarec, L., Kermarrec, N., Marchiol-Fournigault, C., Gil, M. L., Conjeaud, H. and Fradelizi, D. (1991) Target lysis by human LAK cells is critically dependent upon target binding properties, but LFA-1, LFA-3 and ICAM-1 are not the major adhesion ligands on targets. *Int. J. Cancer*, **47**, 473.
13. Rivoltini, L., Cattoretti, G., Arienti, F., Mastroianni, A., Melani, C., Colombo, M. P. and Parmiani, G. (1991) The high lysability by LAK cells of colon carcinoma target cells resistant to doxorubicin is associated with a high

expression of ICAM-1, LFA-3, NCA and a less-differentiated phenotype. *Int. J. Cancer*, **47**, 746.

14. Hawkyard, S. J., James, K. S., Prescott, S., Jackson, A. M., Ritchie, A. W. S., Smyth, J. F. and Chisholm, G. D. (1991) The effects of recombinant human interferon-gamma on a panel of human bladder cancer cell lines. *J. Urol.*, **145 (5)**, 1078.
15. Dustin, M. L., Rothlein, R., Bhan, A. K., Dinarello, C. A. and Springer, T. A. (1986) Induction by IL^{-1} and interferon-gamma: tissue distribution, biochemistry, and function of a natural adherance molecule (ICAM-1). *J. Exp. Med.*, **137**, 245.
16. Dustin, M. L., Singer, K. H., Tuck, D. T. and Springer, T. A. (1988) Adhesion of T lymphoblasts to epidermal keratinocytes is regulated by interferon-gamma and is mediated by intercellular adhesion molecule-1 (ICAM-1). *J. Exp. Med.*, **167**, 1323.
17. Symington, F. W. and Santos, E. B. (1991) Lysis of human keratinocytes by allogeneic HLA class I-specific cytotoxic T cells. *J. Immunol.*, **146**, 2169.
18. Jackson, A. M. (1993) Cytokines, cell adhesion molecules and bladder cancer immunotherapy. Ph.D. Thesis accepted by the University of Edinburgh.
19. Hawkyard, S. J., Jackson, A. M., James, K. S., Prescott, S., Smyth, J. F. and Chisholm, G. D. (1991) The effects of recombinant human interferon-gamma on the growth of a panel of human bladder cancer cell lines. *J. Urol.*, **147**, 1399.
20. Gronberg, A., Ferm, M. T., Tsai, L. and Kiessling, T. (1989) Interferon is able to reduce tumour-cell susceptibility to human lymphokine activated killer cells. *Cell. Immunol.*, **118**, 10.
21. Gronberg, A., Ferm, M. T., Reynolds, N. J. and Ortaldo, J. R. (1988) IFN-gamma treatment of K562 cells inhibits NK cell triggering and decreases the susceptibility to lysis by cytoplasmic granules from large granular lymphocytes. *J. Immunol.*, **140**, 4397.
22. De Fries, R. U. and Golub, S. H. (1988) Characteristics and mechanism of IFN-gamma induced protection of human tumour cells from lysis by lymphokine activated killer cells. *J. Immunol.*, **140**, 3686.

13. REGULATION OF INTEGRIN EXPRESSION

JYRKI HEINO

From the Department of Medical Biochemistry and the MediCity Research Laboratory, University of Turku, Kiinamyllynkatu 10, FIN-20520 Turku, Finland

INTEGRIN FAMILY OF CELL ADHESION RECEPTORS

Integrin-type receptors mediating cell-matrix or cell-cell adhesion are heterodimers composed of a larger α and a smaller β subunit. At the moment, the amino acid composition of at least 8 β and 14 α subunits has been revealed by cDNA cloning. These transmembrane proteins form at least 20 heterodimeric $\alpha\beta$ receptor combinations. Our studies have been focused on integrins mediating the interaction of cells with connective tissue components (Table 13.1, for reviews see Ruoslahti, 1991; Hemler, 1990; Hynes, 1992).

There are multiple receptors for each matrix molecule and most integrin heterodimers have more than one putative ligand. The "classical fibronectin receptor", $\alpha_5\beta_1$, is known to recognize a sequence of three amino acids, arginine-glycine-aspartic acid (RGD), in fibronectin and also the heterodimers containing α_v subunit bind to the same sequence in their ligands. However, most of the other integrins recognize sequences that do not contain RGD. For example $\alpha_4\beta_1$ binds to CS1 domain in fibronectin. Interestingly, this domain is often missing from a fibronectin molecule due to the differential splicing. Laminin contains domains recognized by only some of the integrin-type laminin receptors. Thus, different fibronectin receptors or different laminin receptors might have their own specific functions.

Intracellular domain of β_1 subunit binds to actin filaments probably via talin and α-actinin molecules. Thus, integrins connect extracellular matrix to cytoskeleton. Another important function of the integrin heterodimers is that they are likely to participate in the signal transduction pathways mediating

Table 13.1 Integrin heterodimers binding to matrix molecules

Collagen receptors
$\alpha_1\beta_1$, $\alpha_2\beta_1$, $\alpha_3\beta_1$
Fibronectin receptors
$\alpha_3\beta_1$, $\alpha_4\beta 1$, $\alpha_4\beta_7$, $\alpha_5\beta_1$, $\alpha_v\beta_1$, $\alpha_v\beta_3$, $\alpha_v\beta_6$, $\alpha_{IIb}\beta_3$
Laminin receptors
$\alpha_1\beta_1$, $\alpha_2\beta_1$, $\alpha_3\beta_1$, $\alpha_6\beta_1$, $\alpha_6\beta_4$, $\alpha_7\beta_1$, $\alpha_v\beta_3$
Vitronectin receptors
$\alpha_v\beta_1$, $\alpha_v\beta_3$, $\alpha_{IIb}\beta_3$, $\alpha_v\beta_5$

information about the structure of the matrix into the cells (for review see Schwartz, 1992). The exact molecular mechanism of signal transduction via integrins is still incompletely known. However, the fact that the cell surface integrin pattern does not only regulate the adhesion properties of a cell but can also significantly influence cell behaviour increases the interest to study the factors regulating integrin expression. Especially, the integrin-mediated regulation of cellular enzymes degrading interstitial connective tissue (Werb *et al.* 1989) and basement membrane components (Larjava *et al.* 1993a) has a putative importance in cancer biology and inflammation.

REGULATION OF INTEGRIN EXPRESSION BY CYTOKINES

Transforming growth factor-β1 (TGF-β1) is a prototype molecule in a large family of growth and differentiation factors (for review see Massagué, 1991). The main reservoirs of TGF-β1 are platelets and bone matrix, but many cell types are known to produce TGF-β1 *in vivo* and *in vitro*. TGFs-β control cell growth and differentiation but they are also important regulators of extracellular matrix accumulation. In general TGFs-β increase the production of matrix molecules and decrease their degradation, although there are some tissue specific exceptions from this main rule. TGF-β1 was also the first cytokine reported to regulate integrin expression (Ignotz and Massagué, 1987). The first report showed that TGF-β1, and also TGF-β2, can in many cell lines (including 3T3 cells, human fibroblasts and NRK cells) increase the synthesis and the cell surface expression of β_1 integrin subunit (Ignotz and Massagué, 1987). Later on it was shown that also α_5 subunit was up-regulated by TGF-β1 in fibroblasts (Roberts *et al.* 1988). Furthermore, we have shown that actually the expression on all β_1-associated α subunits (α_1, α_2, α_3, α_4, α_5 and α_6) (Heino *et al.* 1989) and also α_v, α_L, β_3 (Ignotz *et al.* 1989), β_5 and β_6 (Sheppard *et al.* 1992) subunits can be controlled by TGF-β1. Many different cell lines were used in these *in vitro* studies and some general conclusions can be made. First, the effect of TGF-β1 is cell line specific. There are cell lines expressing TGF-β receptors and other biological responses to TGF-β but not showing any significant changes in integrin expression. Second, TGF-β1 is able to regulate the synthesis of all integrin subunits studied at least in some cell lines. In most cases TGF-β1 up-regulates integrin expression. Third, TGF-β1 has not been reported to turn on the expression of an integrin subunit not expressed in a non-stimulated cell.

In some cell types, for example in MG-63 human osteosarcoma cells, TGF-β1 completely changes the cell surface integrin pattern (Heino and Massagué, 1989). The main β_1-integrin heterodimer in MG-63 cells is known to be $\alpha_3\beta_1$, a receptor for laminin, fibronectin, and collagens. Other α subunits, namely α_2 and α_5 are present in much smaller quantities. TGF-β1 decreases the expression of α_3 subunit (about 60%) and concomitantly increases the synthesis of α_2 (about 8-fold) and α_5 (about 3-fold). This suggests independent regulation of different integrin subunits.

Metabolic labellings of cultured cells have revealed the presence of intracellular precursor forms of integrin subunits. N-linked oligosaccharides in these peptides are sensitive to endoglycosidase H -digestion, and therefore presumed to be of high-mannose type. Using pulse chase strategy, it is possible to follow the maturation of integrin subunits as glycoproteins and their concomitant movement from endoplasmic reticulum via Golgi complex to cell surface. Earlier experiments especially with fibroblasts showed that the maturation rate of metabolically labelled pool of β_1 subunits is exceptionally slow ($t_{1/2}$ about 10 hours), whereas the maturation of the pool of α subunits is faster ($t_{1/2}$ about 2 hours) (Ignotz and Massagué, 1987; Akiyama and Yamada, 1987). Interestingly, TGF-β accelerated the maturation of β_1 pool but not the pool of α subunits (Ignotz and Massagué, 1987). We have studied this phenomenon (Heino *et al.* 1989) and made a hypothesis presented in Fig. 13.1. Fibroblasts seem to produce β_1 subunit in large excess over α

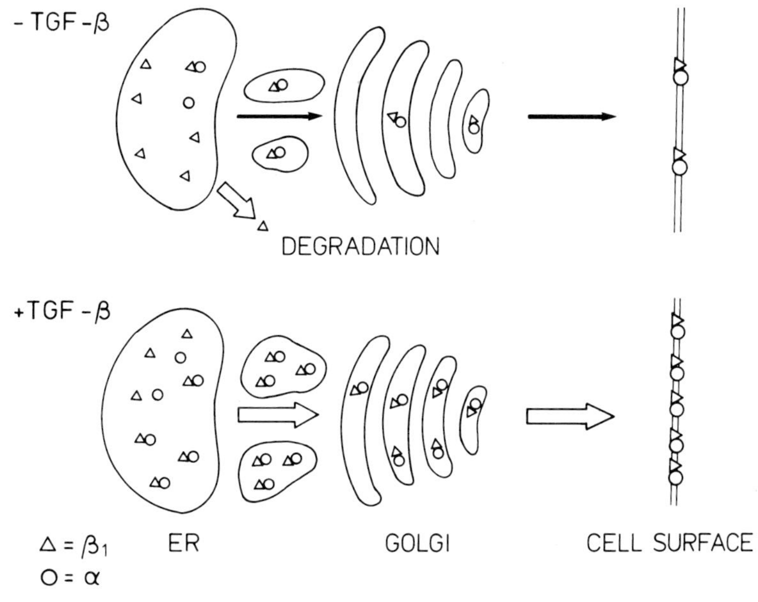

Figure 13.1 The maturation of β_1 integrin subunit is regulated by transforming growth factor-β (TGF-β). Pre-β_1 subunits are produced in large excess to endoplasmic reticulum (ER). Only the ones forming a complex with pre-α subunits can go through the maturation process to the cell surface. TGF-β increases mainly pre-α synthesis leading to increased $\alpha\beta$ complex formation, to increased maturation of the pre-β_1 pool and to the increased cell surface expression of β_1 subunit.

subunits. Furthermore, only a part of pre-β_1 subunits actually maturate, whereas a larger portion of pre-β_1 subunits go to a degradation pathway. The fact that in the immunoprecipitation experiments done with anti-β_1 antibodies also pre-α subunits can be coprecipitated in a complex with pre-β_1 suggests strongly that $\alpha\beta$ complex is formed before their maturation as glycoproteins. Endoglycosidase-treatments suggested that the complex formation takes place already in endoplasmic reticulum. Thus, the maturation of pre-β_1 subunits after $\alpha\beta$ complex formation must take the same time as the maturation of pre-α subunits. However, the large intracellular pool of unlabelled pre-β_1 subunits delays the complex formation of metabolically labelled pre-β_1 subunits. TGF-β increases mainly the synthesis of pre-α subunits and therefore increases the formation of $\alpha\beta$ complexes leading to increased maturation of the radiolabelled pre-β_1 pool. The important conclusion is that the cell surface expression of integrin heterodimers seems to be dependent on the production of pre-α subunits rather than on pre-β_1 synthesis. Thus, a factor changing only the synthesis of β_1 subunit does not necessarily change its cell surface expression. In a recent paper, Sheppard *et al.* (1992) have shown that the regulation of α_v containing heterodimers is very similar. α_v is produced in excess and its maturation process as well as surface expression is dependent on the number of β subunits (β_3, β_5 and β_6) available.

The effect of TGFs-β on multiple cell lines has been studied by cell adhesion experiments. These studies have shown increased ability of the TGF-β treated cells to bind fibronectin (Ignotz and Massagué, 1987; Allen-Hoffmann *et al.* 1988) or collagen (Pignatelli and Bodmer, 1989) and decreased cell adhesion to laminin (Heino and Massagué, 1989). These effects are cell line specific. It is also tempting to speculate that some of the cytokine-related changes in cell behaviour are due to altered expression of cell adhesion receptors. Similarly, some of the effects of TGF-β on cell growth and differentiation might be partially mediated via TGF-β induced collagen accumulation to the cell layer (Nugent and Newman, 1989; Heino and Massagué, 1990).

In addition to TGFs-β also other cytokines have been reported to regulate integrin expression. Interleukin-1β (IL-1β) and tumour necrosis factor-α (TNF-α) are two inflammatory mediators, also known to regulate connective tissue metabolism. In our experiments IL-1β increased α_1 integrin expression in human skin fibroblasts and, moreover, it turned on α_1 expression in MG-63 cells, which normally do not express this α subunit in detectable amounts (Santala and Heino, 1991). The effects of TNF-α were very similar. However, much larger concentration of TNF-α was needed. IL-1β was effective in the concentrations of 0.5–5 units/ml. These are concentrations seen at the site of chronic inflammation, for example in the rheumatoid synovial fluid (Brennan *et al.* 1991). Defilippi *et al.* (1991a) have reported that TNF-α turns on α_1 expression also in human endothelial cells. However, in these cells IL-1β did not affect α_1 expression, whereas both TNF-α and

IL-1β could decrease the expression of $\alpha_6\beta_1$ laminin receptor (Defilippi *et al.* 1992).

In our experiments IL-1β also increased MG-63 cell adhesion to laminin and E1 (Elastase 1) fragment of laminin. This increase was suggested to be due to increased α_1 expression, because the expression of other putative laminin receptors stayed constant (α_v) or slightly decreased (α_3) and, more importantly, because the increased cell attachment to laminin could be inhibited by anti-α_1 integrin monoclonal antibody (Santala and Heino, unpublished results).

In human endothelial cells TNF-α in combination with interferon-γ (IFN-γ) can down-regulate the expression of β_3 integrin subunit (Defilippi *et al.* 1991b). Together TNF-α and IFN-γ also reduced the mobility of endothelial cells. Previously IFN-γ has been reported to increase human fibroblast adhesion on collagen (Clark *et al.* 1989). However, it is not known whether this is caused by increased expression or changed function of integrin-type receptors.

$\alpha_1\beta_1$ is one of the main laminin receptors in neuronal cells. PC12 pheochromocytoma cells exposed to nerve growth factor (NGF) can differentiate to sympathetic neurones and extend neurites on laminin. At the same time NGF increases several fold their α_1 integrin expression (Rossino *et al.* 1990).

The significance of the cytokine-related changes in integrin expression *in vivo* is still incompletely known. We have studied integrin expression in rheumatoid synovium, especially in the synovial lining cell layer. The expression of α_6 integrin subunit was significantly decreased when lining cell layer was hyperplastic (Nikkari *et al.* 1993). Experiments are ongoing to evaluate whether this shift in integrin expression is due to inflammation-related factors present in synovium.

The healing process of human mucosal wounds is another *in vivo* model where we have recently shown marked changes in integrin expression (Larjava *et al.* 1993b). Mucosal keratinocytes express, while in the resting stage, $\alpha_2\beta_1$, $\alpha_3\beta_1$ and $\alpha_6\beta_4$ heterodimers. Our observation was that in migrating keratinocytes, at least two new subunits, namely α_5 and α_v are induced and this induction lasts as long as the basement membrane is formed again underneath the cells. We suggest the putative involvement of cytokines also in this process.

In cell differentiation (Adams and Watt, 1990) and malignant transformation (Plantefaber and Hynes, 1989) the cell surface integrin pattern often changes. Furthermore, differentiation promoting factors, like retinoids, have effects on integrin expression (Defilippi *et al.* 1991a, Rossino *et al.* 1991). Plantefaber and Hynes (1989) have reported that viral transformation of rat1, NRK, and Nil8 cells down-regulates $\alpha_5\beta_1$ and two other integrin receptors, whereas $\alpha_3\beta_1$ expression is retained. However, this is not a general rule, because in other transformation models α_5 expression can also stay constant (Akiyama *et al.* 1990; Ylänne and Virtanen, 1989). Thus, in spite of the fact that altered cell adhesion properties might be essential for the invasion of tumours, the changes in integrin expression in different cancer cells are very variable. Although cytokines and malignant transformation regulate the same integrin subunits, at least in osteogenic sarcoma cells, the molecular mechanism is not identical (Santala *et al.* 1994).

MOLECULAR MECHANISMS CONTROLLING INTEGRIN GENE EXPRESSION

TGF-β1 -related increase in integrin protein synthesis is due to elevated cellular levels of the corresponding mRNAs (Ignotz and Massagué, 1987; Heino *et al.* 1989; Ignotz *et al.* 1989; Heino and Massagué, 1989). Similarly, down-regulation of α_3 integrin expression by TGF-β1 in MG-63 cells correlates with the similar change in mRNA levels (Heino and Massagué, 1989). Similarly, down-regulation of α_3 integrin expression by TGF-β1 in MG-63 cells correlates with the similar change in mRNA levels (Heino and Massagué, 1989). Based on the existing knowledge about TGF-β action on other genes, the transcriptional regulatory mechanism is the most probable. However, the regulation of integrin mRNA half-lives has not been excluded. TGFs-β can use both mechanisms when controlling collagen gene expression (Penttinen *et al.* 1988). The effect of TGFs-β on cell cycle is associated with dephosphorylation of the retinoblastoma gene product. Regulation of matrix production by TGFs-β is, however, independent from this regulatory pathway (Laiho *et al.* 1991).

Molecular cloning of the promoter regions of the α_4, α_5 and β_1 integrin genes (Rosen *et al.* 1991; Birkenmeier *et al.* 1991; Cervella *et al.* 1993) will increase the knowledge about the regulation of integrin expression. Interestingly, the sequence analysis of integrin promoters has revealed several putative binding sites for transcripton factors, including AP-1 consensus sequences. TGFs-β regulate the expression of jun and fos protooncogenes (Pertovaara *et al.* 1989) and, furthermore, AP-1 sites are essential for activation TGF-β1 gene by itself (Kim *et al.* 1990).

Defilippi *et al.* (1991b) reported the decreased synthesis of β_3 integrin subunit in endothelial cells treated with TNF-α in combination IFN-γ. However, they were not able to see changes in the corresponding mRNA levels. This suggests that also the translational regulation of integrin expression might be possible.

CONCLUSIONS

Integrins are a large family of cell adhesion receptors. Cellular integrin pattern changes during wound healing, chronic inflammation, differentiation processes, and malignant transformation. Cell-matrix interaction is an important regulator of cell behaviour, and changes in integrin expression are putatively of great biological importance.

In cell culture experiments integrin expression can be modulated by growth factors and inflammmatory cytokines. These factors are also likely to be involved in *in vivo* regulation of cell adhesion apparatus. However, very little is known about the role of these mechanisms in the pathophysiology of human diseases.

Acknowledgements

Experiments performed in J. H.'s laboratory in Finland were supported by grants from the Academy of Finland, the Sigrid Juselius Foundation and the Finnish Cancer Organizations. I want to thank Dr. Hannu Larjava for fruitful discussions helping the preparation of this manuscript.

REFERENCES

1. Adams, J. C. and Watt, F. M. (1990) Changes in keratinocyte adhesion during terminal differentiation: Reduction in fibronectin binding precedes $\alpha_5\beta_1$ integrin loss from the cell surface. *Cell,* **63**, 425–435.
2. Akiyama, S. K. and Yamada, K. M. (1987) Biosynthesis and acquisition of biological activity of the fibronectin receptor. *J. Biol. Chem.*, **262**, 17536–17542.
3. Akiyama, S. K., Larjava, H. and Yamada, K. M. (1990) Differences in the biosynthesis and localization of the fibronectin receptor in normal and transformed cultrued human cells. *Cancer Res.*, **50**, 1601–1607.
4. Allen-Hoffmann, B. L., Crankshaw, C. L. and Mosher, D. F. (1988) Transforming growth factor β increases cell surface binding and assembly of exogenous (plasma) fibronectin by normal human fibroblasts. *Mol. Cell. Biol.*, **8**, 4234–4242.
5. Birkenmeier, T. M., McQuillan, J. J., Boedeker, E. D., Argraves, S. W., Ruoslahti, E. and Dean, D. C. (1991) The $\alpha_5\beta_1$ fibronectin receptor. Characterization of the α_5 gene promoter. *J. Biol. Chem.*, **266**, 20544–20549.
6. Brennan, F. M., Field, M., Chu, C. Q., Feldmann, M. and Maini, R. N. (1991) Cytokine expression in rheumatoid arthritis. *Br. J. Rheumatol.*, **30**, 76–80.
7. Cervella, P., Silengo, L., Pastore, C., Altruda, F. (1993) Human β_1-Integrin Gene expression is regulated by two promoter regions. *J. Biol. Chem.*, **268**, 5148–5155.
8. Clark, J. G., Dedon, T. F., Wayner, E. A. and Carter, W. G. (1989) Effects of interferon-τ on expression of cell surface receptors for collagen and deposition of newly synthesized collagen by cultured human lung fibroblasts. *J. Clin. Invest.*, **83**, 1505–1511.
9. Defilippi, P., van Hinsbergh, V., Bertolotto, P., Rossino, P., Silengo, L. and Tarone, G. (1991a) Differential distribution and modulation of expression of α_1/β_1 integrin in human endothelial cells. *J. Cell Biol.*, **114**, 855–863.
10. Defilippi, P., Truffa, G., Stefanuto, G., Altruda, F., Silengo, L. and Tarone, G. (1991b) Tumor necrosis factor-α and interferon-τ modulate the expression of the

vitronectin receptor (integrin β_3) in human endothelial cells. *J. Biol. Chem.*, **266**, 7638–7645.

11. Defilippi, P., Silengo, L. and Tarone, G. (1992) $\alpha 6\beta 1$ integrin (laminin receptor) is down-regulated by tumor necrosis factor-α and interleukin-1β in human endothelial cells. *J. Biol. Chem.*, **267**, 18303–18307.
12. Heino, J., Ignotz, R. A., Hemler, M. E., Crouse, C. and Massagué, J. (1989) Regulation of cell adhesion receptors by transforming growth factor-β. Concomitant regulation of integrins that share a common β_1 subunit. *J. Biol. Chem.*, **264**, 380–388.
13. Heino, J. and Massagué, J. (1989) Transforming growth factor-β swithes the pattern of integrins expressed in MG-63 human osteosarcoma cells and causes a selective loss of cell adhesion to laminin. *J. Biol. Chem.*, **264**, 21806–21811.
14. Heino, J. and Massagué, J. (1990) Cell adhesion to collagen and decreased myogenic gene expression implicated in the control of myogenesis by TGF-β. *J. Biol. Chem.*, **265**, 10181–10184.
15. Hemler, M. E. (1990) VLA proteins in the integrin family: structures, functions, and their role on leukocytes. *Annu. Rev. Immunol.*, **8**, 365–400.
16. Hynes, R. (1992) Integrins: versatility, modulation, and signaling in cell adhesion. *Cell*, **69**, 11–25.
17. Ignotz, R. A. and Massagué, J. (1987) Cell adhesion receptors as targets for transforming growth factor-β action. *Cell*, **51**, 189–197.
18. Ignotz, R. A., Heino, J. and Massagué, J. (1989) Regulation of cell adhesion receptors by transforming growth factor-β. Regulation of vitronectin receptor and LFA-1. *J. Biol. Chem.*, **264**, 389–392.
19. Kim, S. J., Angel, P., Lafyatis, R., Hattori, K., Kim, K. Y., Sporn, M. B., Karin, M., Roberts, A. B. (1990) Autoinduction of transforming growth factor beta 1 is mediated by the AP-1 complex. *Mol. Cell-Biol.* **10**, 1492–1497.
20. Laiho, M., Rönnstrand, L., Heino, J., DeCaprio, J. A., Ludlow, J. W., Livingston, D. M. and Massagué, J. (1991) Control of JunB and extracellular protein expression by transforming growth factor-β1 is independent of simian virus 40 T antigen-sensitive growth-inhibitory events. *Mol. Cell. Biol.*, **11**, 972–978.
21. Larjava, H., Lyons, J. G., Salo, T., Mäkelä, M., Koivisto, L., Birkedal-Hansen, H., Akiyama, S. K., Yamada, K. M. and Heino, J. (1993a) Anti-integrin antibodies induce type IV collagenase expression in keratinocytes. *J. Cell. Physiol.* (in press).
22. Larjava, H., Salo, T., Haapasalmi, K., Kramer, R. and Heino, J. (1993b) Expression of integrins and basement membrane components by wound keratinocytes. *J. Clin. Invest.* (in press).
23. Massagué, J. (1991) The transforming growth factor-β family. *Annu. Rev. Cell Biol.*, **6**, 597–641.
24. Nikkari, L., Aho, H., Yli-Jama, T., Larjava, H., Jalkanen, M. and Heino, J. (1993) Expression of integrin family of cell adhesion receptors in rheumatoid synovium. *Am. J. Pathol.*, **142**, 1019–1027.
25. Nugent, M. A. and Newman, M. J. (1989) Inhibition of normal rat kidney cell growth by transforming growth factor-β is mediated by collagen. *J. Biol. Chem.*, **264**, 18060–18067.
26. Penttinen, R. P., Koyayashi, S. and Bornstein, P. (1988) Transforming growth factor-β increases mRNA for matrix proteins both in the presence and in the absence of changes in mRNA stability. *Proc. Natl. Acad. Sci. USA*, **85**, 1105–1108.

27. Pertovaara, L., Sistonen, L., Bos, T. J., Vogt, P. K., Keski-Oja, J. and Alitalo, K. (1989) Enhanced jun gene expression is an early genomic response to transforming growth factor β stimulation. *Mol. Cell Biol.*, **9**, 1255–1262.
28. Pignatelli, M. and Bodmer, W. F. (1989) Integrin-receptor-mediated differentiation and growth inhibition are enhanced by transforming growth factor-β in colorectal tumour cells grown in collagen gel. *Int. J. Cancer*, **44**, 518–523.
29. Plantefaber, L. C. and Hynes, R. O. (1989) Changes in integrin receptors on oncogenically transformed cells. *Cell*, **56**, 281–290.
30. Roberts, C. J., Birkenmeier, T. M., McQuillan, J. J., Akiyama, S. K., Yamada, S. S., Chen, W., Yamada, K. M. and McDonald, J. A. (1988) Transforming growth factor-β stimulates the expression of fibronectin and of both subunits of the human fibronectin receptor by cultured human lung fibroblasts. *J. Biol. Chem.*, **263**, 4586–4592.
31. Rosen, G. D., Birkenmeier, T. M. and Dean, D. C. (1991) Characterization of the α_4 integrin gene promoter. *Proc. Natl. Acad, Sci. USA*, **88**, 4094–4098.
32. Rossino, P., Gavazzi, I., Timpl, R., Aumailley, M., Abbadini, M., Giancotti, F., Silengo, L., Marchisio, P. C. and Tarone, G. (1990) Nerve growth factor induces increased expression of a laminin-binding integrin in rat pheochromocytoma PC12 cells. *Exp. Cell Res.*, **189**, 100–108.
33. Rossino, P., Defilippi, P., Silengo, L. and Tarone, G. (1991) Up-regulation of the integrin α_1/β_1 in human neuroblastome cells differentiated by retinoic acid: correlation with increased neurite outgrowth response to laminin. *Cell Regulation*, **2**, 1021–1033.
34. Ruoslahti E. (1991) Integrins. *J. Clin. Invest.*, **87**, 1–5.
35. Santala, P. and Heino, J. (1991) Regulation of integrin-type cell adhesion receptors by cytokines. *J. Biol. Chem.*, **266**, 23505–23509.
36. Santala, P., Larjava, H., Nissinen, L., Riikonen, T., Määttä, A. and Heino, J. (1994) Suppressed collagen synthesis and induced expression of $\alpha 2\beta 1$ integrin-type collagen receptor in tumorigenic osteosarcoma cells. *J. Biol. Chem.*, (in press).
37. Schwartz, M. A. (1992) Transmembrane signalling by integrins. *Trends in Cell Biol.*, **2**, 304–308.
38. Sheppard, D., Cohen, D. S., Wang, A. and Busk, M. (1992) Transforming growth factor-β differntially regulates expression of integrin subunits in guinea pig airway epithelial cells. *J. Biol. Chem.*, **267**, 17409–17414.
39. Werb, Z., Tremble, P. M., Behrendtsen, O., Crowley, E. and Damsky, C. H. (1989) Signal transduction through the fibronectin receptor induces collagenase and stromelysin gene expression. *J. Cell Biol.*, **109**, 877–889.
40. Ylänne, J. and Virtanen I. (1989) The M_r 140,000 fibronectin receptor complex in normal and virus-transformed human fibroblasts and in fibrosarcoma cells: identical localization and function. *Int. J. Cancer*, **43**, 1126–1136.

14. THE DISTRIBUTION OF INTEGRIN AND NON-INTEGRIN RECEPTORS AND THEIR LIGANDS IN HUMAN THYMUS

SUZANNE M. WATT[1], J. ALERO THOMAS[2], SIDNEY CHANG[3], JOHN DAVIES[4], MICHAEL A. HORTON[4], ANDREW J. EDWARDS[5] and JIRI BARTEK[6]*

[1]*MRC Molecular Haematology Unit, Institute of Molecular Medicine, Oxford, UK*
[2]*London School of Hygiene and Tropical Medicine, London, UK*
[3]*Royal College of Surgeons, London, UK*
[4]*Haemopoiesis* [5]*FACS and* [6]*Epithelial Cell Biology Laboratories, Imperial Cancer Research Fund, London, UK*
**Research Institute of Clinical and Experimental Oncology, Brno, Czechoslovakia*

The development of specific T lymphocyte subpopulations in the thymus depends on the selective interaction of T cell precursors with microenvironmental cells of mesenchymal, epithelial, endothelial and haemopoietic origin. Knowledge of the distribution of integrin and non-integrin adhesion receptors on these cells *in vivo* and *in vitro* is essential to our understanding of the mechanisms involved. To partially achieve this, human thymic epithelial cell strains representing subcapsular or outer cortical epithelium were generated with a defective retroviral vector containing the SV40T antigen and subsequently analysed for their adhesion receptor phenotype.

The epithelial nature of the cell strains was confirmed by the presence of keratins 8 and 18, by desmosomes and tonofilaments. All cell strains, whether of fetal of post-natal derivation, express Thy1 and HLA Class 1 at high levels suggesting a subcapsular or outer cortical epithelial origin. These transformed human thymic epithelial cell strains express the VLA $\beta 1$ receptor and α chains of VLA-2, VLA-3 and VLA-6. The cells are also positive for LFA-3 and ICAM-1 and weakly express $\beta 3$, $\beta 4$ and VNRα. The distribution of integrin and other receptor molecules on thymic epithelial cells in tissue sections generally corresponds to the phenotypic profile on cultured thymic epithelium. Differential levels of expression of VLA-$\beta 1$ and $\beta 2$ chains as well as VLA-4, VLA-6 and LFA-1α(L) on human thymocyte subsets were observed in three colour flow cytometric analyses. These studies and those examining the immunohistochemical localization of extracellular matrix molecules indicate that both the distribution of receptor-ligand pairs

Correspondence. Dr. S. M. Watt, The MRC Molecular Haematology Unit, Institute of Molecular Medicine, The John Radcliffe Hospital, Headington, Oxford, OX3 9DU, UK.

and the level of expression of adhesion molecules may influence T cell development within the thymus.

INTRODUCTION

The generation of immunocompetent T cells in the thymus depends on a functional microenvironment consisting of haemopoietic, epithelial, myoid, endothelial, and mesenchymal cells.[1] The importance of the epithelium in this process is suggested by the findings that an epithelial rudiment will support multi-clonal develoment of single T cell precursors in thymic organ culture[2] and thymic epithelial hyperplasia may lead to over-proliferation of thymocytes.[3,4] Although six different epithelial cell subsets have been identified throughout the subcapsular, cortical and medullary regions,[5] the function of the individual types of thymic epithelium in regulating normal T cell development is not completely understood.

Thymocyte-epithelial cell interactions are thought to be mediated by molecules which include both the integrins and members of the immunoglobulin superfamily of cell surface receptors, which are also involved in leucocyte adhesion to other cells and/or to extracellular matrix (ECM) proteins.[6–8] Integrin receptor subfamilies each contain heterodimeric molecules consisting of common β and unique α subunits that recognize distinct cellular or ECM ligands.[6–8] Of the eight β subunits so far described, $\beta1$, $\beta2$ and $\beta3$ define the VLA antigen, leucocyte integrin (Leu-CAM) and cytoadhesin subfamilies respectively.[6,8] Some of these receptors have been implicated in thymocyte development.[9,10] However the contribution that specific thymic epithelial cells make to this process has been hindered by their heterogeneity and by the difficulty in isolating homogeneous cell populations.[1,5,11,12] To understand the function of thymic epithelium in this process, the expression of integrin and non-integrin receptors and their ligands was studied on thymic tissue sections and their phenotypic patterns compared with those of isolated thymocytes and SV40 transformed thymic epithelial cell strains.

MATERIALS AND METHODS

Cells and Tissues

Human fetal and post-natal thymus tissue was obtained and processed for single cell suspensions or for tissue section analysis as described earlier.[12,13] Thymic epithelial cell strains, SM3, SM4, SM5 (fetal) and SM6 (neonatal), were derived from fetal and post-natal human thymus by infection with LJ-SVLT amphotrophic virus carrying the SV40 large T antigen and the neomycin resistance genes.[12] Epithelial cell strains SM1 and SM2 derived from human fetal thymus by cotransfection with SV40 ori$^-$ virus and the neomycin resistance gene have been described previously.[11]

Antibodies

Mouse and rat monoclonal antibodies (Mabs) used in the analyses are described in references 11–13. Polyclonal rabbit and monoclonal (FN3 and FN4) anti-human fibronectin, collagen type IV and laminin were also used for section staining.[14]

Immunocytochemical and Electron Microscopic Analyses

Phenotyping studies on tissue sections were performed with the peroxidase anti-peroxidase technique (PAP), the samples being counterstained in Mayer's haemalum, as described previously.[12] Adherent cell monolayers were fixed in acetone at 4°C or acetone-methanol (1:1) at −20°C for 10 minutes and were then sequentially incubated with the appropriate first and second layer antibodies.[13] The reaction was developed either by the PAP technique or with fluorescein isothiocyanate (FITC)-conjugated rabbit $F(ab)_2$ anti mouse or anti-rat immunoglobulin (Ig; Dakopatts, Copenhagen, Denmark). Electron microscopy was performed on fixed adherent epithelial monolayers as described by Watt *et al.*[12,13]

Flow Cytometric Analysis

Thymocytes were isolated on Ficoll-hypaque (Pharmacia Fine Chemicals, Uppsala, Sweden; density <1.077 g/cc). Single cell suspensions of thymic epithelial cell strains (grown as adherent monolayers) were obtained by addition of 2 mM EDTA in PBS. Cells were labelled with each monoclonal antibody and with FITC-conjugated rabbit $F(ab)_2$ anti-mouse or anti-rat immunoglobulin (Ig; Dakopatts, Copenhagen, Denmark). For three colour analysis, cells were also initially labelled with the appropriate monoclonal antibody followed by FITC-rabbit $F(ab)_2$ anti-mouse or anti-rat Ig (Dakopatts, Copenhagen, Denmark) and subsequently blocked with normal mouse Ig. For the second and third colours, CD4-biotin, streptavidin-Tandem (Southern Biotechnology Associates, USA) and CD8-phycoerythrin (PE) (Becton-Dickinson, USA) were then added and viable cells were analysed on a FACScan.

Immunoprecipitation and Immunoblotting

Thymic epithelial cell strains were labelled with 1 mCi ^{35}S-methionine (Amersham, UK), the solubilized cell extracts were immunoprecipitated and the isolated molecules analysed by polyacrylamide gel electrophoresis and fluorography.[14] Cytoskeletal proteins were analysed by immunoblotting.[12]

Growth *in vivo* and *in vitro*

The epithelial cell strains were analysed for their growth in methocel and in nu/nu mice as described by Watt *et al.*[12]

RESULTS

The Establishment of Human Thymic Epithelial Cell Strains

Infection of primary epithelial cell cultures with the LJ-SVLT retrovirus and subsequent selection of cells for their G418 resistance led to the establishment of two sets of cell strains.[12] One clone (SM6) from post-natal thymus and three (SM3, SM4, SM5) from fetal thymus were analysed in detail and compared with the fetal thymic cells strains, SM1 and SM2, generated by SV40ori- transfection.[11] All cell strains expressed nuclear SV40 large T and the p53 protein (Table 14.1 and Fig. 14.1 a and b) The cells differed in their morphological characteristics having cuboidal, fusiform or spindle-like appearances (see reference 13). An example of the morphology of two of the SV40 transformed cell strains is presented in Fig. 14.1 (c, post-natal SM6 cells; d, fetal SM3 cells).

Immunocytochemistry and electron microscopy were used to analyse all the cell strains. The presence of keratins (Table 14.1 and Figs 14.1f and 1g), desmosomes (Fig. 14.1e) and tonofilaments was confirmed in all cases. All the cell strains stained homogeneously with keratins 8 and 18 (Table 14.1 and Fig. 14.1f). Unlike SM1 and SM2, some of the SM3, SM4, SM5 and SM6 cells expressed keratin 7 (Table 14.1 and Fig. 14.1g), as was confirmed by

Table 14.1 Phenotype of Human Epithelial Cell Strains

Epitope	*Antibody*	*Cell Strains*					
		SM1	*SM2*	*SM3*	*SM4*	*SM5*	*SM6*
SV40 T	Pab 419/ Pab 423	+ +	+ +	+ +	+ +	+ +	+ +
p53	Pab 421	+ +	+ +	+ +	+ +	+ +	+ +
Desmoplakin 1&2	DP 1&2	nd	nd	+	+	+	+
Desmoglein	DG3.10	nd	nd	+	+	+	+
Desmin	DE-B-5	nd	nd	–	nd	–	–
Keratin 7	C18	–	–	% +	+	+	+ w
Keratin 8	C4.3	+ w	+ +	+ + / +	+ + / +	+ + / +	+ w
Keratin 18	CO4	+ +	+ +	+ +	+ +	+ +	+ +
Thy1	19/22	+ +	+ +	+ +	+ +	+ +	+ +
HLA Class1	W6/32	+ +	+ +	+ +	+ +	+ +	+ +
	TE3B	+ w	+ w	+ w	+ w	+ w	+ w
GQ gangliosides	A2B5	+ w	+ w	+ w	+ w	+ w	+ w

Details of negative staining are provided in references 11 to 13.
nd = not determined; + = positive; – = negative; w = weak; % = percentage of cells staining.

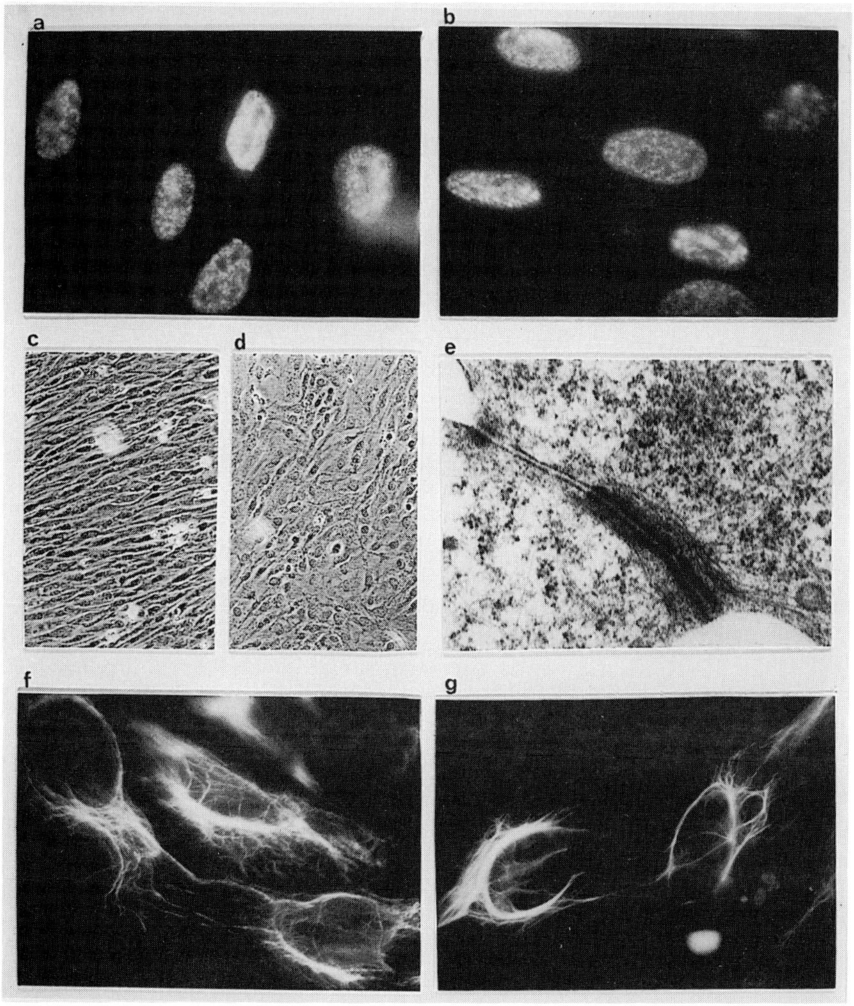

Figure 14.1 SV40 Transformed Epithelial Cell Strains. SV40 epithelial cells from neonatal (SM6; a,b,e,f) or fetal (SM3: b,d,g) thymus. a and b show staining for SV40 large T and p53 respectively. Cell morphology(c,d), desmosomes(e) and staining for keratin 18 (f) and 7 (g) are also shown.

immunoblotting.[13] Occasional cells within SM3, SM4 and SM5 were keratin 19 positive, but keratin 14 was absent from all clones. Detailed comparative studies using fresh frozen samples of normal thymus showed phenotypic similarity between the transformed cells and the outer thymic cortical epithelium (Thomas *et al.*, manuscript in preparation). When analysed by immunofluorescence, all cell strains labelled with antibodies to desmoplakins 1 and 2 and to desmoglein but were non-reactive with an anti-desmin antibody to myoid cells (Table 14.1). The cells were also analysed with a panel of monoclonal antibodies to macrophages, haemopoietic, mesenchymal and lynphoid cells and were found to be negative.[12] In contrast, Thy1 and HLA Class1 were strongly expressed, while the antigens identified with TE3B and A2B5 were present in low amounts (Table 14.1). HLA Class II was generally not present on the clones unless these were treated with γ-interferon. All the cell strains had an extended lifespan *in vitro*, but no cells escaped the degenerative phase. None of the cell strains formed tumours in nu/nu mice and none grew in methocel.

Expression of Adhesion Receptors on Isolated Cells

(a) Epithelial Cell Strains

The positive results for integrin staining were similar for all cell strains and are summarized in Table 14.2. Complete results are presented in Watt *et al.*[13] of the four β subunits examined, $\beta 1$ was clearly expressed by the thymic epithelial cell strains. FACScan analyses of the epithelial cell strains revealed the presence of VLA-3 in significant amounts, with weak expression of VLA-2, VLA-6 and $\beta 4$. In both fetal and post-natal thymic epithelial strains, the VNRαv, $\beta 3$ and the complexed VNRαv$\beta 3$ forms of the vitronectin receptor were identified with the LM142, VI-PL2, and 23C6 Mabs respectively. Expression of the Ig Superfamily of adhesion molecules was also examined. The transformed human thymic epithelial cell strains were LFA-3 and ICAM-1 positive. A limited immunoprecipitation experiment confirmed the presence of $\beta 1$, the α subunits of VLA-2 and VLA-3 integrin chains, ICAM-1, low levels of VNRαv and the absence of the α subunit of VLA-1 on the thymic epithelial strains.

(b) Differential levels of integrin expression on thymocyte subpopulations

The $\beta 1$ and $\beta 2$ subunits, VLA-4, VLA-6, and LFA-1α(L) were expressed in significant amounts, but the majority of thymocytes lacked other cytoadhesion molecules belonging to the integrin superfamily. Isolated thymocytes were divided into subsets based on CD4 and CD8 binding and the relative levels of expression of adhesion receptors determined by three colour flow cytometric analyses. Representative histograms showing the median fluorescence intensity of $\beta 1$, VLA-4 LFA-1α(L) and VLA-6 on these subsets are presented in figure 14.2. VLA-$\beta 1$ and VLA-4 antibodies were tested and consistently showed a higher median level of fluorescence on the $CD4^+8^+$ subset when compared to both the $CD4^-8^+$ and $CD4^+8^-$ subpopulations (Fig. 14.2 b and c). Interestingly, the $CD4^+CD8^-$ subset showed some heterogeneity with a minor, brighter subpopulation, the significance of which requires further examination. In contrast to VLA-4 and $\beta 1$, the expression

Table 14.2 Positive Expression of Integrins in Human Fetal and Postnatal Thymus

Antigen	*Mab*	*Tissue Reactivity in Sections*						*Single Cells*	
			Epithelium			*Thymocytes*			
		SC	*C*	*M*	*HC*	*C*	*M*	*T*	*TEC*
INTEGRIN $\beta 1$	TS2/16	–	+/–	+	+/–	+	+	+	+
	AJ2	+	+	+	+	+	+	+	+
	K20	+	+	+	+	+(v)	+(v)	+	+
	A-IA5	+	–	+	+	+(v)	+(v)	+	+
	DH12	+	+	+	+	+(v)	+(v)	+	+
INTEGRIN $\beta 4$	439–89	–	–	–	–	–	–	–	$+^{w}$
VLA-2	Gil4#	+	+	+	+	–	–	–	+
	12F1#	+	–	–	–	–	–	–	+
VLA-3	J143#	+	–	+	$+^{w}$	–	–	–	+
VLA-4	HP2/1, B-5G10*	–	–	–	–	+	+	+	–
VLA-6	GoH-3	+	?	+	+	+	+	+	+
INTEGRIN $\beta 2$	MHM23								
	60.3	–	–	–	–	+	+	+	–
LFA-1α	CIMT	–	–	–	–	+	+	+	–
INTEGRIN $\beta 3$	V1-PL2 }	–	–	–	–	–	–	–	$+^{w}$
	Y2/51 }								
	CLB/THRI** }	–	–	–	–	–	–	–	$+^{w}$
VNR	13C2	$+^{w}$(v)	–	–	–	–	–	–	$+^{w}$
	23C6	–	–	–	+	–	–	–	$+^{w}$
	LM142	+(v)	–	%+	+	–	–	–	+

Abbreviations:

SC	= Subcapsular	v	= variable levels of staining between different sections although all positive
C	= Cortex	**	= Tested only on fetal thymic sections
M	= Medulla	#	= Tested only on post-natal thymus
HC	= Hassall's Cortex	*	= Minor proportion of cell negative
VNR	= Vitronectin Receptor	+	= Most cells positive
Mab	= Monoclonal Antibody	–	= No staining compared to a negative control
?	= Staining uncertain	%+	= A proportion of cells staining
T	= thymocytes	w	= Weak staining
TEC	= Thymic epithelial cell strains(SM1 to SM6)	+/–	= Some sections showing positive staining and others showing negative staining

of VLA-6 and LFA-1α(L), was similar or higher on the single positive populations than on the double positive (Figure 14.2 d), with the $CD4^{+}8^{-}$ cells staining marginally more intensely than the $CD4^{-}8^{+}$ cells. The majority of cells in the $CD4^{-}8^{-}$ subset were $\beta 1+$, VLA – 4 +, VLA – 6 + and LFA-1(αL) +, with staining for VLA-5 being observed at low levels on a proportion of these cells. Higher median fluorescence intensities were obtained for VLA-5, VLA-6 and LFA-1(αL) expression on the $CD4^{-}8^{-}$

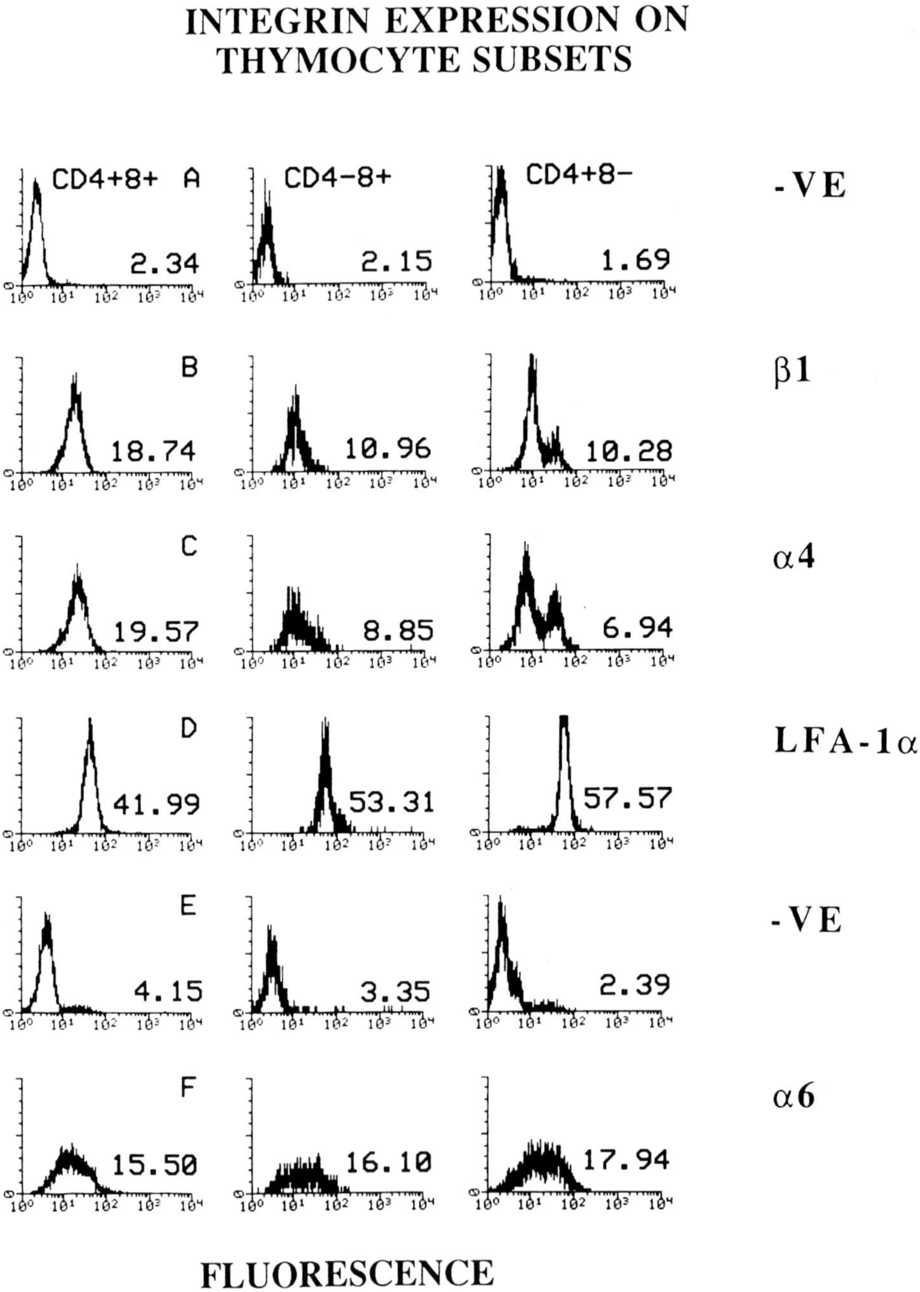

Figure 14.2 Three Colour Analyses of Thymocyte Subsets. Thymocytes were labelled with the appropriate anti-integrin Mab together with FITC anti-mouse or anti-rat Ig. Available binding sites were blocked with normal mouse Ig and the cells subsequently stained with CD-4-biotin together with streptavidin-Tandem and CD8-PE. Representative histograms of selected CD4/CD8 subsets show the median fluorescence values for each integrin molecule. The graphs show negative controls with (A) FITC-anti-mouse or (E) anti-rat Ig; (B) A-IA5, (C) P4G9 for VLA-4; (D) Mab38 for LFA-1; (F) GoH3 for VLA-6.

subset than for cells expressing CD4 and/or CD8, while the median fluorescence for VLA – 4 on the $CD4^-8^-$ subset was similar to that observed on the $CD4^+8^+$ population (data not shown). Without functional maturation studies, it is not possible to determine conclusively whether these $CD4^-8^-$ cells represent thymocyte precursors destined for subsequent T cell differentiation.

Expression of Adhesion Receptors and their Ligands *in Vivo*

A summary of the staining distribution of adhesion receptors and their ligands on the thymic epithelial network *in situ* is presented in Table 14.2 and in Fig. 14.3. The results for isolated single cell analyses coincided with the section staining for the integrin molecules in most cases. The majority of thymocytes *in vivo* were $\beta 1$, $\beta 2$, VLA-4, VLA-6 and LFA-1α(L) positive, with all cortical and a proportion of medullary thymocytes expressing VLA-4, when assessed with the B5G10 Mab (Fig. 14.3a) VLA-6 as defined by the Mab, GoH3, reacted in frozen sections with subcapsular and medullary thymocytes and in the cortex showed positive staining on a proportion of thymocytes but only in perivascular locations (Fig. 14.3b). A similar pattern of antigen expression to the cell strains (Table 14.2) was observed for the epithelium in sections, except that LFA-3 staining was found on the medullary epithelium only, and ICAM-1 was absent from subcapsular or outer cortical epithelium. ICAM-1 was, however, present on cells with morphological characteristics of thymic nurse cells (Fig. 14.3c). With the Mabs, J143 and Gi14 (Fig. 14.3d) to VLA-3 and VLA-2 respectively, staining of thymic epithelium but not of thymocytes was observed in post-natal thymic sections.

The ECM ligands, (laminin, fibronectin and type IV collagen) showed a similar pattern of expression in all samples, with a continuous band of intense homogeneous staining surrounding the entire thymic tissue and forming fine extensions into the outer regions of the cortex. Similar linear staining was present around blood vessels in the extra-thymic fibrous stroma of the interlobular septa and in the cortex. An example of collagen type IV staining is shown in Fig. 14.3f. A close correlation between the distribution of ECM proteins and blood vessel endothelium stained with Factor VIII related antigen. In addition to this common staining pattern, fibronectin also showed a reticular pattern of staining throughout the cortex which was reminiscent of thymic cortical epithelium. Fibronectin was not expressed on the medullary cells, but was detected on Hassall's corpuscles which also stained for laminin. Monoclonal antibodies, FN3 and FN4, to the whole fibronectin molecule and to the 45kD fragment respectively showed similar patterns of staining to those observed with polyclonal anti-sera. Combined immunocytochemistry with anti-sera to cytokeratins and to ECM proteins, however, revealed that the distribution of ECM proteins was not co-ordinated with cytokeratin positive cells but was present on closely adjacent structures.

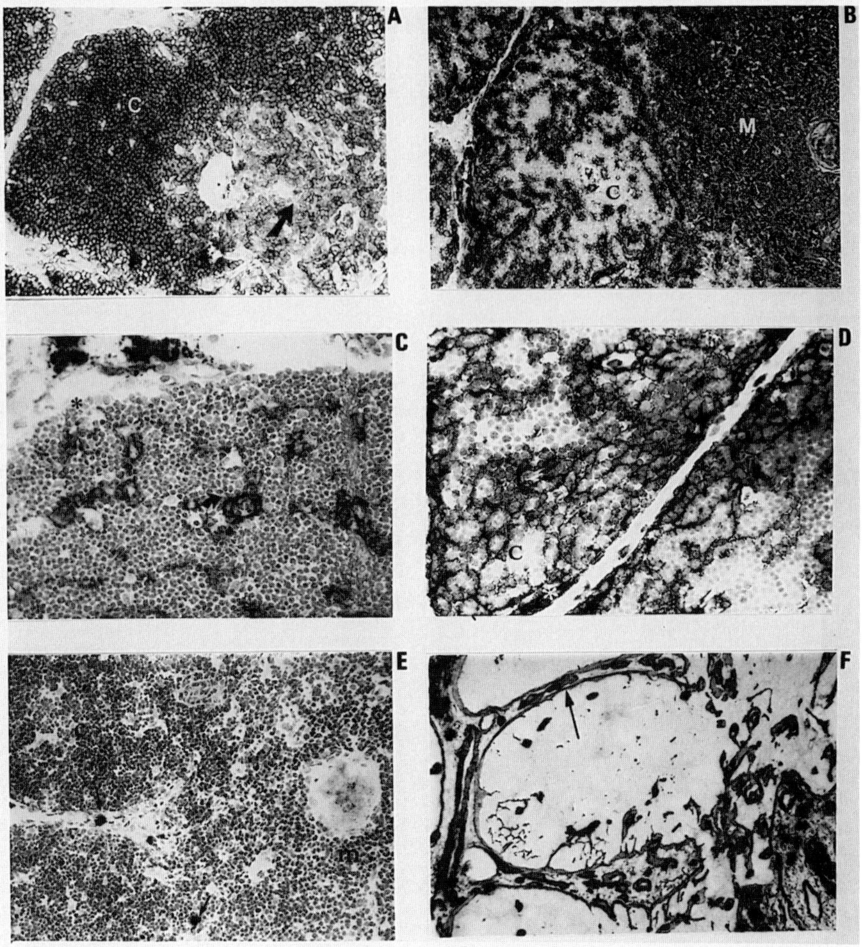

Figure 14.3 Immunohistochemical Staining of Thymocytes in Frozen Sections from Human Post-natal Thymus.

(a) Positive staining on cortical thymocytes (C) with the B5G10 Mab whereas fewer cells are B5G10 (VLA-4) positive in the medulla (→). [Magnification = ×250]

(b) Strong staining for VLA-6, defined by Mab GoH3, is demonstrated on a proportion of cortical thymocytes (C) which show a restricted distribution in perivascular areas. All cells in the thymic medulla (M) are VLA-6 positive. [Magnification = ×400]

(c) ICAM-1 determinants defined by the Mab, R.R1/1, are selectively expressed on cortical (C) and medullary epithelial cells. A structure reminiscent of a thymic nurse cell is also positive (→) Subcapsular epithelial cells are negative (*). [Magnification = ×400]

(d) VLA-2 determinants outline the fine reticular network of post-natal human thymus. Positive staining on subcapsular (*) and cortical (C) epithelium is demonstrated with the Gi14 Mab. [Magnification = ×400]

(e) Negative control lacking the primary antibody only and showing a cryostat section of post-natal human thymus. There is no staining of thymic cells in cortex (C) or medulla (M). Cells containing non-specific endogenous peroxidase activity are present in the interlobular septum and at the corticomedullary junction (←) [Magnification = ×400]

(f) Collagen Type IV expression on post-natal thymus distinctly outlines the basement membrane (→) surrounding the thymic parenchyma and thymic vasculature. [Magnification = ×250].

The peripheral membrane-like structure formed by ECM proteins was in close apposition to and occasionally enclosed subcapsular epithelial cells and also showed a diffuse intercellular distribution of fibronectin in the cortex (data not shown).

DISCUSSION

Our studies indicate that the majority of human thymocytes express significant levels of VLA-4, VLA-6, LFA-1α(L) and the β1 and β2 molecules but negligible amounts of VLA-1, VLA-2, VLA-3, VLA-5 and the β3, β4, IIb/IIIa, VNRαv, αM and αX subunits. Division of thymocytes into four subpopulations based on differential expression of the CD4 and CD8 defined cell surface molecules revealed different levels of expression of adhesion receptors. The immature CD4$^-$8$^-$ subset was VLA-β1, VLA-4, VLA-5, VLA-6 and LFA-1α(L) positive. Since the CD4$^-$8$^-$ cells are not thymocyte precursors exclusively, the importance of these integrin receptors in the migration and entry of precursors into the human thymus cannot be determined without detailed functional analyses. The cortical CD4$^+$8$^+$ thymocytes constitute the major proportion of cells in the thymus and are thought to develop from CD4$^-$8$^-$ thymocytes.[14] This CD4$^+$8$^+$ cortical population expresses significant amounts of β1, VLA-4, VLA-6 and LFA-1α(L). VLA-6 (α6β1) expression, defined by the GoH3 Mab, on cortical thymocytes *in situ* appeared to be higher around the perivascular regions which were positive for the known α6β1 ligand, laminin.

Medullary thymocytes can be divided into CD4$^+$8$^-$ and CD4$^-$8$^+$ subsets, thereby sharing a phenotype in common with peripheral T cells and representing the most mature components in the thymus.[14] In addition, the CD4$^-$8$^-$ subset may differentiate into CD4$^+$8$^+$ cells via immature CD4$^-$8$^+$ cells. while CD4^{dull+}8$^-$ cells contain T cell precursors with the potential to produce $\gamma\delta$ and $\alpha\beta$ T cell receptor (TCR) bearing subsets.[15,16] Both single positive subsets expressed β1, VLA-4, VLA-6 and LFA-1α(L). In all experiments, β1 and VLA-4 expression was higher on the CD4$^+$8$^+$ subset than on the majority of CD4$^+$8$^-$ or CD4$^-$8$^+$ cells, whereas VLA-6 and LFA-1α(L) showed similar or higher median fluorescence intensities on the CD4$^+$8$^-$ or CD4$^-$8$^+$ thymocyte populations. The CD4$^+$8$^-$ subset generally showed a biphasic pattern of β1 and VLA-4 expression, the significance of which requires further examination. Studies on thymic sections agreed with the analyses on isolated thymocyte suspensions, with two exceptions involving small percentages of cells. With the B5G10 Mab to VLA-4, a minor proportion of medullary thymocytes appeared negative in sections, while, with the Mabs to VLA-5, all sections appeared negative although the CD4$^-$8$^-$ subset was positive by FACS analysis.

The thymic epithelium both *in vivo* and *in vitro* differentially expressed VLA-β1, VLA-2, VLA-3, VLA-6, VNRαv, ICAM-1 and LFA-3 molecules.

In contrast to the studies of Zutter and Santoro.[17] our results show VLA-2 on subcapsular epithelium with the 12F1 Mab, but with the Gil4 Mab the whole epithelial network was stained. Both antibodies reacted with blood vessel endothlium. VLA-3 was detected on both the subcapsular and medullary epithelium. Ruco *et al.*[18] have subsequently confirmed these findings for VLA-2 and VLA-3 expression. The VNRαv subunit, which is present on subcapsular and on a proportion of medullary epithelial cells, has the capacity to associate with $\beta 1$, $\beta 3$, $\beta 5$ or $\beta 6$ integrins. Since very low levels of $\beta 3$ integrins and the VNRαv$\beta 3$ complex (detected with the 23C6 Mab[19]) were found on subcapsular or medullary epithelium *in situ*, the possibility arises that the VNRαv subunit may preferentially form an association with $\beta 1$, $\beta 5$ or $\beta 6$ integrins in this epithelium. Of particular interest was the apparent lack of ICAM-1 and LFA-3 on the subcapsular epithelium when tissue sections were examined, although these appeared upregulated on transformed thymic epithelial cell strains. These findings confirm those of Smith and Thomas[20] and suggest that CD2/LFA-3 and ICAM-1/LFA-1 interactions are not important in the initial migration and entry of precursors into the thymus nor in the association of these cells with subcapsular epithelium. This would also agree with the studies of Kyewski *et al.*[21] in which CD2 Mabs do not inhibit entry or expansion of T cell precursors in mouse thymic epithelial rudiments *in vitro*. Recent research has shown that LFA-1 and CD2 have additional ligands.[22–26] It would be of interest to know the distribution of these ligands on the thymic epithelia.

Our studies on human post-natal thymus show that, in sections, fibronectin, collagen type IV and laminin are concentrated around the capsule and blood vessel endothelium and extend into the outer cortical region of the thymus. While fibronectin was also present throughout the cortex and on Hassall's corpuscles, collagen type IV was absent in these areas. In its simplest form, it is therefore possible that both the immature $CD4^{dull+}8^-$, $CD4^-8^+$ and $CD4^-8^-$ subsets as well as the more mature $CD4^+8^+$ population might interact with fibronectin via VLA-4 ($\alpha 4\beta 1$). The $CD4^-8^-$ population may also use VLA-5 ($\alpha 5\beta 1$) and VLA-6 ($\alpha 6\beta 1$) for fibronectin and laminin interactions, respectively, during entry into the thymus and at the initial stages of thymocyte development. Expression of VLA-4 and VLA-6 on medullary thymocytes did not correlate with fibronectin or laminin distribution around the medullary epithelium, suggesting that if these receptors are important in this region of the thymus, then they probably associate with different ligands. Subsequent studies by Sawada *et al.*[27] have indicated that VLA-4 is more important for adhesion of cortical thymocytes to fibronectin and thymic stromal cells than it is for the adhesion of medullary thymocytes. Adhesion receptors, such as VLA-4 ($\alpha 4\beta 1$ or $\alpha 4\beta 7$), have the potential to interact with more than one ligand[28–30] and may preferentially use different ligands either in different tissues[31,32] or when a particular adhesion receptor is selectively activated. Previous studies have indicated that immature precursor cells which invade the capsule and proliferate require functional fibronectin and laminin receptors.[20,33]

The presence of VLA-2 on subcapsular and cortical epithelium, of VLA-3 and VNRαv on the subcapsular cortex and of both VLA-2 and VLA-3 on blood vessel endothelium corresponds to the presence of at least some of their associated ECM ligands in these regions of the thymus. VLA-2 and VLA-3 have also been implicated in cell-cell adhesion[34,35] and may function in holding the epithelial network together in those regions of the thymus that apparently lack ECM ligands. VLA-3 has recently been shown to bind to an epithelial ligand, epiligrin,[36] and the function of VLA-2 appears to depend on its activation state.[37]

In conclusion, we could envisage that VLA-4 and VLA-5 on immature thymocytes both interact with different regions of the fibronectin molecule, namely with the CS-1[38] and RGD sequences[8] respectively. These fibronectin molecules could be bound closely to the subcapsular epithelium via VLA-2 and VLA-3 on epithelium, thereby maintaining a three-dimensional network within which thymocyte development can proceed in association with other receptors and with soluble factors. Further regulation may occur by expression of different fibronectin splice variants at various locations in the thymus. In addition, VLA-2, VLA-3 and VLA-6 on the subcapsular epithelium may present laminin to immature thymocytes which would then bind this component via their VLA-6 adhesion receptors. The exact mechanisms of these interactions in the thymus require further investigation. However, the availability of defined thymic epithelial cell strains, thymocyte subsets and the SCID-hu mouse model make such studies feasible.

Acknowledgements

The authors would like to thank Drs F. Katz, L. Wong, N. Al-Mahdi and G. Morgan for thymus material and Drs N. Hogg, M. Hemler, L. Old and T. Springer for antibodies. This work has been published in detail in references 12 and 13.

Acknowledgements for Research Support

This work is supported by the Medical Research Council and the Imperial Cancer Research Fund of Great Britain.

REFERENCES

1. Haynes, B. F. (1986) The role of the thymic microenvironment in promotion of early stages of human T cell maturation. *Clin. Res.*, **34**, 422.
2. Ready, A. R., Jenkinson, E. J., Kingston, R. and Owen, J. T. T. (1984) Successful transplantation across major histocompatibility barrier of deoxyguanosine-treated embryonic thymus expressing class II antigens. *Nature*, **310**, 231.

3. Ruther, U., Muller, W., Sumida, T., Tokuhisa, T., Rajewsky, K. and Wagner, E. F. (1988) *c-fos* expression interferes with thymus development in transgenic mice. *Cell*, **53**, 847.
4. Spanopoulou, E., Early, A., Elliott, J., Crispe, I. N., Ladyman, H., Ritter, M., Watt, S. Grosveld, F. and Kioussis, D. (1989) Complex lymphoid and epithelial thymic tumours in *Thy1-myc* transgenic mice. *Nature*, **342**, 185.
5. Osculati, F., Balercia, G. and Mathe, G. (1988) Human thymic epithelium in culture: an experimental model for the study of thymic microenvironment. *Biomed. and Pharmacother.*, **42**, 395.
6. Bierer, B. E. and Burakoff, S. J. (1988) T cell adhesion molecules. *FASEB J.*, **2**, 2584.
7. Yong, F. and Khwaja, A. (1990) Leucocyte cellular adhesion molelcules. *Blood Reviews*, **4**, 211.
8. Hemler, M. E. (1990) VLA proteins in the integrin family: structures, functions, and their role on leukocytes. *Annu. Rev. Immunol.*, **8**, 365.
9. Savagner, P., Imhof, B. A., Yamada, K. M. and Thiery, J.-P. (1986) Homing of hemopoietic precursor cells to the embryonic thymus: characterization of an invasive mechanism induced by chemotactic peptides. *J. Cell Biol.*, **103**, 2715.
10. Cardarelli, P. M., Crispe, I. N. and Pierschbacher, M. D. (1988) Preferential expression of fibronectin receptors on immature thymocytes. *J. Cell Biol.*, **106**, 2183.
11. Mizutani, S., Watt, S. M., Robertson, D., Hussein, S., Healy, L. E., Furley, A. J. and Greaves, M. F. (1987) Cloning of human thymic subcapsular cortex epithelial cells with T-lymphocyte binding sites and hemopoietic growth factor activity. *Proc. Natl. Acad. Sci. USA*, **84**, 4999.
12. Watt, S. M., Thomas, J. A., Murdoch, S. J., Kearney, L., Chang, S. and Bartek, J. (1991) Human thymic epithelial cells are frequently transformed retroviral vectors encoding simian virus 40. *Cell. Immunol.*, **138**, 456.
13. Watt, S. M., Thomas, J. A., Edwards, A. J., Murdoch, S. J. and Horton, M. A. (1991) Adhesion receptors are differentially expressed on developing thymocytes and epithelium in human thymus. *Exp. Hematol.*, **20**, 1101.
14. Adkins, B., Mueller, C., Okada, C. Y., Reichert, R. A., Weissman, I. L. and Spangrude, G. J. (1987) Early events in T cell maturation. *Ann. Rev. Immunol.*, **5**, 325.
15. MacDonald, H. R., Budd, R. C. and Howe, R. C. (1988) A CD3- subset of CD4 – 8 + thymocytes: a rapidly cycling intermediate in the generation of CD4 + 8 + cells. *Eur. J. Immunol.*, **18**, 519.
16. Hugo, P., Waanders, G. A., Scollay, R., Shortman, K. and Boyd, R. (1990) Ontogeny of a novel CD4 + CD8 – CD3 – thymocyte subpopulation: a comparison with CD4 – CD8 + CD3 – thymocytes. *Int. Immunol.*, **2**, 209.
17. Zutter, M. M. and Santoro, S. A. (1990) Widespread histologic distribution of the $\alpha_2\beta_1$ integrin cell-surface collagen receptor. *Am. J. Path.*, **137**, 113.
18. Ruco, L. P., Paradiso, P., Pittiglio, M., Diordoro, M. G., Gearing, A. J., Mainiero, F., Gismondi, A., Santoni, A. and Baroni, C. D. (1993) Tissue distribution of very late activation antigens-1/6 and very late activation ligands in the normal human thymus and in thymoma. *Am. J. Pathol.*, **142**, 765.
19. Horton, M. J. (1990) Current status review. Vitronectin receptor: tissue specific expression or adaptation to culture? *Int. J. Exp. Pathol.*, **71**, 741.
20. Smith, M. E. F. and Thomas, J. A. (1990) Cellular expression of lymphocyte

function associated antigens and the intercellular adhesion molecule-1 in normal tissue. *J. Clin. Pathol.*, **43**, 893.

21. Kyewski, B., Jenkinson, E. J., Kingston, R., Altevogt, P., Owen, M. J. and Owen, J. J. T. (1989) Expression of CD2 during the differentiation of mouse thymocytes. *Eur. J. Immunol.*, **19**, 951.
22. Xu, H., Tong, I. L., De-Fougerolles, A. R. and Springer, T. A. (1992) Isolation, characterization and expression of mouse ICAM-2 complementary and genomic DNA. *J. Immunol.*, **149**, 2650.
23. Vazeux, R., Hoffman, P. A., Tomita, J. K., Dickinson, E. S., Jasman, R. L., St. John, T. and Gallatin, W. M. (1992) Cloning and characterization of a new intercellular adhesion molecule ICAM-R. *Nature*, **360**, 485.
24. Fawcett, J., Holness, C. L., Needham, L. A., Turley, H., Gatter, K. C., Mason, D. Y. and Simmons, D. L. (1992) Molelcular cloning of ICAM-3, a third ligand for LFA-1, constitutively expressed on resting leukocytes. *Nature*, **360**, 481.
25. Arulanandam, A. R. N., Moingeon, P., Concino, M. F., Recny, M. A., Kato, K., Yagita, H., Koyasu, S. and Reinherz, E. L. (1993) A soluble multimeric recombinant CD2 protein identifies CD48 as a low affinity ligand for human CD2: divergence of CD2 ligands during the evolution of humans and mice. *J. Exp. Med.*, **177**, 1439.
26. Deckert, M., Kubar, J., Zoccola, D., Bernad-Pomier, G., Angelisova, P., Horejsi, V. and Bernard, A. (1992) CD59 molecule: a second ligand for CD2 in T cell adhesion. *Eur. J. Immunol.*, **22**, 2943.
27. Sawada, M., Nagamine, J., Takeda, K., Utsumi, K., Kosugi, A., Tatsumi, Y., Hamaoka, T., Miyake, K., Nakajima, K. and Watanabe, T. *et al.* (1992) Expression of VLA-4 on thymocytes. Maturation stage-associated transition and its correlation with their capacity to adhere to thymic stromal cells. *J. Immunol.*, **149**, 3517.
28. Shimizu, Y., van Seventer, G. A., Horgan, K. J. and Shaw, S. (1990) Roles of adhesion molecules in T-cell recognition: fundamental similarities between four integrins on resting human T cells (LFA-1, VLA-4, VLA-5, VLA-6) in expression, binding and costimulation. *Immunol. Rev.*, **114**, 109.
29. Hemler, M. E., Elices, M. J., Parker, C. and Takada, Y. (1990) Structure of the integrin VLA-4 and its cell-cell and cell-matrix adhesion functions. *Immunol. Rev.*, **114**, 45.
30. Kassner, P. D., Teixido, J., Chan, B. M., Parker, C. M. and Hemler, M. E. (1992) Analyses of VLA-4 structure and function. *Adv. Exp. Med. Biol.*, **323**, 163.
31. Miyake, K., Weissman, I. L., Greenberger, J. S. and Kincade, P. W. (1991) Evidence for a role of the integrin VLA-4 in lympho-hemopoiesis. *J. Exp. Med.*, **173**, 599.
32. Guan, J.-L. and Hynes, R. O. (1990) Lymphoid cells recognise an alternatively spliced segment of fibronectin via the integrin receptor $\alpha_4\beta_1$. *Cell*, **60**, 53.
33. Utsumi, K., Motoyuki, S., Seiji, N., Nagamine, J., Sakata, T., Iwagami, S., Kita, Y., Teraoka, H., Hirano, H., Ogata, M., Hamaoka, T. and Fujiwara, H. (1991) Adhesion of immature thymocytes to thymic stromal cells through fibronectin molecules and its significance for the induction of thymocyte differentiation. *Proc. Natl. Acad. Sci. USA*, **88**, 5685.
34. Kaufmann, R., Frosch, D., Westphal, C., Weber, C. and Klein, C. (1989) Integrin VLA-3: Ultrastructural localization at cell-cell contact sites of human cell cultures. *J. Cell Biol.*, **109**, 1807.

35. Carter, W. G., Kaur, P., Gil, S. G., Gahr, P. J. and Wayner, E. A. (1990) Distinct functions for integrins $\alpha_3\beta_1$ in focal adhesions and $\alpha_6\beta_4$/bullous pemphigoid antigen in a new stable anchoring contact (SAC) of keratinocytes: relation to hemidesmosomes. *J. Cell Biol.*, **111**, 3141.
36. Domloge-Hultsch, N., Gammon, W. R., Briggaman, R. A., Gil, S. G., Carter, W. G. and Yancey, K. B. (1992) Epiligrin, the major human keratinocyte integrin ligand, is a target in both an acquired autoimmune and an inherited subepidermal blistering skin disease. *J. Clin. Invest.*, **90**, 1628.
37. Chan, B. M. and Hemler, M. E. (1993) Multiple functional forms of the integrin VLA-2 can be derived from a single alpha 2 cDNA clone: interconversion of forms induced by an anti-beta 1 antibody. *J. Cell. Biol.*, **120**, 537.
38. Nojima, Y., Humphries, M. J., Mould, A. P., Komoriya, A., Yamada, K. M., Schlossman, S. F. and Morimoto, C. (1990) VLA-4 mediates CD3-dependent $CD4^+$ T cell activation via the CS1 alternatively spliced domain of fibronectin. *J. Exp. Med.*, **172**, 1185.

15. EXPRESSION AND FUNCTION OF CD66 DEFINED MEMBERS OF THE CARCINOEMBRYONIC ANTIGEN FAMILY

SUZANNE M. WATT[1], DAVID L. SIMMONS[2], JON FAWCETT[2], ANA TEIXEIRA[3], ELMA TCHILIAN[4], FRITZ GRUNERT[5], GERHARDT NAGEL[5], HERMAN WALDMANN[6] and GRACIELLA SALA-NEWBY[7]

[1]*The MRC Molecular Haematology Unit, Institute of Molecular Medicine, The John Radcliffe Hospital, Oxford, UK*
[2]*Cell Adhesion Laboratory, Imperial Cancer Research Fund, Institute of Molecular Medicine, Oxford, UK*
[3]*Medical Oncology Unit, Imperial Cancer Research Fund, London, UK*
[4]*Tumour Immunology Unit, Imperial Cancer Research Fund, London, UK*
[5]*Institut fur Immunobiologie, Albert-Ludwigs Universitat, Freiburg, Germany*
[6]*Department of Pathology, University of Cambridge, Cambridge, UK*
[7]*Department of Medical Biochemistry, University of Wales College of Medicine, Cardiff, UK*

The monoclonal antibody (MoAb), YTH71.3.2, has been classified as CD66. Two other monoclonals, CE6/2D3.1 and YPC2/12.1, share a common cellular specificity, reacting with colonic epithelium and cells of the granulocytic lineage, but not with the majority of haemopoietic progenitor cells from human bone marrow. Both YTH71.3.2 and CE6/2D3.1 recognise a similar epitope, therefore allowing CE6/2D3.1 to be classified as CD66. However, this epitope is distinct from that recognised by YPC2/12.1. All three antibodies react with molecules of 50–55kD, 80–100kD and 130–200kD from a partially purified CEA extract of colonic adenocarcinomas and with a 90–130kD component on HL-60 cells. Expression cloning studies, together with the analysis of HeLa or CHO cell transfectants of CEA, biliary glycoprotein (BGPc), NCA and CGM6 (NCA-95) cDNAs, indicate that CD66 and YPC2/12.1 bind to CEA, BGPc and NCA to varying extents, while YPC2/12.1 also binds NCA-95.

Preliminary adhesion assays indicate that cell lines representing colonic epithelial and neutrophilic lineages adhere to Cos transfectants expressing BGPc. Furthermore, CHO cells expressing BGPc appear to aggregate homotypically in a temperature dependent manner, suggesting that splicing of the cytoplasmic tail of BGP does not affect its initial adhesion properties. Although some studies indicate that CD66 defined molecules express carbohydrate moieties identified by CD15, analyses of soluble forms of BGP and of CHO transfectants expressing BGPc show no significant binding of the Leu-M1 or CSLex antibodies to the Le^x and sialyl Le^x determinants respectively of the CD15 defined antigen. This undoubtedly reflects the absence of the correct fucosyl transferases for Le^x and sialyl Le^x expression in these cell lines. Using soluble domain variants of BGPc, our results show that the CD66 MoAbs, YTH71.3.2 and CLB/gran 10, bind to the V domain of BGPc and that this domain will mediate homotypic adhesion to CHO-BGPc transfectants in a temperature dependent manner. The soluble constructs were also used for epitope mapping of 25 MoAbs. Fourteen of these mapped to the N-terminal domain, whereas the remainder required the presence of the IgC2-set domains.

Correspondence. S. M. Watt, The MRC Molecular Haematology Unit, Institute of Molecular Medicine, The John Radcliffe Hospital, Headington, Oxford, OX3 9DU, UK.

INTRODUCTION

The carcinoembryonic antigen (CEA) family now comprises at least 29 different CEA-related cDNAs or genes, which are tandemly arranged on the long arm of chromosome 19 and are highly homologous at both the nucleotide and amino acid levels.[1] Further heterogeneity of this family is achieved by differential glycosylation and splicing of some of its members.[1] The cDNAs isolated to date encompass both membrane linked and secreted forms of CEA-related antigens and this has allowed their division into two subgroups: the membrane bound CEA subfamily and the secreted pregnancy specific glycoprotein (PSG) subfamily.[1] The relationship between cloned cDNA sequences and multiple forms of CEA-related proteins identified by immunoblotting and immunoprecipitation is, however, not always clear.

All CEA-related proteins so far described are members of the immunoglobulin (Ig) superfamily, with the extracellular region having a structure consisting of a leader peptide, an IgV-like domain and a variable number of or no IgC-like domains.[1,2] For the CEA subfamily, a short hydrophobic carboxy-terminal region may be post-translationally modified to allow membrane anchoring via PI-linkage[3,4] or may represent a transmembrane domain, which is followed by a cytoplasmic tail of variable length.[5–7] The former PI-linked proteins include CEA itself and the non-specific cross reacting antigens (NCAs), while the latter include the differentially spliced biliary glycoprotein (BGP) and CGM1 molecules.[1,2–7]

A number of CEA-related molecules have been located in human granulocytes.[2,6–13] These include the differentially glycosylated NCA molecule, which exhibits molecular weights of 90 and 50kD, NCA-like proteins with molecular weights of 160kD (NCA-160), 95kD (NCA-95), 80kD and 26kD, and the PSG 5 glycoprotein. The monoclonal antibody, YTH71.3.2, has been clustered as CD66.[13–15] In this article, we define the molecules identified by CD66 and CD66-like monoclonal antibodies and discuss their expression and possible functions within the haemopoietic system.

MATERIALS AND METHODS

Cell Preparation

Human bone marrow leucocytes (density <1.114g/ml) were obtained as described earlier.[13] Peripheral blood neutrophils were enriched by centrifugation on a discontinuous gradient of Ficoll-hypaque with densities of 1.077 and 1.114 g/ml and collected from the 1.114 g/ml interface. Residual erythrocytes in this fraction were lysed at room temperature for 15 minutes with isotonic ammonium chloride.[13] The HL-60, Molt 4, Daudi, Raji, KG1, LS174T and HT29 cell lines were maintained in RPMI-1640 medium with 10% (w/v) fetal calf serum (FCS) and diluted to 2×10^5 cells per ml with fresh medium on the day before analysis. Single cell suspensions of the colon carcinoma cell lines, LS174T and HT29, and of CHO transfectants were obtained by trypsinization or with 2mM EDTA in phosphate buffered saline (0.02M phosphate buffer containing 0.145M sodium chloride, pH 7.4; PBS) prior to use. HL-60 cells (2×10^5/ml) were induced to mature neutrophils by the addition of luM retinoic acid for 7 days.[13]

Monoclonal Antibodies

The rat monoclonal antibodies, YTH71.3.2, YPC2/12.1 and CE6/2D3.1, were produced as detailed earlier.[13] The CD33 MoAb was obtained from Seralab, Crawley, UK. The 3B10 antibody, a mouse monoclonal antibody that reacts with the CEA family[16] was kindly provided by Dr H. Durbin. The LeuMl, CSLex and CD3 antibodies were obtained from Becton-Dickinson, San Jose, CA, USA, while the CD34 antibodies, ICH-3, HPCA-2 were obtained from Becton-Dickinson, San Jose, CA, USA or Immunotech, Marseilles, France and Qbend 10 and CD7 were prepared as described.[17,18] All remaining CD66 and CD66-like MoAbs were obtained from the IVth and Vth Leucocyte Culture Conference Workshops.[14] The latter were kindly provided by Professor Keith Skubitz, University of Minnesota, Minneapolis, USA. The antibodies were used as culture supernatants, ascites or purified on Protein G-Sepharose (Pharmacia, Uppsala, Sweden).[13–17]

Cell Staining, Flow Cytometry and Clonogenic Assays

Adherent cell monolayers and 5 micron cryostat sections of normal colon and colonic adenocarcinomas were used unfixed or fixed in acetone or p-formaldehyde and stained with antibodies.[13,19] Bone marrow mononuclear cells, neutrophils and the cell lines were stained for flow cytometric analysis and sorting as described.[13] Sorted cells were then analysed for their morphology and haemopoietic progenitor cell content.[20] Clonogenic cells which formed colonies containing more than 50 cells were scored on days 7 to 14 as non-erythroid clones and included macrophage (M), granulocyte (G) and mixed granulocyte-macrophage (GM) cell types and as pure erythroid (BFU-E) and mixed erythroid (CFU-Mix) clones. The latter contained at least three cell types, one of which was erythroid. Non-erythroid (G/M) and erythroid (CFU-E) clusters were scored at day 7 and contained up to 50 cells.

CEA and Cell Membrane Analyses

Partially purified CEA preparations (kindly donated by Dr G. T. Rogers) were isolated from liver metastases of colon carcinomas by perchloric acid extraction and fractionation on Sepharose 6B and on Con A Sepharose.[13] Crude cell membrane samples were prepared, electrophoresed on 7.5%

polyacrylamide gels (SDS-PAGE), blotted on nitrocellulose and probed with CD66 and CD66-like antibodies as described.[13]

Expression Cloning and Sequencing

cDNA libraries prepared from the HT29 colon carcinoma cell line, from colonic carinomas and from the promyelocytic cell line, HL-60, all in the πHM3 vector were transfected into Cos cells using dextran sulphate.[21] Cells expressing CD66 defined antigens were selected by three cycles of panning with the YTH71.3.2, YPC2/12.1 and CE6/2D3.1 antibodies followed by protoplast fusion.[6,21,22] A single cDNA clone was obtained and was completely sequenced using ^{35}S-methionine labelling together with the Sequenase kit and the Sequagel sequencing system (National Diagnostics, Aylesbury, UK) in accordance with the manufacturer's instructions.

Production of Soluble BGPc Domain Variants

(a) The pIG vector

Details of the pIG vector containing the Fc genomic fragment of human IgG1 will be provided elsewhere [Fawcett *et al.* manuscript in preparation;[6]]. In brief, the IgGl gene segment was amplified from 250ng of human genomic DNA from K562 cells by the polymerase chain reaction (PCR) using 5′ and 3′ oligonucleotide primer sets which were:

GENBANK HUMIGCC4 sequence:
Bam HI
5′TCGGATCC GGAGGGAGGG TGTCTGCTGG AAGCAGG3′ and
ramp at 2000–1970:
Not1
5′TGATCGCGGCCGC TGCCTCCCTC ATGCCACTCA GGCCTCAG3′

and were designed to include BamHI and Notl sites as shown. The PCR product was digested with HindIII/Notl, cloned into Bluescript (pBluescript II SK) and sequenced on the forward strand only using sequence specific primers at 200bp intervals. No differences from the Genbank sequence were seen. The single BamHl site in pCDM8 was removed by mung bean nuclease digestion and religation. The Fc genomic fragment was moved as a HindIII/Notl fragment from Bluescript and ligated into HindIII/Notl digested BamHl-minus pCDM8. This brings in additional cloning sites at the 5′ extracellular domain (ECD) cloning site: HindIII-EcoRV-EcoRI-PstI-SmaI-BamHI.

(b) Soluble BGPc constructs

The cDNA sequences encoding the different ECDs, V, VAIBI and VAIBIA2 of the BGPc molecule and the entire ECD of the CD33 defined molecules were amplified by PCR using 94°C for 10 minutes, followed by 4 cycles of 94°C for 0.5 minutes, 55°C for 1.5 minutes and 72°C for 1 minute and 24 cycles of amplification at 94°C for 0.5 minutes, 55°C for 0.5 minutes and 72°C for 1 minute, both with Taq polymerase. The primers complementary to 5′ and 3′ sequences of the πHM3 vector and the coding region of BCPc ECD variants and the CD33 ECD respectively were:

(i) *πH3M or pCDM8 Primer (at 2140)*
5′ GAG AAC CCA CTG CTT AAC TGG 3′

(ii) *BGPc(N) Reverse Primer (at 435–412)*
BspE1 EcoR1 Consensus Splice Donor BGpc(N) Reverse Complement Sequence
5′ CTG ATC CGG AGA ATT CCT TAC CTG T GGG CAG CTC CGG GTA TAC ATG GAA 3′

(iii) *BGPc(NA1B1) Reverse Primer (at 957–934)*
BspE1 EcoR1 Consensus Splice Donor BGpc(B1) Reverse Complement Sequence
5′ CTG ATC CGG AGA ATT CCT TAC CTG T AGT GAC TAT GAT CGT CTT GAC TGT 3′

(iv) *BGPc(NA1BA2) Reverse Primer (at 1269–1246)*
BspE1 EcoR1 Consensus Splice Donor BGpc(N) Reverse Complement Sequence
5′ CTG ATC CGG AGA ATT CCT TAC CTG T ATT TTC TTG TGG TAG AGC ATT ATA 3′

(v) *CD33 Reverse Primer (at 789–768)*
BspE1 Consensus Splice Donor BGpc(N) Reverse Complement Sequence
5′ GAT CAG ATC TAC TTA CCT TAC CTG T ATG AAC CAC TCC TGC TCT GGT 3′

These were designed to include restriction enzyme sites as shown. For the production of BGPc-Fc domain constructs, PCR products of the BGPc ECDs and the pIG vector were digested with HindIII and EcoRI. The CD33 ECD PCR products and the pIG vector digested with HindIII/BgIII and HindIII/BamH1 respectively. The digests were purified at 0.8% (wt/vol) agarose gels prior to ligation with T4 ligase. Since splice donor sites were just upstream of the 3′ enzyme site, the ECD sequence of BGPc and CD33 were than spliced in frame into the IgG1 hinge with the resulting chimaeric molecules dimerizing via this hinge. After MC1061 bacterial transformation with these constructs, plasmid preparations of the BGPc-Fc domain constructs and the CD33-Fc were purified on CsC1 gradients. The BGPc(NA1B1A2), BGPc(NA1B1) and BGPc(N) sequences were determined as described above using primers 200bp apart. The former two PCR products were identical to the appropriate sequences of the cloned BGPc ECD, while the BGPc(N) sequence showed identity for the first 430bp of the total 435bp ECD sequence used. The Fc constructs were transfected into Cos 7 cells using DEAE-Dextran[23] and secreted proteins collected on days 3 to 7 of culture. The secreted proteins were used unlabelled or metabolically labelled with ^{35}S-methionine, then purified on Protein A-Sepharose (Pharmacia Fine Chemicals, Uppsala, Sweden), eluted with 0.1M glycine HCI (pH3) or with 4M imidazole and analysed by SDS polyacrylamide gel electrophoresis (SDS-PAGE) on 7.5% (wt/vol) gels under reducing and non-reducing conditions. The unlabelled proteins were blotted onto Hybond C extra and probed with CD66 (YTH71.3.2) using the ECL system from Amersham (Amersham, Bucks., UK) as described by Watt *et al.*[6] Both the protein A-Sepharose purification and the ECL blotting procedures were carried out according to the manufacturers' instructions.

HeLa and CHO Cell Transfectants

cDNAs encoding CEA, NCA and NCA-95 were cloned into the pBEH-pUC9 expression vector. HeLa cells transfected with these constructs together with the pSV2neo plasmid (which confers neomycin resistance) were selected in G418 to give stable HeLa-CEA, HeLa-NCA and HeLa-CGM6 (NCA-95) cell lines expressing the respective proteins.[13] BGPc[6] and the CD33[24] defined cDNAs (both in the πHM3 vector), CD34[25] cDNA (in the pCDM8 vector) or the pSV2neo plasmid were used to produce stable transfectants. Plasmids containing the BGPc cDNA was linearized with SfiI. Plasmids containing the CD33 and CD34 defined cDNAs were linearized with HindIII. The pSV2neo construct was linearized with BamH1. Linearized plasmids to the cell surface proteins together with pSV2neo were electroporated into CHO cells in suspension in DMEM using a Biorad electroporator at 400V and 500 to 960uF. After selection in 0.5mg/ml G418 and three cycles of sorting with the appropriate antibodies, stable CHO-BGPc, CHO-CD33 and CHO-CD34 transfectants were obtained. HeLa cells transfected with the pSV2neo plasmid (HeLa-Neo) and non-transfected CHO cells were used as negative controls. Cos 7 cells were transfected with cDNAs encoding BGPc and VCAM-1[22] or mock transfected using DEAE-Dextran[6,21,22] and used 3 days later.

Binding and Adhesion Assays

Cells (2×10^4 to 4×10^5) and CEA and BGP-Fc domain variant proteins (10ug/well) were dispersed in or bound to flexible microtitre plates prior to addition of the appropriate antibody.[6,26] After washing the sample, the binding was monitored by addition of ^{125}I-labelled sheep anti-rat Ig or by the sequential addition of biotinylated goat anti-mouse or rat Ig, avidin-horse radish peroxidase and the p-nitrophenyl phosphate substrate (all from Sigma Chemical Co., St Louis, Miss, USA). Samples were counted in a γ-counter or analysed on an Elisa reader at 405nm.[27] Details of cell-cell adhesion assays for CHO and Cos cell transfectants are given in references 6, 22 and 28.

For binding of transfectants to soluble constructs, 25 well Sterilin plates were coated with rabbit anti-human Ig (Sigma Chemical Co., St Louis, Miss., USA) prior to incubation with different soluble domain Fc constructs as described.[6,26] ^{3}H-thymidine 5% (vol/vol) dialysed FCS with or without 5mM EDTA or in medium containing Ca++ and/or Mg++ at 5mM concentrations[26] were allowed to adhere to the plates at 4°C or 37°C for 1 hour. The plates were washed twice with the same medium and the adherent cells solubilised with 1% (wt/vol) sodium dodecyl sulphate (SDS), placed into 3 mls of ecoscint scintillant and counted on a β counter.[6,26]

RESULTS

(a) Cellular Expression of CD66 Defined Antigens

YTH71.3.2, YPC2/12.1 and CE6/2D3.1 all bound to colonic epithelium, to the colonic carcinoma cell lines, LS174T and HT29, to peripheral blood neutrophils and the promyelocytic HL-60 cell line, but not to erythroid (K562), monocytic (U937) or lymphoid (Daudi, Raji, Molt 4) cell lines nor to isolated peripheral blood erythrocytes (of A, B or O blood groups), monocytes or lymphocytes. Since the colonic cell lines express CEA-related products, the antibodies were tested for their reactivity with a partially purified CEA preparation and showed positive binding. Details of this are presented in reference 13.

YTH71.3.2 and YPC2/12.1 were examined in detail by flow sorting for their reativity with clonogenic and morphologically defined cells in normal human bone marrow. CD7 was used as a control antibody that reacts with T lymphocytes but not with neutrophilic nor clonogenic cells (Fig. 15.1). With both YTH71.3.2 and YPC2/12.1, cells of the neutrophil series were present in the positive fraction (Fig. 15.1). These included cells from the late myeloblast or early promyelocyte stages to the mature neutrophils. Monocytes and lymphocytes were generally negative. Reactivity with eosinophils was inconclusive because of the low content of these cells in the samples examined. Most of the recovered cells initiating day 7 erythroid clusters and

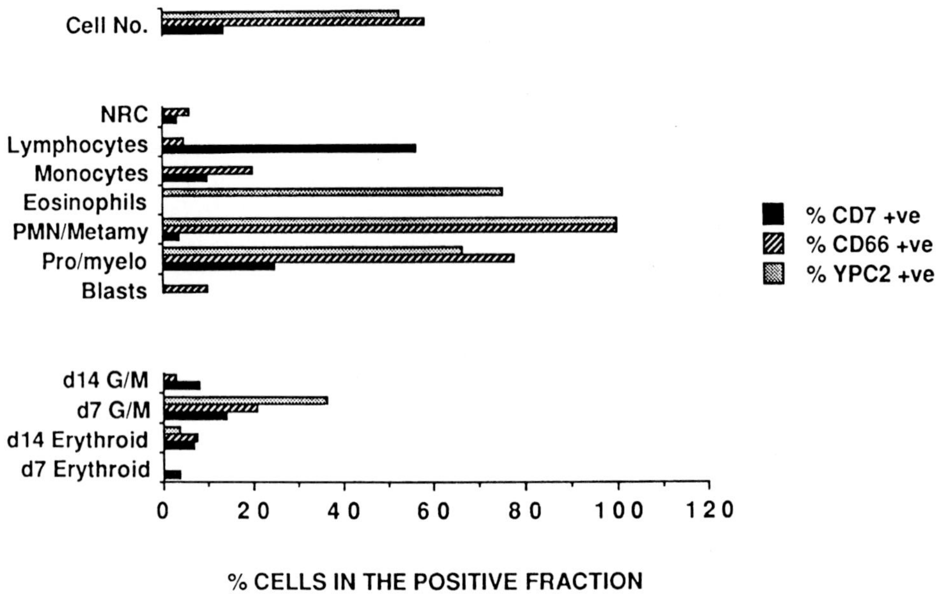

Figure 15.1 Reactivity of Human Bone Marrow Cells with CD66 and CD66-like Antibodies. Leucocytes (density <1.114g/ml) from normal human bone marrow were labelled with the YPC2/12.1, YTH71.3.2 or CD7 antibodies. The latter was used as a control antibody that binds T lymphocytes but not to the haemopoietic progenitors measured nor to cells of the neutrophil series. After addition of FITC-anti-rat Ig, cells where separated by flow cytometry into positive and negative fractions, (negative) a threshold defined using cells stained with an irrelevant antibody and the fluoresceinated second layer reagent as shown in [11]. The sorted cells were cytocentrifuged for morphological analysis or they were plated in methocel for progenitor cell analysis. The recovery of cells in the positive is presented and was calculated as a percentage of cells recovered when compared to the combined positive and negative fractions. Cells analysed include promyelocytes/myelocytes (pro/myelo), neutrophils/band, forms/metamyelocytes (PMN/Metamy), nucleated red cells (NRC), d7 erythroid clusters (d7 CFU-E), d7 G/M clusters and colonies (d7 G/M), d14 erythroid colonies (d14 BFU-E and CFU-Mix) and D14 G/M colonies (d14 G/M).

day 14 erythroid, myeloid and multipotential colonies occurred in the negative fraction. However, a substantial proportion of cells initiating day 7 non-erythroid clusters and colonies co-purified with the neutrophilic cells and were presumably restricted to this lineage (Fig. 15.1). Quantitative cell sorting studies indicated that the epitopes detected by CD66 and CD66-like antibodies increase in their level of expression as cells mature from promyelocytes to mature neutrophils. This was confirmed by an increase in the mean level of YTH71.3.2, YPC2/12.1 and CE6/2D3.1 binding on mature neutrophils when these were induced from the immature promyelocytic cell line, HL-60, with retinoic acid.[13]

(b) YTH71.3.2 and CE6/2D31 Identify Similar Epitopes

Competitive binding experiments were used to test the epitope specifity of each antibody. YPC2/12.1 and YTH71.3.2 were iodinated and assayed for their binding to a crude CEA preparation coated on to microtitre plates. Displacement of this binding was assessed by the addition of unlabelled YPC2/12.1, YTH71.3.2 and CE6/2D3.1 The binding of labelled YTH71.3.2 was displaced by cold CE6/2D3.1 or YTH71.3.2 (both giving 90% inhibition of binding) but not with YPC2/12.1 and vice versa [data not shown]. This indicates that YPC2/12.1 recognizes a different epitope to that detected with YTH71.3.2 and CE6/2D3.1 and that YTH71.3.2 and CE6/2D3.1 share common epitopes or epitopes that are sufficiently close to cause steric hindrance in binding.

(c) CD66 Identifies the CEA Family

(i) Immunoblotting

Extracts of the LS-174T and HL-60 cell lines [gel 1] and of the partially purified CEA [gel 2] were analyzed for their binding of the CD66 antibodies; the data being presented in Fig. 15.2. The YPC2/12.1 (track B), CE6/2D3.1 (track C) and YTH71.3.2 (track D), but not H58 to CEA itself (track A), recognized a 90–130kD component from HL-60 cell membrane preparations [Fig. 15.2; gel 1]. With the LS174T cell extract [Fig. 15.2; gel 1], the YPC2/12.1 antibody recognized antigens in the molecular weight regions of 160–200kD and 90–130kD (track H); YTH71.3.2 bound the 90–130kD molecule (track I); CE6/2D3.1 bound preferentially to the lower molecular weight component (track J), while its binding to the 160–200kD molecule varied from one experiment to another; and H58 labelled the 160–200kD band only (track G). With the partially purified CEA preparation [Fig. 15.2; gel 2], YPC2/12.1 detected 160–200kD, 90–100kD and 50–55kD bands (track A); CE6/2D3.1 bound weakly to the 160–200kD molecules, but preferentially detected the lower molecular weight components (track B), while YTH71.3.2 bound only to the 90–100kD and 50–55kD molecules (track C). In contrast, H58 reacted strongly with the 160–200kD component (track D). Affinity chromatography of the CEA extract showed that the CD66 and YPC2/12.1 antibodies identified components in the 100–200kD, 80kD and 50–55 kD molecular weight regions in all cases [data not shown]. YTH71.3.2 therefore binds to the higher molecular weight species even though this binding was not detected by immunoblotting. This suggests that protein folding is important for the detection of some antigenic sites.

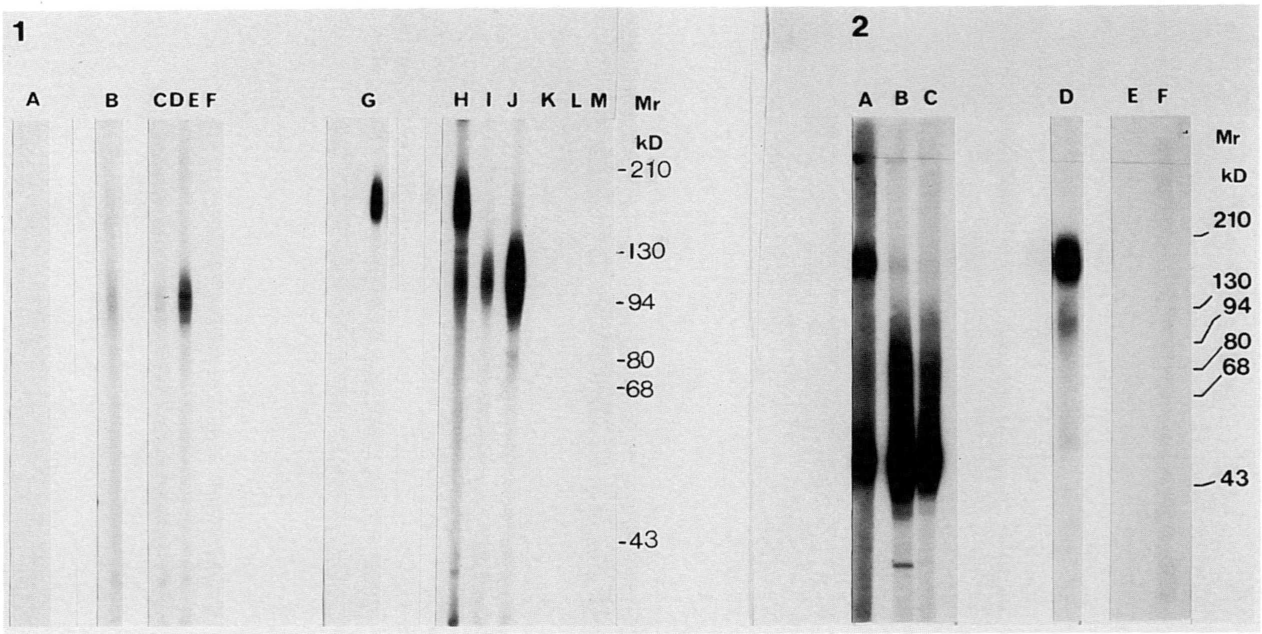

Figure 15.2 Immunoblotting of HL-60, LS174T and a CEA preparation using CD66 and CD66-like Mabs. The HL-60 (gel 1: A-F) and LS174T (gel 1: G-M) cell membrane and CEA (gel 2: A-F) preparations were solubilized, run on 7.5% SDS-polyacylamide gels under reducing (gel 2) or non-reducing (gel 1) conditions, transferred on to nitrocellulose, blotted with the antibodies indicated below and developed with ^{125}I-anti-mouse or rat Ig prior to autoradiography. Bands of similar molecular weight were obtained under reducing and non-reducing conditions. The molecular weight markers were myosin, 210kD; β-galactosidase, 130kD; Phosphorylase b, 94kD; human transferrin, 80kD; bovine serum albumin, 68kD; ovalbumin, 43kD. The antibodies used for immunoblotting are for gel 1: tracks A, G - H58 (anti-CEA); tracks B, H - YPC2/12.1; tracks C, I - YTH71.3.2; tracks D, J -CE6/2D3.1; tracks E, K -rabbit anti-mouse Ig; tracks F, L - sheep anti-rat Ig; and for gel 2: Tracks (A) YPC2/12.1; (B) CE6/2D3.1; (C) YTH71.3.2; (D) H58; (E) rabbit anti-mouse Ig; (F) sheep anti-rat Ig. (After Watt *et al.*[13])

(ii) Expression Cloning and Analysis of Transfectants

Using the cloning procedure of Aruffo and Seed,[21] our studies indicated that the CD66 and YPC2/12.1 antibodies all identified the BGPc molecule. Full sequence analysis is presented elsewhere.[6]

HeLa or CHO cells stably expressing CEA (HeLa-CEA), NCA (HeLa-NCA), NCA-95 (HeLa-CGM6) and BGPc (CHO-BGPc) molecules were labelled with YPC2/12.1, YTH71.3.2 and CE6/2D3.1 and analysed on the FACScan. The relative mean fluorescence values for this binding are shown in Figs 15.3 and 15.4. CE6/2D3.1 and YTH71.3.2 recognize the HeLa-NCA and CHO-BGPc transfectants strongly and HeLa-CEA weakly. CE6/2D3.1 binds CEA more avidly than does YTH71.3.2 confirming the immunoblotting experiments. YPC2/12.1 identifies determinants present on all four CEA, NCA, NCA-95 and BGPc species expressed by the transfectants. No binding to the HeLa-Neo, CHO and CHO-CD34 negative controls was observed. Additional studies[15] indicate that the CD66 MoAbs also recognize the transmembrane CEA-related CGM1 molecules.

(d) Adhesion Properties of BGPc

Previous studies have indicated that BGPa has cadherin-like properties, mediating homotypic adhesion in a Ca^{++} and temperature manner.[27] BGPc is a splice variant of BGPa, containing the same ECD structure but with the cytoplasmic tail reduced in size.[6] Our preliminary studies indicate that CHO-BGPc cells will aggregate in the presence of Ca^{++} at 37°C (Fig. 15.5). This aggregation was not inhibited with the YTH71.3.2 or YPC2/12.1 antibodies. Cos cells transiently expressing BGPc or VCAM-1 (as a possible control) were analysed for their adhesion to a variety of cell lines representing T (Molt 4), B (Ramos), myeloid (HL-60, U937), erythroid (K562) and colonic epithelial (HT29) cells. All cell lines failed to adhere except HT29, HL-60 and possibly U937 cells (Fig. 15.6).

In order to show which domain of BGPc might mediate homotypic adhesion to CHO-BGPc transfectants, soluble constructs carrying either the N-terminal domain only or the N-terminal domain plus two or three IgC2-set domains were analysed for binding to CHO-BGPc transfectants. Figure 15.7 shows that at 37°C in the presence of cations, all constructs adhered to CHO-BCPc transfectants but not to CHO-CD34 cells.

The CD33-Fc protein was used as a negative internal control for specificity of the CHO-BGPc adhesion. The transfectants did not bind to this molecule nor to BSA. Adhesion to fibronectin [FN] of all the transfectants was however observed and this was particularly marked at 37°C in the presence of cations (Fig. 15.7). Thus the N-terminal IgV like domain is necessary for homotypic adhesion.

Stocks *et al.*[28] have shown that the NCA-160 molecule expresses CD15 defined determinants, which have adhesive properties. Initial studies show

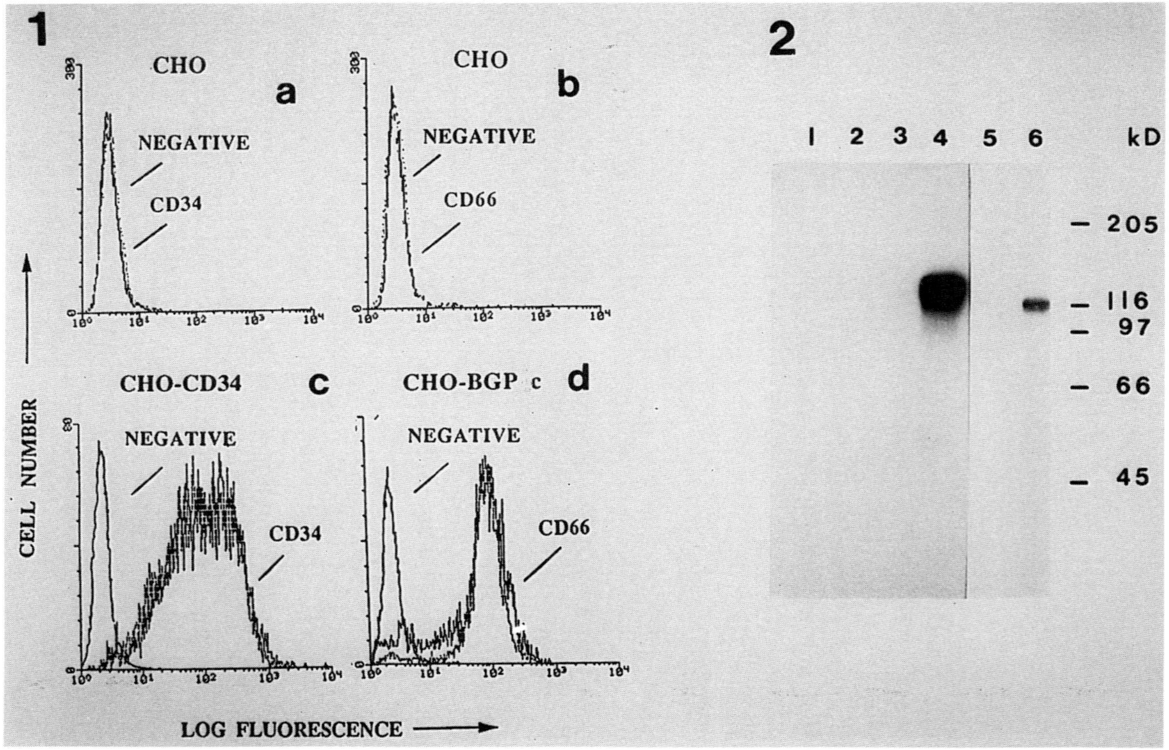

Figure 15.3 Production and Analysis of CHO Transfectants. (1) CHO cells were transfected with BGPc (CHO-BGPc) and cDNA representing the CD34 defined antigen (CHO-CD34) prior to three or four cycles of flow sorting with the appropriate MoAbs (see Materials and Methods). Overlapping fluorescence histograms revealed that detachment of the adherent cells with 2mM EDTA in PBS or by mild·trypsinization did not affect the positive staining (1 c and d). No staining of non-transfected CHO cells was observed with either CD34 (1 a) or CD66 (1 b) MoAbs. Negative controls represented cells stained with the second layer reagent only. (2) The CHO-BGPc (tracks 3 and 4) and CHO-CD34 (tracks 5 and 6) transfectants and the CHO cells (tracks 1 and 2) were lysed, analysed by SDS-PAGE and the proteins transferred to nitocellulose and immunoblotted using the ECL system with CD66 (YTH71.3.2; tracks 1 and 4), with CD34 (ICH3; tracks 2 and 6) or without the first layer antibody (tracks 3 and 5). Molecular weight markers were myosin, 250kD; β-galactoside, 116kD; phosphorylase b, 97kD; bovine serum albumin, 66kD ovalbumin, 45kD; carbonic anhydrase, 29kD.

A

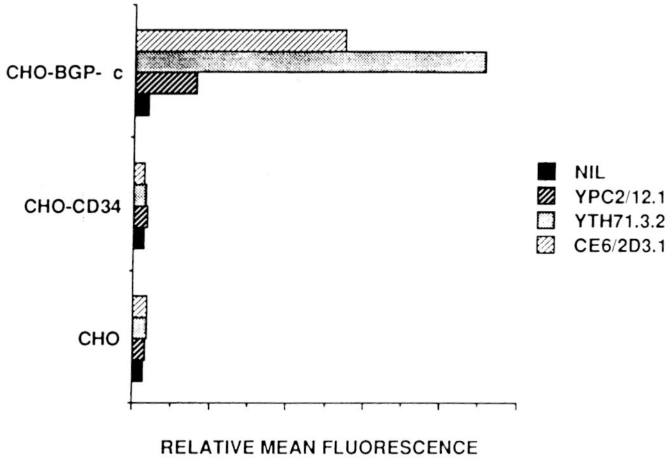

B

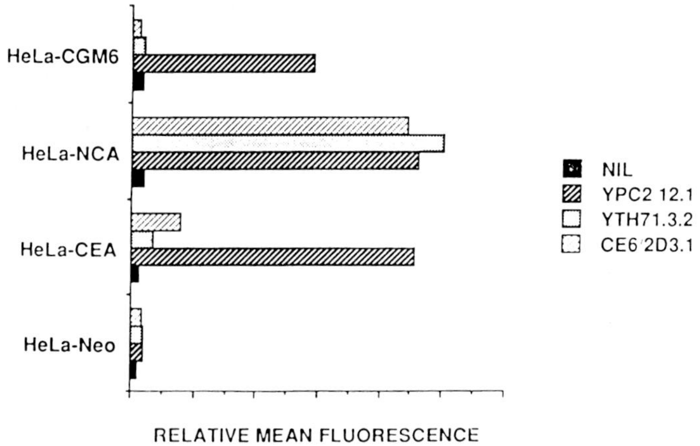

Figure 15.4 CD66 and CD66-like Antibody binding to stable HeLa and CHO Transfectants. CHO cells were transfected with the BGPc cDNA (CHO-BGPc) and HeLa cells with cDNAs encoding CEA (HeLa-CEA), NCA (HeLa-NCA) and NCA-95 (HeLa-CGM6). Untreated CHO cells, CHO transfected with cDNA of the CD34 defined antigen (CHO-CD34) and HeLa cells transfected with pSV2neo plasmid (HeLa-Neo) were used as negative controls. Cells were stained without the first Mab (Nil) or with YTH71.3.2, CE6/2D3.1 and YPC2/12.1 followed by the fluorescent second antibody as described in Materials and Methods. The mean fluorescence is plotted for each analysis.

that CHO-BGPc and the BGP-Fc molecule (Fig. 15.8 a and b) do not bind CD15 MoAbs to the Le^x determinant (Fig. 15.8 c and d) in the presence or

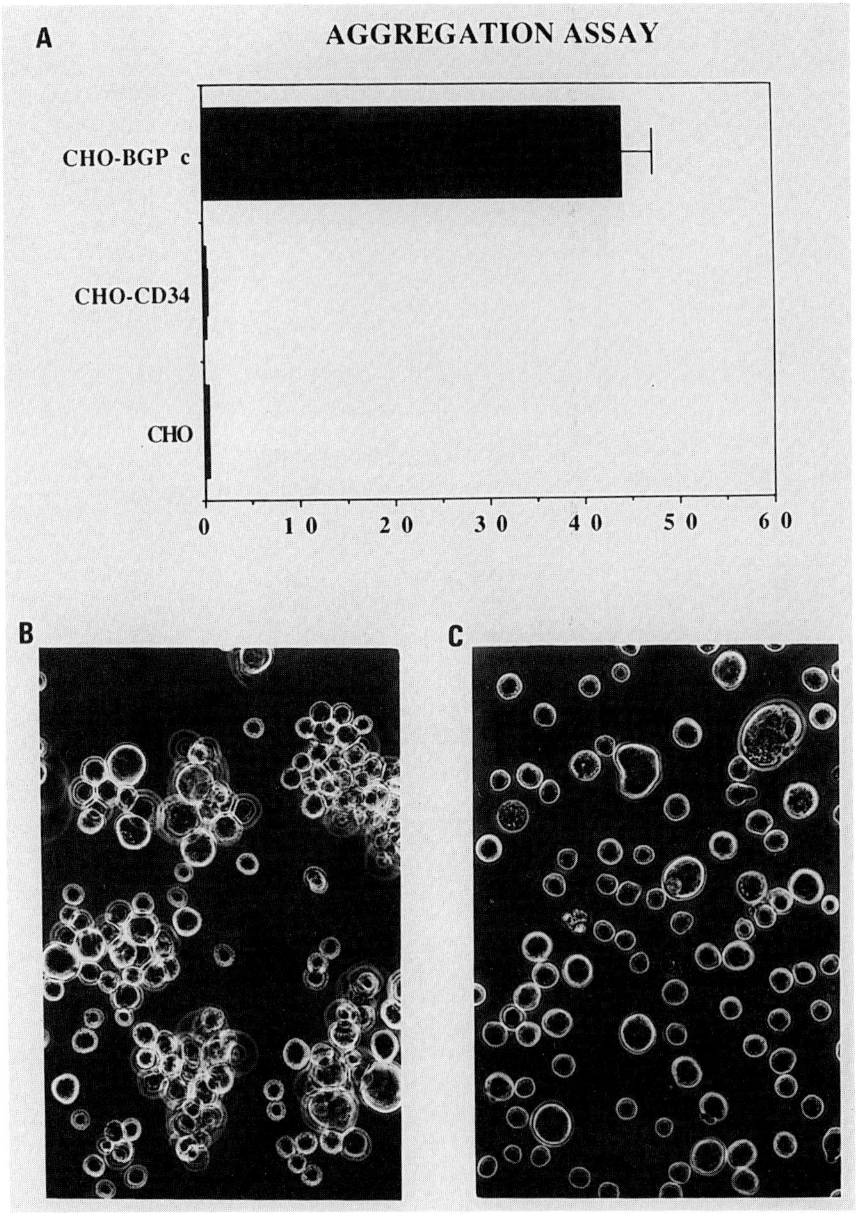

Figure 15.5 Aggregation of CHO-BGPc Transfectants. CHO, CHO-CD34 and CHO-BGPc transfectants (10^6 cells/ml) were incubated for 1 hour at 37° with agitation in DMEM medium containing Ca + + and Mg + + . Single cells remaining were counted and subtracted from the total number of cells used in order to provide an estimate of the number of aggregated cells. A: the percentages of cells forming aggregates were determined from 6 replicate samples. Values are means ± SD; B: CHO-BGPc cell aggregates; C: non-transfected CHO cells do not aggregate under the same conditions.

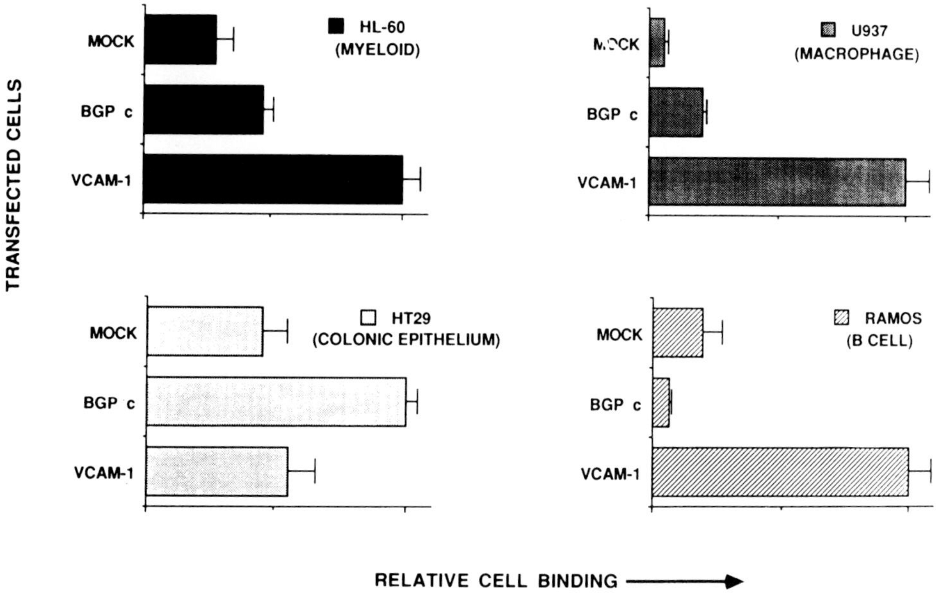

Figure 15.6 Adhesion of Cell Lines to Cos Cell Transfectants The cell lines indicated were labelled with ^{3}H-thymidine and examined for their ability to adhere to Cos cells transfected with cDNAs encoding BGPc and VCAM-1 or without cDNA (Mock transfection). The maximum level of binding for each line is normalized to 100%. Values are the mean ± SD of 4 replicates.

absence of neuraminidase. MoAbs to the sialyl Le^{x} determinant were also negative (data not shown).

The soluble domain variants were also examined for their ability to bind to CD66 MoAbs. Figure 15.9 shows that while the CD66 MoAbs, YTH71.3.2 and CLB/gran 10 bind to all constructs and is therefore recognizing the N-terminal domain, YPC2/12.1 appears to require additional domains for optimal binding. Twenty five MoAbs from the Vth Leukocyte Culture Conference were also analyzed for binding to the soluble domain constructs. The results are shown in Fig. 15.10. Fourteen of the MoAbs reacted with all three soluble constructs and therefore recognized the N-terminal domain of BGPc (Fig. 15.10A). Eleven of the MoAbs required the presence of the IgC2 set domains (Fig. 15.10B). The data do not show whether these MoAbs react with the A1 or B1 domains specifically or require both domains for binding.

DISCUSSION

There are now known to be at least 12 CEA-related genes/proteins

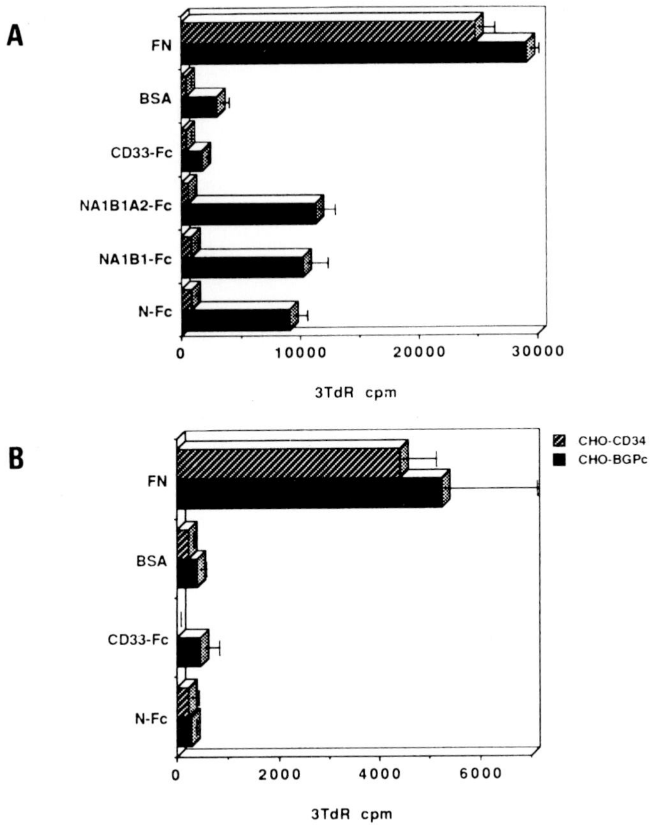

Figure 15.7 Adhesion of Soluble BGPc Domain Variants to CHO-BGPc Transfectants. BGPc domain-Fc [N, N1A1B1, NA1B1B2] and CD33-Fc proteins were attached to 25 well plates via a rabbit anti-human Ig bridge (see Materials and Methods). CHO-BGPc and CHO-CD34 transfectants labelled with ^{3}H-thymidine [^{3}TdR] were allowed to bind to these proteins for 1 hour and the adherent cells solubilised and counted in a β scintillation counter. Wells coated with BSA served as negative controls, and those coated with fibronectin [FN] as positive controls for binding. Binding was carried out at 37°C [A,C] or at 4°C [B]. The data show the mean cpm ± SD of triplicate results in one experiment. Similar results were obtained in 3 independent experiments.

expressed in human granulocytes [reviewed in 1,2,6,12,13,15]. However, the relationship between cloned cDNA sequences and multiple forms of CEA-related proteins identified by immunoblotting and immuoprecipitation is not always clear. This paper shows that the CD66 master antibody, YTH71.3.2, and two others like it, CE6/2D3.1 and YPC2/12.1, react with CEA-related molecules. Both YTH71.3.2 and CE6/2D3.1 recognize the same or closely linked epitopes that are distinct from those identified by YPC2/12.1. Within

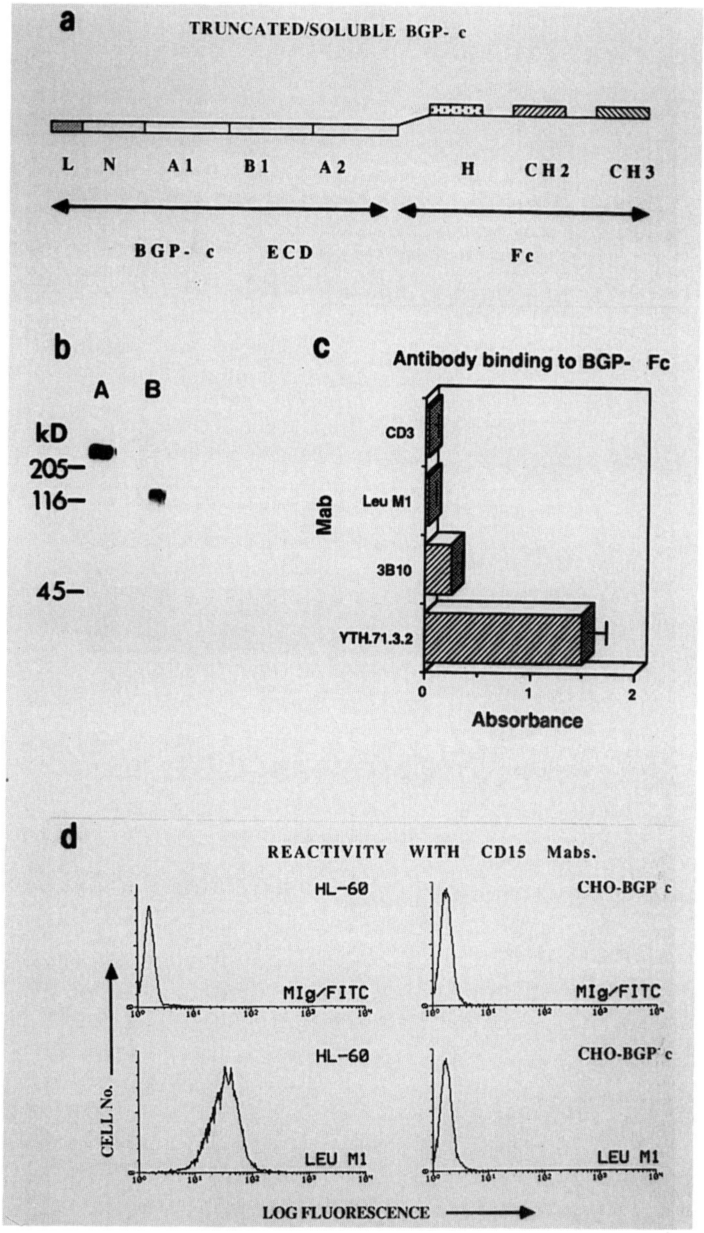

Figure 15.8 Soluble BGP-Fc and the CHO-BGPc Transfectant do not express the Le^x Determinant. (a) Schematic representation of the soluble or truncated form of BGPc. The extracellular domain (ECD) and BGPc containing the leader (L), N-terminal IgV-like domain (N) and three IgC like domains (A1, B1, A2) was linked to an Fc construct of human IgG1 containing hinge (H), CH2 and CH3 domains. (b) Purified BGP-Fc proteins were analysed by SDS-PAGE under non-reducing (A) and reducing (B) conditions and immunoblotted with YTH71.3.2 (CD66) using the ECL system. Molecular weight markers are given in Fig. 3. (c) The BGP-Fc (LeuM1) protein was attached to 96 well plates and analysed for binding of CD66 (YTH71.3.2), CD66-like (3B10) and CD15 (LeuM1) Mabs. CD3 was used as a negative control. The reaction was developed with biotinylated anti-mouse or rat Ig, alkaline phosphatase conjugated-avidin and the p-nitrophenyl phosphate substrate and the absorbance determined at 405 nm. Values represent the mean ± SD of 4 readings. (d) Fluorescence histograms of HL-60 and CHO-BGPc cells labelled with CD15 (LeuM1) and FITC-anti-mouse Ig. Negative controls were labelled with the second layer only.

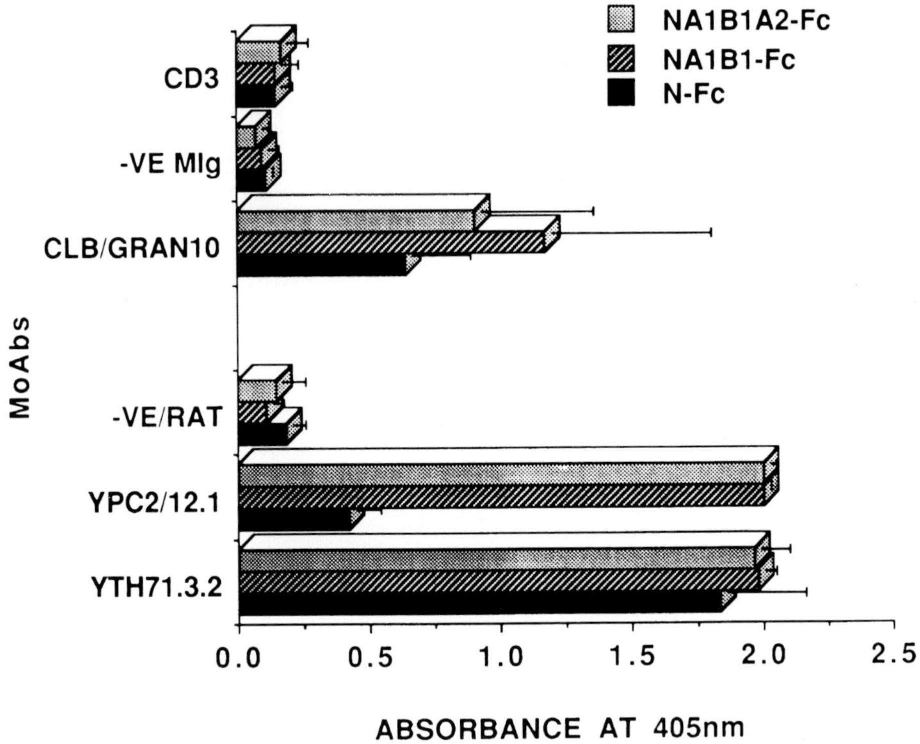

Figure 15.9 CD66 Identifies the N-Terminal Damain of BGPc. The different soluble Fc domain proteins of BGPc: [N], [NA1B1] and [NA1B1A2] were attached to microtiter wells. The MoAbs YTH71.3.2, YPC2/12.1 and CLB/gran 10 were allowed to bind, the wells washed and assayed using an Elisa based alkaline phosphatase reaction described in the Materials and Methods. CD3 or no first layer MoAb (-ve rat/mouse) were used as negative controls. The absorbance at 405 nm indicates the relative level of binding of each MoAb to each soluble protein.

the haemopoietic lineage, all three MoAbs react with cells to the mature neutrophil series from the promyelocytic or late myeloid progenitor cell to the mature neutrophil stages. Binding to monocytes was low or negligible. We have also shown that YPC2/12.1 recognizes at least four CEA-related molecules, CEA, NCA, NCA-95 (CGM6) and BGP, while CD66 identifies NCA, BCP and CEA. Van der Schoot *et al.*[12] have confirmed our findings regarding binding of their CD66 MoAbs to NCA and CEA. They also detect a 130–180kD phosphorylated protein (now known to be a BGP molecule) on mature neutrophils. Our recent findings show that CD66 also identifies the CGM1 family on neutrophils.[15] Since splice variants of the CGM1 molecules possess only an N-terminal IgV-like domain, it is apparent that the CD66

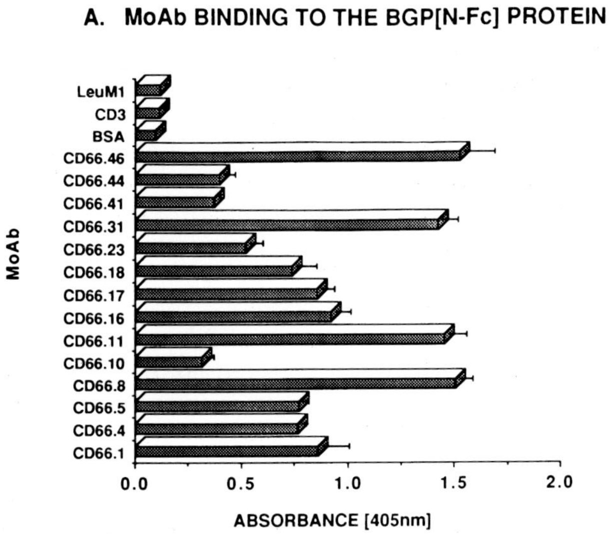

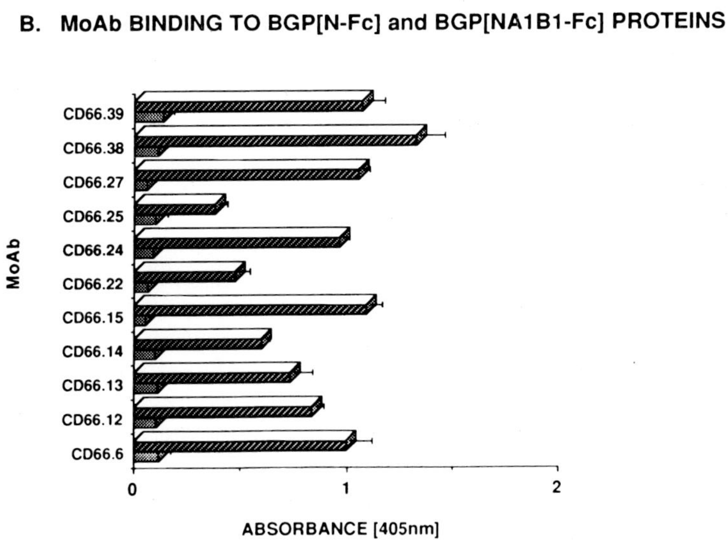

Figure 15.10 Reactivity of CD66 clustered MoAbs with soluble BGPc Domain Proteins. 25 MoAbs submitted to the CD66 workshop of the Vth Leucocyte Culture Conference were analyzed for binding to soluble BGPc domain proteins by an ELISA assay as described in Fig. 15.9. BSA was included in place of the BGPc-Fc proteins as a negative control, while Cd3 and LeuM1 were used as negative control MoAbs. A: Binding of MoAbs to the BGPc [N-Fc] protein; B: Binding of MoAbs to the BGPc [N-Fc] (■) and BGP [NA1B1-Fc] (□) proteins.

MoAbs, YTH71.3.2 and CLB/gran 10, must react with this domain. We have confirmed this by showing positive binding of these MoAbs to a soluble BGP molecule carrying only the N-terminal region. It was of interest that the YPC2/12.1 MoAb, however, appeared to require additional BGPc domains for its molecular recognition. A number of other MoAbs submitted to the Vth Leucocyte Culture Conference also shared these characteristics.

The distribution of CEA-related molecules in neutrophilic cells may provide insight into their functions during myeloid development. Differentiation within the granulocytic compartment involves programmed changes in the adhesion of these cells to stromal cells and extracellular matrix (ECM) molecules in bone marrow.[29] Such haemopoietic-stromal cell/matrix interactions are thought to bring cells of the neutrophil lineage into contact with growth factors or inhibitiors, that are secreted locally, concentrated by heparin binding components of the ECM or are presented as transmembrane components by stromal cells.[29–34] Maturation of neutrophils results in dissociation of these cells from stromal elements and their entry into the peripheral blood and tissues. This latter process and the subsequent activation of neurophils in the periphery involves migration between and adhesion to blood vessel endothelium. Some studies have suggested that increased sialylation of cell surface carbohydrates may result in the release of granulocytes into the circulation and allow their adhesion to endothelium.[35–39]

The CD66 defined molecules described here are all members of the Ig superfamily, are involved in homotypic and/or heterotypic adhesion and all, except CEA itself, are expressed in the granulocyte lineage.[1] Differences in the levels of cell surface expression of NCA and NCA-95[40] have been described during neutrophilic maturation. Although BGP has recently been identified in granulocytes by cDNA cloning[2] and by peptide sequence analysis of the isolated NCA-160 protein, changes in the level of expression of BGP proteins during neutrophilic differentiation have not been clearly defined. This is complicated by the number of BGP splice variants. The BGP family includes, amongst its members, a set of transmembrane proteins that are differentially spliced in the extracellular and/or the cytoplasmic domains. Recently at least 8 transmembrane BGP molecules produced by alternative splicing have been identified.[6,15] These all carry the N-terminal IgV like domain and have none [BGPx, BGPx[1]], two [BGPb, BGPd], or these [BGPa, BGPc] IgC2 set domains on the third IgC domain is replaced with a 31 amino acid non IgC2 set domain [BGPy and BGPz]. These variants are also differentially spliced in their cytoplasmic tails. The expression at the protein level and the role of the different BGP splice variants has not been elucidated either *in vivo* or *in vitro*. However, it may be that the third IgC like domains mediate additional heterotypic adhesion functions[7] such as those observed with inserted sequences in N-CAM splice variants[41,42] and in the FGF receptor.[43] Thus, the existence of different splice variants suggests that these molecules may be specifically expressed at different stages of cell maturation, in different tissues or during neoplastic development, or that they may be involved in heterotypic binding to as yet unidentified ligands. In the latter case, each isoform would function according to differences in their cytoplasmic and extracellular domains. While NCA and NCA-95 molecules appear to mediate adhesion in a Ca^{++} and temperature independent manner,[1] BGPa has been reported to act as a cadherin by mediating homotypic adhesion at 37°C in the presence of Ca^{++}.[27] Splicing of the cytoplasmic tail of BGPa by 53bp results in the production of BGPc, in which the 3′ nucleotide sequence is out of frame with the BGPa sequence. In this article, we have shown that splicing of the cytoplasmic tail of BGPa does not appear to affect the ability of CHO-BGPc transfectants to aggregate. Furthermore, our results show that homotypic adhesion of CHO-BGPc transfectants to immobilized BGPc[N]-Fc, BGPc[NA1B1-Fc, or BGPc[NA1B1A2]-Fc proteins is temperature dependent. More recent experiments suggest that cations promote this adhesion.[25]

These experiments do not explain the function of CEA-related molecules during myeloid development. The role of BGP in heterotypic adhesion is unknown. Some recent studies suggest that the NCA-160 molecule (a BGP molecule) on mature neutrophils has attached carbohydrate moieties identified by CD15 Mabs.[28] The Le^x and/or sialyl Le^x oligosaccharides identified by CD15 Mabs are amongst those molecules implicated in adhesion of neutrophils to selectins (ELAM-I and CD62) on activated endothelium.[36–39] Although sialyl Le^x is expressed by a number of cell types, these cells do not necessarily all share the same adhesiveness or avidity of binding to the selectins.[36] It has therefore recently been suggested[36] that Le^x or sialyl Le^x moieties must be presented by the appropriate cell surface molecule, in the correct orientation or conformation, or must be structurally modified for high affinity adhesion to occur. It is of interest (i) that NCA-160 (or BGP) expressed Le^x determinants,[28,38] (ii) that BGP and the selectins require Ca^{++} for cell adhesion[27,36–39] and (iii) that the CEA family may function as accessory adhesion molecules by controlling the binding activity of type 1 collagen receptors to ECM components.[44] We therefore examined the possibility that the Le^x or sLe^x determinants defined by the CD15 MoAbs, LeuM1 and $CSLe^x$,[45] are attached to BGPc. Our studies show that this oligosaccharide is neither present or accessible on CHO-BGPc transfectants nor on a soluble BGPc molecule secreted from Cos cells. These results may be explained by incomplete glycosylation of BGPc in our system.

It has also been suggested that CEA-related molecules may act as ecto-ATPases[46] with the sinusoidal ecto-ATPase acting as a ligand for P2-purinergic receptors and the canalicular ecto-ATPase forming a retrieval system for adenosine recovery or function in bacterial attachment.[47] Whether the CEA family also functions to enhance granulocyte/stromal cell or ECM interactions in bone marrow or possibly via the additional IgC2 domains or whether changes in glycosylation of these molecules during neutrophil maturation weakens such interactions requires more detailed studies.

Acknowledgements

This work is supported by the Medical Research Council, the Imperial Cancer Research Fund, an EC concerted action programme on human haematopoietic stem cells and an EC Cancer Fellowship.

REFERENCES

1. Thompson, J. A., Grunert, F. and Zimmerman, W. (1991) The carcinoembryonic antigen gene family: Molecular biology and clinical perspectives. *J. Clin. Lab. Analysis*, **5**, 344.
2. Kuroki, M., Arakawa, F., Matsuo, Y., Oikawa, S., Misumi, Y., Nakazato, H. and Matsuoka, Y. (1991) Molecular cloning of nonspecific cross-reacting antigens in human granulocytes. *J. Biol. Chem.*, **266**, 11810.
3. Kolbinger, F., Schwartz, K., Brombacher, E., von Kleist, S. and Grunert, F. (1989) Expression of an NCA cDNA in NIH/3T3 cells yields a 110K glycoprotein, which is anchored into the membrane via glycosyl phosphatidyl inositol. *Biochem. Biophys. Res. Commun.*, **161**, 1126.
4. Hefta, S. A., Hefta, L. J. F., Lee, T. D., Paxton, R. J. and Shively, J. E. (1988) Carcinoembryonic antigen is anchored to membranes by covalent attachment to a glycosylphosphatidylinositol moiety: Identification of the ethanolamine linkage site. *Proc. Natl. Acad. Sci. USA*, **85**, 4648.
5. Barnett, T. R., Kretschmer, A., Austen, D. A., Goebel, S. J., Hart, J. T., Elting, J. J. and Karmarch, M. E. (1989) Carcinoembryonic antigens: Alternative splicing accounts for the multiple mRNAs that code for novel members of the carcinoembryonic antigen family. *J. Cell Biol.*, **108**, 267.
6. Watt, S. M., Fawcett, J., Teixeira, A. M., Gschmeissner, S. E., Murdoch, S. J., Hajibagheri, N. M. A. and Simmons, D. L. (1993) CD66 identifies the biliary glycoprotein (BGP) adhesion molecule: Cloning, expression and adhesion functions of the BGPc splice variant. *Blood*, (Submitted).
7. Barnett, T. R., Drake, L. and Pickle, W. (1993) Human biliary glycoprotien gene: Characterization of a family of novel alternative spliced RNAs and their expressed proteins. *Mol. Cell Biol.*, **13**, 1273.
8. Buchegger, F., Schreyer, M., Carrel, S. and Mach, J.-P. (1984) Monoclonal antibodies identify a CEA crossreacting antigen of 95kD (NCA-95) distinct in antigenicity and tissue distribution from the previously described NCA of 55kD. *Int. J. Cancer*, **33**, 643.
9. Audette, M., Buchegger, F., Schreyer, M. and Mach, J-P. (1987) Monoclonal antibody against carcinoembryonic antigen (CEA) identifies two forms of crossreacting antigens of molecular weight 90,000 and 160,000 in normal granulocytes. *Mol. Immunol.*, **24**, 1177.
10. Kuroki, M., Matsuo, Y., Kuroki, M. and Matsuoka, Y. (1990) Nonspecific cross-reacting antigen (NCA) expressed by human granulocytes: Six Species with different peptide sizes and membrane anchoring forms. *Biochem. Biophys. Res. Commun.*, **166**, 701.

11. Oikawa, S., Inuzuka, C., Kuroki, M., Matsuoka, Y., Kosaki, G. and Nakazato, H. (1989) Pregnancy-specific-β1-glycoprotein, a CEA gene family member, expressed in a human promyelocytic leukemia cell line, HL-60: Structures of protein, mRNA and gene. *Biochem. Biophys. Res. Commun.*, **163**, 1021.
12. Van der Schoot, C.E., Kuijpers, T. W., Nagel, G., Grunert, F., Daams, M. and von dem Borne, A. E. G. Kr. (1991) The granulocytic CD66 and CD67 antigens belong to the carcinoembryonic antigen family. *Blood*, (In Press).
13. Watt, S. M., Sala-Newby, G., Hoang, T., Gilmore, D. J., Grunert, F., Nagel, G., Murdoch, S. J., Tchilian, E., Lennox, E. S. and Waldmann, H. (1991) CD66 identifies a neutrophil-specific epitope within the hematopoietic system that is expressed by members of the carcinoembryonic antigen family of adhesion molecules. *Blood*, **78**, 63.
14. Van der Schoot, C. E., Daams, M., von dem Borne, A. E. G. Kr, Skubitz, K. M., Skubitz, A. P. N., van Agthoven, A., Braully, H., Romagne, F., Lanini, S., Kniep, B., Civin, C. I. Fackler, M. J., Chorvath, B., Duraj, J., Bazil, V., Horejsi, V., Hildreth, J., Hyman, J. and Tetteroo, P. A. T. (1989) Biochemical analysis of the myeloid panel. In White Cell Differentiation Antigens in Leucocyte Typing IV, edited by Knapp, W., Dorken, B., Gelks, W. R., Rieber, E. P., Schmidt, R. E., Stein, H., von dem Borne, A. E. G. Oxford University Press, Oxford, p. 887.
15. Nagel, G., Grunert, F., Kuijpers, T. W., Watt, S. M., Thompson, J. and Zimmermann, W. (1993) Genomic organization, splice variants and expression of CGM1, a CD66-related member of the carcinoembryonic antigen gene family. *Eur. J. Biochem.*, **214**, 27.
16. Richman, P. I. and Bodmer, W. F. (1987) Monoclonal antibodies to human colorectal epithelium: markers for differentiation and tumour characterization. *Int. J. Cancer.*, **15**, 317.
17. Watt, S. M., Karhi, K., Gatter, K., Furley, A. J. W., Katz, F. E., Healy, L. E., Altass, L. J., Bradley, N. J., Sutherland, D. R., Levinsky, R. and Greaves, M. F. (1987) Distribution and epitope analysis of the cell membrane glycoprotein (HPCA-1) associated with human hematopoietic progenitor cells. *Leukemia.*, **1**, 417.
18. Hale, G., Hoang, T. K., Prospero, T., Watt, S. M. and Waldmann, H. (1983) Removal of T cells for bone transplantation. Comparison of rat monoclonal antibodies of different isotypes. *Mol. Biol. Med.*, **1**, 305.
19. Watt, S. M., Thomas, J. A., Edwards, A. J., Murdoch, S. J. and Horton, M. A. (1992) Adhesion receptors are differentially expressed on developing thymocytes and epithelium in human thymus. *Exp. Hematol.*, **20**, 1101.
20. Zhou, Y. Q., Stanley, E. R., Clark, S. C., Hatzfeld, J., Levesque, J. P., Federici, C., Watt, S. M. and Hatzfeld, A. (1988) Interleukin-3 and interleukin-1α allow earlier bone marrow progenitors to respond to human colony-stimulating factor 1. *Blood.*, **72**, 1870.
21. Aruffo, A. and Seed, B. (1987) Molecular cloning of a CD28 cDNA by a high efficiency cos cell expression system. *Proc. Natl. Acad. Sci. USA*, **84**, 8573.
22. Simmons, D. L. and Needham, L. A. (1992) Cloning cell surface molecules using monoclonal antibodies. In Vascular endothelium: Interactions with circulating cells, edited by Gordon S., Elsevier, Holland.
23. Stamenkovic, I., Sgroi, D., Aruffo, A., Sy, M. S. and Anderson, T. (1991) The B lymphocyte adhesion molecule CD22 interacts with leukocyte common antigen CD45RO on T cells and α2–6 sialtransferase, CD75, on B cells. *Cell*, **66**, 1133.

24. Simmons, D. and Seed, B. (1993) Isolation of a cDNA encoding CD33, a differentiation antigen of myeloid progenitor cells. *J. Immunol.*, **141**, 2797.
25. Simmons, D. L., Satterhwaite, A. B., Tenen, D. G. and Seed, B. (1992) Molecular cloning of a cDNA encoding CD34, a sialomucin of human haematopoietic stem cells. *J. Immunol.*, **148**, 267.
26. Teixeira, A. M., Fawcett, J., Simmons, D. L. and Watt, S. M. (1993) The N-domain of the biliary glycoprotein [BGP] adhesion molecule mediates homotypic binding: domain interactions and epitope analysis of BGPc. *Blood*, (submitted).
27. Rojas, M., Fuks, A. and Stanners, C. P. (1990) Biliary glycoprotein, a member of the immunoglobulin supergene family, functions *in vitro* as a Ca^{2+} – dependent intercellular adhesion molecule. *Cell Growth Diff.*, **1**, 527.
28. Stocks, S. C., Albrechtsen, M. and Kerr, M. A. (1990) Expression of the CD15 differentiation antigen (3-fucosyl-N-acety1-lactosamine, Le^x) on putative neutrophil adhesion molecules CR3 and NCA-160. *Biochem. J.*, **268**, 275.
29. Campbell, A. D., Long, M. W. and Wicha, M. S. (1987) Haemonectin, a bone marrow adhesion protein specific for cells of the granulocytic lineage. *Nature*, **329**, 744.
30. Gordon, M. Y., Riley, G. P., Watt, S. M. and Greaves, M. F. (1987) Com-partmentalisation of a human growth factor, GM-CSF, by glycosaminoglycans in the bone marrow microenvironment. *Nature*, **326**, 403.
31. Roberts, R., Gallagher, J., Spooncer, E., Allen, T. D., Bloomfield, F. and Dexter, T. M. (1988) Heparan sulphate bound growth factors: A mechansim for stromal cell mediated haemopoiesis. *Nature*, **332**, 376.
32. Tanaka, Y., Adams, D. H., Hubscher, S., Hirano, H., Siebenlist, U. and Shaw, S. (1993) T-cell adhesion induced by proteoglycan-immobilized cytokine MIP-1 beta. *Nature*, **361**, 15.
33. Nicola, N. A. (1989) Hemopoietic cell growth factors and their receptors. *Annu. Rev. Biochem.*, **58**, 45.
34. Watt, S. M., Williamson, J., Genevier, H., Hatzfeld, A., Fawcett, J., Nesbitt, S. A., Simmons, D. L. and Coombe, D. R. (1993) The heparin binding PECAM-1 adhesion molecule is expressed by CD34 positive haematopoietic precursor cells with early myeloid and B-lymphoid cell phenotypes. *Blood*, (In Press).
35. Baker, M. A., Kanani, A., Brockhausen, I., Schachter, H., Hindenburg, A. and Taub, R. N. (1987) Presence of cytidine 5'monophospho-N-acetylneuraminic acid: Galb1–3Ga1NAc-Ra(2–3)-sialyltransferase in normal human leukocytes and increased activity of the enzyme in granulocytes from chronic myelogenous leukemia patients. *Cancer Res.*, **47**, 2763.
36. Zhou, Q., Moore, K. L., Smith, D. F., Varki, A., McEver, R. P. and Cummings, R. D. (1991) The selectin GMP-140 binds to sialylated, fucosylated lactosaminoglycans on both myeloid and nonmyeloid cells. *J. Cell Biol.*, **115**, 557.
37. Larsen, E., Palabrica, T., Sajer, S., Gilbert, G. E., Wagner, D. D., Furie, B. C. and Furie, B. (1990) PADGEM-dependent adhesion of platelets to monocytes and neutrophils is mediated by a lineage-specific carbohydrate, LNF III (CD15). *Cell*, **63**, 467.
38. Kuijpers, T. W., Hoogerwerg, M., van der Laan, L. J. W., Nagel, G., van der Schoot, C. E., Grunert, F. and Roos, D. (1992) CD66 non-specific cross-reacting antigens are involved in neutrophil adherence to cytokine-activated endothelial cells. *J. Cell Biol.*, **118**, 457.

39. Bevilacqua, M. P., Stengelin, S., Gimbrone, M. A. and Jr, Seed, B. (1989) Endothelial leukocyte adhesion molecule 1: An inducible receptor for neutrophils related to complement regulatory proteins and lectins. *Science*, **243**, 1160.
40. Novak-Hofer, I., Blauenstein, P. and Schubiger, P. A. (1990) Regulation of the cell surface expression of a nonspecific cross-reacting antigen variant during differentiation of HL-60 cells. *Can. Res.*, **50**, 7437.
41. Santoni, M., Barthels, D., Vopper, G., Boned, A., Goridis, C. and Wille, W. (1989) Differential exon usage involving an unusual splicing mechanism generates at least eight types of N-CAM cDNA in mouse brain. *EMBO J.*, **8**, 385.
42. Small, S., Haines, S. and Akeson, R. (1988) Polypeptide variation in an N-CAM extracellular immunoglobulin-like fold is developmentally regulated through alternative splicing. *Neuron*, **1**, 1007.
43. Miki, T., Bottaro, D., Flemmin, C., Smith, C., Burgess, A., Chan, A. and Aaronson, S. (1992) Determination of ligand-binding specificity by alternative splicing: two distinct growth factor receptors encoded by a single gene. *Proc. Natl. Acad. Sci. USA*, **89**, 246.
44. Pignatelli, M., Durbin, H. and Bodmer, W. F. (1990) Carcinoembryonic antigen functions as an accessory adhesion molecule mediating colon epithelial cell-collagen interactions. *Proc. Natl. Acad. Sci. USA*, **87**, 1514.
45. Skacel, P. O., Edwards, A. J., Harrison, C. T. and Watkins, W. M. (1991) Enzymic control of the expression of the X determinant (CD15) in human myeloid cells during maturation: the regulatory role of 6′-sialyltransferase. *Blood*, **78**, 1452.
46. Lin, S.-W. and Guidotti, G. (1989) Cloning and expression of a cDNA coding for a rat liver membrane ecto-ATPase. *J. Biol. Chem.*, **264**, 14408.
47. Leusch, H.-G., Drzeniek, Z., Hefta, S. A., Markos-Pusztai, Z. and Wagener, C. (1991) The putative role of members of the CEA-gene family (CEA, NCA and BGP) as ligands for the bacterial colonization of different human epithelial tissues. *Zbl. Bakt.*, **275**, 118.

16. EXPRESSION OF ADHESION MOLECULES BY GUT EPITHELIAL CELLS – POSSIBLE IMPLICATIONS FOR MUCOSAL IMMUNITY

D. KVALE[1,2] and P. BRANDTZAEG[1]

[1]*Laboratory for Immunohistochemistry and Immunopathology, Institute of Pathology, and* [2]*Medical Dept. A, Rikshospitalet, The National Hospital, University of Oslo, Oslo, Norway*

INTRODUCTION

The mucosal immune system is particularly well developed in the gut which has been estimated to contain at least 80% of all Ig-producing cells of the body (Brandtzaeg *et al.* 1992). Intestinal immune responses must be tightly regulated as some of the enormous load of microbiological and dietary antigens within the gut lumen is constantly penetrating the mucosa.

We have been interested in the immunological properties of intestinal epithelial cells (IEC). Interaction between IEC and the numerous intraepithelial T cells and other leucocytes may be an important event in the regulation of mucosal immune responses. However, the presence of adhesion molecules on IEC has been questioned because of negative observations in chronic inflammatory states *in vivo* (see below). Whether IEC under certain conditions might express adhesion molelcules is of great interest as induction of T cells and mediation of many of their effector functions apparently require interaction with such surface membrane proteins (Springer, 1990). We have therefore studied the presence of adhesion molecules on differentiated colonic epithelial cell lines *in vitro* (Kvale *et al.* 1992) and discuss here our results in relation to present knowledge of immunological functions of IEC and intraepithelial lymphocytes (IEL).

Immunological properties of the intestinal epithelial cell

The major mucosal Ig isotype produced in the gut is J chain-containing polymeric IgA (pIgA). Along with pIgM it is selectively and actively translocated through IEC after binding to secretory component (SC), a pIg receptor glycoprotein expressed basolaterally on the IEC. The ligands for SC are extruded into the gut lumen as secretory IgA and secretory IgM in complex with the cleaved pIg receptor (Brandtzaeg, 1985). We have previously shown that SC expression can be upregulated *in vitro* by

* Correspondence. Dr. D. Kvale, Medical Dept. A, Rikshopitalet, N–0027 Oslo 1, Norway. Phone: 47–22868280 Fax: 47–22868281

interferon-γ (IFN) and tumour necrosis factor-α (TNF) (Sollid *et al.* 1987; Kvale *et al.* 1988a). Moreover, Phillips *et al.* (1990) recently showed that also IL-4 has a similar effect. Upregulation of SC on IEC takes place in various mucosal inflammatory conditions (Brandtzaeg *et al.* 1992).

Another important immunological property of IEC is expression of HLA molecules; epithelial HLA class I products are constitutively present throughout the gut (Brandtzaeg *et al.* 1992) and appear to be regulated by cytokines in a manner similar to SC *in vitro* (Kvale *et al.* 1988b). Normal small intestinal villous epithelium furthermore expresses HLA class II molecules constitutively; in all parts of the gut such expression by IEC is increased during inflammatory processes (Brandtzaeg *et al.* 1992). We have previously shown that epithelial HLA clas II expression requires the presence of IFN *in vitro* and can be additionally increased after co-stimulation with TNF (Kvale *et al.* 1987b). Thus, IEC might theoretically take up and present luminal peptides to T cells in a class II-restricted fashion (Mayrhofer *et al.* 1990); *in vitro* data suggest that IEC are able to stimulate both CD8$^+$ suppressor T cells and CD4$^+$ helper T cells (Bland and Warren, 1986a,b; Mayer and Shlien, 1987; Kaiserlian *et al.* 1989). Furthermore, IEC express nonpolymorphic HLA class I-like CD1d and CD1a molecules in normal and inflammatory mucosa, respectively. These CD1 markers may apparently be involved in the recognition of altered IEC by subpopulations of intraepithelial T cells (Blumberg and Belk, 1993).

Intraepithelial Leucocytes And Epithelial Expression Of Adhesion Molecules *In Vivo*

A striking feature of the gut epithelium is the presence of IEL even in the normal state. Although many types of leucocytes may be found within the intestinal epithelium of humans, the major fraction consists of lymphocytes. Most IEL are T cells bearing the α/β T cell receptor (TCRα/β), CD8$^+$, and CD45RB/R0; this phenotype suggests that they are memory T cells with cytotoxic and/or suppressor function. A minor fraction (2–10%) of human IEL express the TCRγ/δ employing a remarkable predominance of $V_\delta 1/J_\delta 1$ gene product compared with peripheral blood (Halstensen *et al.* 1989; 1990). However, the functions of IEL remain obscure, although there is some experimental evidence suggesting that interactions between CD8$^+$ T cells and IEC may mediate suppression of mucosal hypersensitivity reactions (*i.e.* induction of "oral tolerance") (Mowat, 1987; Bland and Whiting, 1990; Mayer *et al.* 1990; Brandtzage *et al.* 1992).

T cell activation apparently depends on a supportive receptor-mediated stimulus in addition to the TCR-mediated antigen binding (Schwartz, 1990). Notably, adhesion molecules are required for many T cell functions, including antigen-specific recognition by naive cells. The interaction between intercellular adhesion molecule-1 (ICAM-1, CD54) and the integrin lymphocyte function-associated antigen-1 (LFA-1, CD11a/CD18, or $\alpha^L\beta_2$) appears particularly important to this end, whereas interplay between LFA-3/CD2 may be more involved in the amplification of T cell responses (Springer, 1990). It should be noted in this context that human IEL quite frequently express both CD2 (Ebert, 1989) and CD11a (Jarry *et al.* 1990; Cepek *et al.* 1993) adhesion molecule ligands in the normal state. Also, virtually all intraepithelial T cells express the putative immunoregulatory integrin $\alpha^E\beta_7$ recognised by monoclonal antibody (mAb) HML-1; this adhesion molecule seems to be crucial for the affinity of IEL to the epithelium (Cepek *et al.* 1993). In view of the uncertain functions of IEL, adhesion molecules on IEC have therefore recently been in the focus of interest.

ADHESION MOLECULE EXPRESSION BY DIFFERENTIATED GUT CELL LINES *IN VITRO*

Methods

We studied two differentiated human colonic cell lines for the expression of adhesion molecules: HT-29 (subclone m3) with a constitutive high level of SC (Kvale *et al.* 1988c) and Caco-2. The cells were grown in microplates under standard conditions as described in detail elsewhere (Kvale *et al.* 1992).

ICAM-1 and LFA-3 surface membrane expression was measured by cellular enzyme-linked immunosorbent assay (CELISA). Adherent, viable monolayers were labelled as described (Kvale *et al.* 1992) with the following primary mAbs; anti-ICAM-1 (clones RR1/1 and 84H10) (Makgoba *et al.* 1989); anti-LFA-3 (clone TS2/9) (Makgoba *et al.* 1989), or control mAb IgG1. Binding of primary mAb was measured after incubation with rabbit anti-mouse IgG followed by peroxidase-conjugated swine anti-rabbit IgG and development of substrate colour reaction. Total cellular content of ICAM-1 and LFA-3 was similarly quantified by CELISA on cells sequentially fixed with ethanol and periodate-lysine-paraformaldehyde before incubation with primary mAb.

ICAM-1 mRNA was detected and quantified relative to purified total RNA after hybridization of Northern and slot blots with a ^{32}P-labelled DNA antisense oligoprobe comprising the positions 894–928 of the ICAM-1 cDNA (Staunton *et al.* 1988).

Results

Preliminary experiments showed good covariation between surface membrane-associated and total cellular content of ICAM-1 and LFA-3; the CELISA results suggested that the adhesion molecules were mostly expressed on the surface.

Unstimulated viable or fixed HT-29 and Caco-2 cells produced weak CELISA signals for ICAM-1 with optical density (OD) values three to four times the background. Replacement of primary mAbs with control mAb gave OD signals virtually identical to the background. Endogenous peroxidase activity was not detected.

ICAM-1 expression was enhanced after IFN stimulation in both cell lines in a dose-dependent manner (Fig. 16.1). Similar, but less pronounced responses, were obtained after incubation with TNF, IL-1, and to a lesser extent IL-6, the latter in concentrations at 1000 IU/ml (Fig. 16.2). Co-incubation with IFN and TNF (Fig. 16.1), or with IFN and IL-1 or higher doses of IL-6, synergistically increased the ICAM-1 protein levels (Fig. 16.2). A three-fold increase of the ICAM-1 OD signal was seen already after coincubation for 4 h with IFN and TNF; at this time point a strong synergistic effect was present and a near maximal response occurred after 12–24 h (data not shown). Incubation with IL-4 tended to decrease slightly the basal expression of ICAM-1 in both HT-29m3 and Caco-2 cells. Moreover, IL-4 reduced the stimulatory effect of IFN on ICAM-1 (Fig. 16.2), although the extent of inhibition was variable.

LFA-3 was constitutively present in both cell lines with OD values four to six times above the background. Expression of LFA-3 was not significantly changed after stimulation with IFN, TNF, IL-1, IL-4, or IL-6 (Fig. 16.1).

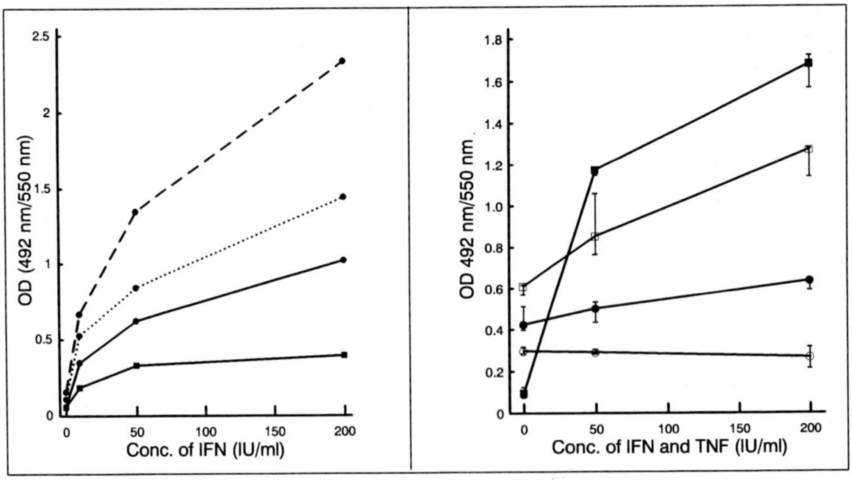

Figure 16.1 Left panel: Cellular ICAM-1 CELISA for HT-29m3 cells after 40 h incubation with IFN (squares) as well as IFN in combination with TNF (circles) at 10 (whole line), 50 (dotted line), and 200 IU/ml (broken line). Each OD value represents the median of duplicate measurements above background (background OD, 0.04). Right Panel: Surface expression of ICAM-1 (squares) and LFA-3 (circles) measured by CELISA in viable, unstimulated HT-29m3 (solid symbols) and Caco-2 cells (open symbols) and after 24 h co-incubation with equal concentrations of IFN and TNF as indicated. Each OD value represents the median of triplicate measurements above background with ranges indicated (background OD, 0.04).

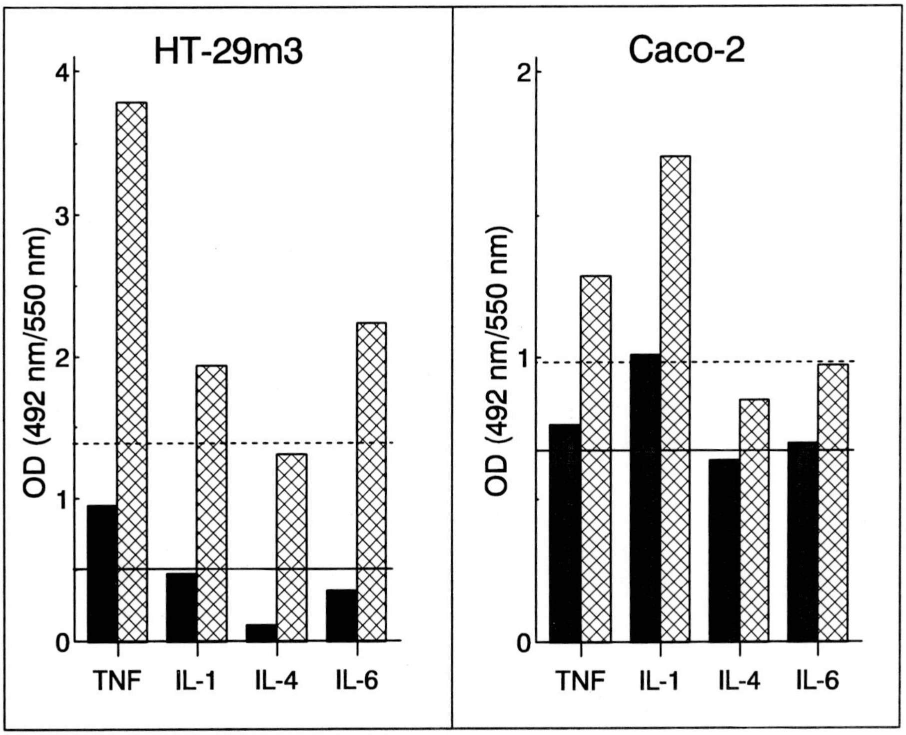

Figure 16.2 Cellular ICAM-1 levels in HT-29m3 (left panel) and Caco-2 cells (right panel) measured by CELISA after 40 h incubation with IFN at 250 IU/ml (broken horizontal line), or with TNF, IL-1, IL-4 and IL-6 at 250 IU/ml in the absence (solid bars) or presence (hatched bars) or IFN (250 IU/ml). ICAM-1 signal in unstimulated cells indicated by horizontal unbroken line. IL-6 enhanced ICAM-1 expression only at higher concentrations than used in this experiment. Each OD value represents the median of duplicate measurements above background.

Northern blots of total RNA from unstimulated and IFN-stimulated HT-29m3 cells hybridized with antisense ICAM-1 probe showed a single, strong band of approximately 3.5 kb and a much weaker band of about 2.5 kb after (Fig. 16.3). A dose-dependent five-fold increase of ICAM-1 mRNA levels was detected after stimulation with IFN for 12 h (Fig. 16.4), whereas the levels of β-actin mRNA remained stable (data not shown).

Discussion

In vivo expression of adhesion molecules can be induced in various human epithelial cells such as keratinocytes and renal tubular cells (Faull and Russ, 1989; Singer, 1990). By contrast, the human intestinal epithelium has usually been found negative (see below). We therefore wanted to examine with

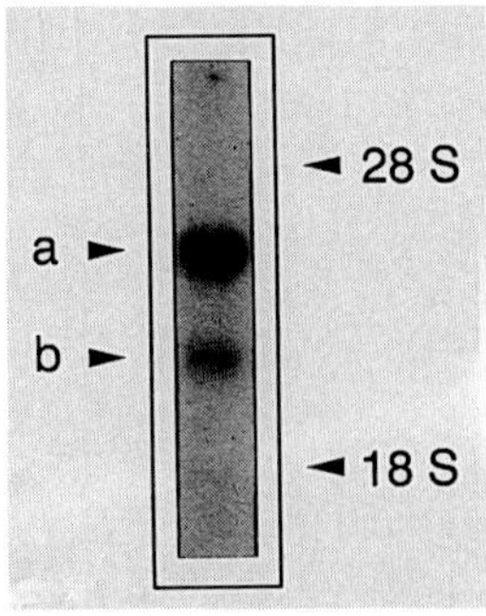

Figure 16.3 Autoradiogram using ICAM-1 antisense probe on Northern blot of total RNA from HT-29m3 from IFN stimulated cells (100 IU/ml, 12 h). The sizes of bands **a** and **b** were about 3.5 kb and 2.5 kb, respectively, estimated relative to the 28S and 18S rRNA bands.

highly sensitive methods whether such expression was detectable on colonic epithelial cell lines stimulated with cytokines *in vitro*. We found constitutive but low levels of LFA-3 and especially ICAM-1 both in Caco-2 and HT-29m3 cells. Moreover, ICAM-1 expression increased after stimulation with IFN, TNF, and IL-1, and with higher concentrations of IL-6. Also, striking synergistic effects were seen already within 4 h after costimulation of both HT-29m3 and Caco-2 cells with IFN, and TNF, IL-1 or IL-6. This rapid cytokine-induced enhancement of ICAM-1 was in keeping with the efffects previously observed for other cell types (Dustin *et al.* 1988; Makgoba *et al.* 1989; Springer, 1990), although ICAM-1 expression by epidermal keratinocytes was apparently not responsive to IL-1 (Makgoba *et al.* 1989; Singer, 1990). Actual presence of ICAM-1 in HT-29m3 cells was further supported by the demonstration of two ICAM-1 mRNA transcripts identical to other cell types (Staunton *et al.* 1988; Simmons *et al.* 1988). Accumulation of specific ICAM-1 mRNA after IFN stimulation has likewise been shown for other cells (Springer, 1990).

The disagreement between our *in vitro* and previous *in vivo* observations with regard to expression of ICAM-1 by IEC might be caused by differences in methodology (sensitivity or interpretation of immunohistochemical staining signals) and nature of test cells. Thus, it remains to be firmly established whether IEC express LFA-3 and ICAM-1 *in situ*. Several immunohistochemical studies have refuted this possibility both in the normal state (Dustin *et al.* 1986) and in various chronic inflammatory diseases (Sturgess *et al.* 1990; Malizia *et al.* 1991; Scheynius and Engstrand, 1991; Smart *et al.* 1991). By contrast, some ICAM-1 positivity were recently observed on colonic crypt cells in acute graft-versus-host disease (Norton *et al.* 1992). Moreover, a preliminary report mentioned that normal unstimulated IEC did express LFA-3 which, however, appeared of no importance for the immunological function in mixed lymphocyte culture assay (Panja *et al.* 1991). Nevertheless,

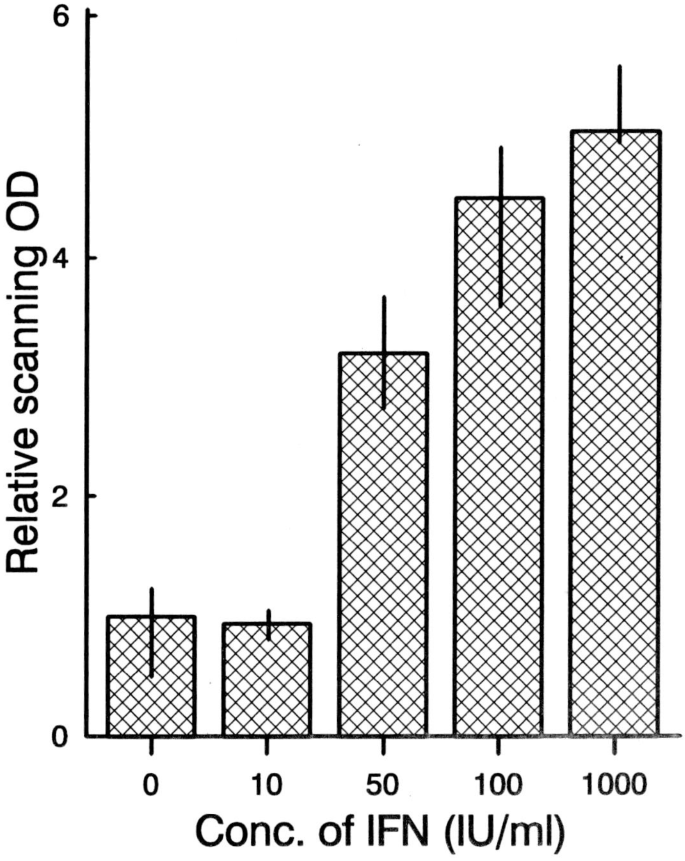

Figure 16.4 ICAM-1 mRNA levels in HT-29m3 cells after stimulation with IFN for 12 h. Data were obtained by densitometric analyses of slot blots hybridized with ICAM-1 probe and represent medians of RNA levels from triplicate cultures with ranges indicated. OD is presented relative to that obtained for untreated cells.

these results did not exclude that adhesion molecules are important if IEC present antigens derived from the gut lumen.

Although our experiments were performed on transformed cell lines, both Caco-2 and HT-29m3 cells share several features with normal IEC. HT-29m3 cells show high constitutive expression of SC (Kvale *et al.* 1988c) like intestinal colonic crypt cells. Also Caco-2 cells are classified as highly differentiated (Chantret *et al.* 1988). Previous reports, moreover, have shown that ICAM-1 expression by transformed cells generally agrees with the behaviour of corresponding normal tissue (Dustin *et al.* 1988). Our observations were recently supported by Kaiserlian *et al.* (1991) who found expression of functional ICAM-1 molecules on stimulated Caco-2 and HT-29 cells, and by Das *et al.* (1993) who found ICAM-1 protein expression in another colonic carcinoma cell line (DLD-1) after IFN-stimulation.

In conculsion, we have provided evidence for low, but constitutive, expression of ICAM-1 and LFA-3 in two intestinal epithelial cell lines and striking upregulation of ICAM-1 protein and mRNA levels by various cytokines except IL-4 . These data are interesting in relation to the putative function of IEL and lamina propria leucocytes juxtaposed to the IEC in health and disease. Perhaps IEC can express ICAM-1 temporarily *in vivo* during the initial phase of mucosal immune activation, but that this apparently crucial adhesion molecule for T cell activation becomes rapidly downregulated by local cytokines, for example IL-4 (Thornhill and Haskard, 1990). This situation would be in keeping with our *in vitro* results and the question as to what extent adhesion molecules are expressed by IEC immediately after immune activation *in situ* should be reevaluated.

Acknowledgements

This work was supported by The Norwegian Cancer Society, The Norwegian Research Council, Torsted's Fund, Anders Jahre's Fund, and Knudsen's Fund. We thank Dr. P. Krajci for valuable assistance with isolation and quantification of RNA.

REFERENCES

1. Bland, P. W. and Warren, L. G. (1986a) Antigen presentation by epithelial cells of the rat small intestine. I. Kinetics, antigen specificity and blocking by anti-Ia antisera. *Immunology,* **58**, 1–7.
2. Bland, P. W. and Warren, L. G. (1986b) Antigen presentation by epithelial cells of the rat small intestine. II. Selective induction of suppressor T cells. *Immunology*, **58**, 9–14.
3. Bland, P. W. and Whiting, C. V. (1990) Intestinal processing of soluble protein antigen, in *Advances in Mucosal Immunology*, edited by MacDonald T. T., Challacombe, S. J., Bland, P. W, Stokes, C. R., Heatley, R. V. and Mowat, A. M., Kluwer Academic Publishers, London, pp. 17–22.
4. Blumberg, R. S. and Belk, S. P. (1993) Role of CD1 in human IEL-epithelial cell interactions. *Mucosal Immunology Update,* **1**, 6.
5. Bradtzaeg, P. (1985) Role of J chain and secretory component in receptor-mediated glandular and hepatic transport of immunoglobulins in man. *Scand. J. Immunol.*, **22**, 111–146.
6. Brandtzaeg, P., Halstensen, T. S., Huitfeldt, H. S., Krajci, P., Kvale, D., Scott, H. and Thrane, P. S. (1992) Epithelial expression of HLA, secretory component (poly-Ig receptor), and adhesion molecules in the human alimentary tract. *Ann. NY Acad. Sci.,* **664**, 157–179.

7. Cepek, K. L., Parker, C. M., Madara, J. L. and Brenner, M. B. (1993) Integrin $\alpha^E\beta_7$ mediates adhesion of T lymphocytes to epithelial cells. *J. Immunol.*, **150**, 3459–3470.
8. Chantret, I., Barbat, A., Dussaulx, E., Brattain, M. G. and Zweibaum, A. (1988) Epithelial polarity, villin expression, an enterocytic differentiation of cultured human colonic carcinoma cell: A survey of twenty cell lines. *Cancer Res.*, **48**, 1936–1942.
9. Das, K. M., Squillante, L. and Robertson, F. M. (1993) Amplified expression of intercellular adhesion molecule-1 (ICAM-1) and M(r) 40K protein by DLD-1 colon tumor cells with interferon-gamma. *Cell. Immunol.*, **147**, 215–221.
10. Dustin, M. L., Roethlein, R., Bhan, A. K., Dinarillo, C. A. and Springer, T. A. (1986) Induction by IL-1 and interferon-γ: Tissue distribution, biochemistry, and function of a natural adherence molecule (ICAM-1). *J. Immunol.*, **137**, 245–254.
11. Dustin, M. L., Staunton, D. E. and Springer, T. A. (1988) Supergene families meet in the immune system. *Immunol. Today*, **9**, 213–215.
12. Ebert, E. C. (1989) Proliferative response of human intraepithelial lymphocytes to various T cell stimuli. *Gastroenterology*, **97**, 1372–1381.
13. Faull, R. J. and Russ, G. R. (1989) Tubular expression of intercellular adhesion molecule-1 during renal allograft rejection. *Transplantation*, **48**, 226–230.
14. Halstensen, T. S., Scott, H. and Brandtzaeg, P. (1989) Intraepithelial T cells of the TcRγ/δ + CD8 – and $V_\delta 1/J_\delta 1$ + phenotypes are increased in coeliac disease. *Scand. J. Immunol.*, **30**, 665–672.
15. Halstensen, T. S., Scott, S. and Brandtzaeg, P. (1990) Human CD8 + intraepithelial T lymphocytes are mainly CD45RA-RB + and show increased co-expression of CD45R0 in celiac disease. *Eur. J. Immunol.*, **20**, 1825–1830.
16. Jarry, A., Cerf-Bensussan, N., Brousse, N., Selz, F. and Guy-Grand, D. (1990) Subsets of CD3 + (T cell receptor α/β or γ/δ) and CD3 – lymphocytes isolated from normal human gut epithelium display phenotypical features different from their counterparts in peripheral blood. *Eur. J. Immunol.*, **20**, 1097–1103.
17. Kaiserlian, D. and Revillard, J. P. (1989) Murine enterocytes can present soluble antigen to specific class II-restricted CD4 + T cells. *Eur. J. Immunol.*, **19**, 1513–1516.
18. Kaiserlian, D., Rigal, D., Abello, J. and Revillard, J. P. (1991) Expression, function regulation of the intercellular adhesion molecule-1 (ICAM-1) on human intestinal epithelial cell lines. *Eur. J. Immunol.*, **21**, 2415–2421.
19. Kvale, D., Løvhaug, D., Sollid, L. M. and Brandtzaeg, P. (1988a) Tumor necrosis factor-α upregulates expression of secretory component, the epithelial receptor for polymeric immunoglobulins. *J. Immunol.*, **140**, 3086–3089.
20. Kvale, D., Brandtzaeg, P. and Løvhaug, D. (1988b) Up-regulation of the expression of secretory component and HLA molecules in a human colonic cell line by tumour necrosis factor-α and γ-interferon. *Scand. J. Immunol.*, **28**, 351–357.
21. Kvale, D., Bartek, J., Sollid, L. M. and Brandtzaeg, P. (1988c) Rapid selection of cultured cells with increased expression of a membrane marker (secretory component). *Int. J. Cancer*, **42**, 638–641.
22. Kvale, D., Krajci, P. and Brandtzaeg, P. (1992) Expression and regulation of adhesion molecules ICAM-1 (CD54) and LFA-3 (CD58) in human intestinal cell lines. *Scand. J. Immunol.*, **35**, 669–676.

23. Makgoba, M. W., Sanders, M. E and Shaw, S. (1989) The CD-2-LFA-3 and LFA-1-ICAM pathways: relevance to T-cell recognition. *Immunol. Today*, **10**, 417–422.
24. Malizia, G., Calabrese, A., Cottone, M., Raimondo, M., Trejdosiewicz, L. K., Smart, C. J., Oliva, L. and Pagliaro, L. (1991) Expression of leukocyte adhesion molecules in inflammatory bowel disease. *Gastroenterology*, **100**, 150–159.
25. Mayer, L. and Shlien, R. (1987) Evidence for function of Ia molecules on gut epithelial cells in man. *J. Exp. Med.*, **166**, 1471–1483.
26. Mayer, L., Siden, E., Becker, S. and Eisenhardt, D. (1990) Antigen handling in the intestine mediated by normal enterocytes, in *Advances in Mucosal Immunology*, edited by MacDonald, T. T., Challacombe, S. J., Bland, P. W., Stokes, C. R., Heatley, R. V. and Mowat, A. M. Kluwer Academic Publishers, London, pp. 23–28.
27. Mayrhofer, G., Spargo, L. D. J., Davidson, G. P. and Huang, Z-H. (1990) Ultrastructural distribution of class II MHC molecules in mucosal antigen presenting cells and their relationship to antigen uptake and processing in *Advances in Mucosal Immunology*, edited by MacDonald, T. T., Challacombe, S. J., Bland, A. W., Stokes, C. R., Heatley, R. V. and Mowat, A. M. Kluwer Academic Publishers, London, pp. 29–33.
28. Mowat, A. M. (1987) The regulation of immune responses to dietary protein antigens. *Immunol. Today*, **8**, 93–98.
29. Norton, J., Sloane, J. P., Al-Saffar, N. and Haskard, D. D. (1992) Adhesion molecules expression in acute graft-versus-host disease. *Clin. Exp. Immunol.*, **87**, 231–236.
30. Panja, A., Salamon, P. and Mayer, L. (1991) Adhesion molecules are expressed on intestinal epithelial cells but do not participate in T cell interactions. Abstract No. 1418, *Digestive Disease Week*, New Orleans, 1991. Programs and Abstracts, p. A-356.
31. Phillips, J. O., Everson, M. P. Moldoveanu, Z., Lue, C. and Mestecky, J. (1990) Synergistic effect of IL-4 and IFN-γ on the expression of polymeric Ig-receptor (secretory component) and IgA binding to human epithelial cells. *J. Immunol.*, **145**, 1740–1744.
32. Scheynius, A. and Engstrand, L. (1991) Gastric epithelial cells in *Helicobacter pylori*-associated gastritis express HLA-DR but not ICAM-1. *Scand. J. Immunol.*, **33**, 237–241.
33. Schwartz, R. H. (1990) A cell culture model for T lymphocyte clonal anergy. *Science*, **248**, 1349–1356.
34. Simmons, D., Makgoba, M. W. and Seed, B. (1988) ICAM, an adhesion ligand of LFA-1, is homologous to the neural cell adhesion molecule NCAM. *Nature*, **331**, 624–627.
35. Singer, K. H. (1990) Interactions between epithelial cells and T lymphocytes: Role of adhesion molecules. *J. Leukocyte Biol.*, **48**, 367–374.
36. Smart, C. J., Calabrese, A., Oakes, D. J., Howdle, P. D. and Trejdosiewicz, L. K. (1991) Expression of LFA-1 β_2 integrin (CD11α/CD18) and ICAM-1 (CD54) in normal and coeliac small bowel disease. *Scand. J. Immunol.*, **34**, 299–306.
37. Sollid, L. M., Kvale, D., Brandtzaeg, P., Markussen, G. and Thorsby, E. (1987) Interferon-γ enhances expression of secretory component, the epithelial receptor for polymeric immunoglobulins. *J. Immunol.*, **138**, 4303–4306.

38. Springer, T. A. (1990) Adhesion receptors of the immune system. *Nature*, **346**, 425–434.
39. Staunton, D. E., Marlin, S. D., Stratowa, C., Dustin, M. L. and Springer, T. A. (1988) Primary structure of ICAM-1 demonstrates interaction between members of the immunoglobulin and intergrin supergene families. *Cell*, **52**, 925–933.
40. Sturgess, R. P., McCartney, J. C. Makgoba, M. W., Hund, C. H., Haskand, D. H. and Ciclitra, P. J. (1990) Diferential upregulation of intercellular adhesion molecule-1 in coeliac disease. *Clin. Exp. Immunol.*, **82**, 489–492.
41. Thornhill, M. H. and Haskard, D. O. (1990) IL-4 regulates endothelial cell activation by IL-1, tumor necrosis factor, or IFN-γ. *J. Immunol.*, **145**, 865–872.

17. CHARACTERISATION OF A MOLECULE INVOLVED IN CELL ACTIVATION AND ADHESION

GILL MARLAND, JANET ALLEN* and LYNDA PARTRIDGE

Krebs Institute for Biomolecular Research, Department of Molecular Biology and Biotechnology, University of Sheffield, P.O. Box 594, Firth Court, Western Bank, Sheffield S10 2UH, UK

**Department of Physiology, University of Cambridge, Downing Street, Cambridge CB2 73G, UK*

INTRODUCTION

Molecules which are involved in cell-cell or cell-matrix interactions are of crucial importance in the immune response. Recently, there has also been a great deal of interest in the role of adhesion molecules in tumour progression and metastasis. We previously described the properties of a monoclonal antibody, JW7, which recognised an apparently novel adhesion molecule which was expressed on leukocytes (Partridge *et al.* 1989; Partridge *et al.* 1990). We have now cloned and sequenced cDNA corresponding to the JW7 antigen. The nucleotide sequence is identical to that encoding the ME491 antigen which has been described as a marker for the early stages of melanoma (Hotta *et al.* 1988). More recently the ME491 antigen has been shown to be identical to CD63, a platelet protein which is expressed at the cell surface following platelet activation (Metzelaar *et al.* 1991). Our findings may have interesting implications for the role of this molecule on normal and malignant cells.

RESULTS AND DISCUSSION

The JW7 monoclonal antibody was raised against adherent human monocytes. The antigen is expressed on the surface and in the cytoplasm of all normal peripheral blood leukocytes and a large number of leukocyte cell lines (Partridge *et al.* 1990). Analysis of frozen tissue sections has shown that antibody strongly stains B cells in the mantle zone of lymphoid tissue and various macrophage populations.

Expression of the JW7 antigen appears to be related to cell activation. Increased staining is observed on T lymphocytes stimulated with phytohaemagglutinin and monocytes stimulated with phorbol ester. Macrophages in reactive lymph nodes and in sections of gut from patients with inflammatory bowel disease are also strongly stained (Partridge *et al.* 1990).

The JW7 antibody has been shown to inhibit the adhesion of peripheral blood monocytes to plastic surfaces (Dransfield, 1987). Intercellular adhesion reactions are also inhibited as demonstrated by its effects on phorbol ester-induced aggregation (Fig. 17.1). JW7 inhibits phorbol ester-induced homotypic adhesion of peripheral blood T lymphocytes, cells of a B cell line, JY, and of U937 cells. The kinetics of cluster inhibition differ from that obtained with antibodies directed against LFA-1 or ICAM-1. These antibodies inhibit aggregation completely, whereas in the presence of JW7, cluster formation is initiated but reverses after 18-24 hours. These results suggest that the JW7 antigen plays a role in the maintenance of intercellular adhesion.

A cDNA library derived from human placenta was constructed in the expression vector πH3M. Clones containing cDNA corresponding to the JW7 antigen were selected using a COS cell positive selection system (Seed *et al.* 1987) and amplified in *E. coli.* The nucleotide sequence of the insert (approximately 0.9 kb) was determined and compared with those in the GenBank and UEMBL data banks. The sequence of the JW7 antigen was found to be identical with that of the ME491 antigen, previously described as a marker for melanoma (Fig. 17.2).

We have demonstrated by flow cytometry that JW7 gives the same reactivity profile as a panel of anti-CD63 antibodies on melanoma cell lines, leukocytes and monocytic cell lines (data not shown). JW7 also stains melanoma cells in frozen tissue sections, while two anti-CD63 antibodies react with a similar antigen expressed by both melanoma and monocytic cell lines, as shown by Western blotting (Fig. 17.3).

Hydrophobicity analysis has indicated that CD63 contains four transmembrane domains (Fig. 17.4). This structure is characteristic of the transmembrane 4 (TM4) superfamily of plasma membrane antigens, of which CD63 is a member. There are currently twelve members of this family, including CD9, CD37 and CD53. They are predominantly associated with leukocytes and tumour cells, although two notable exceptions are Sm23 and Sj23, which are present on the parasitic helminths *Schistosoma mansoni* and *S. japonicum* respectively. As yet no function has been firmly ascribed to any of the TM4 family members, although their structure suggests that they may be involved in signal transduction.

CONCLUSION

We have shown that the JW7 antigen is identical to CD63 and to the ME491 antigen. This is a member of a recently described superfamily of proteins. CD63 has been described as a platelet protein which is expressed at the cell surface platelet activation (Metzelaar *et al.* 1991). The ME491 antigen has been detected on melanoma cells but not on normal melanocytes (Hotta *et al.* 1988). Since its expression decreases as melanoma becomes more

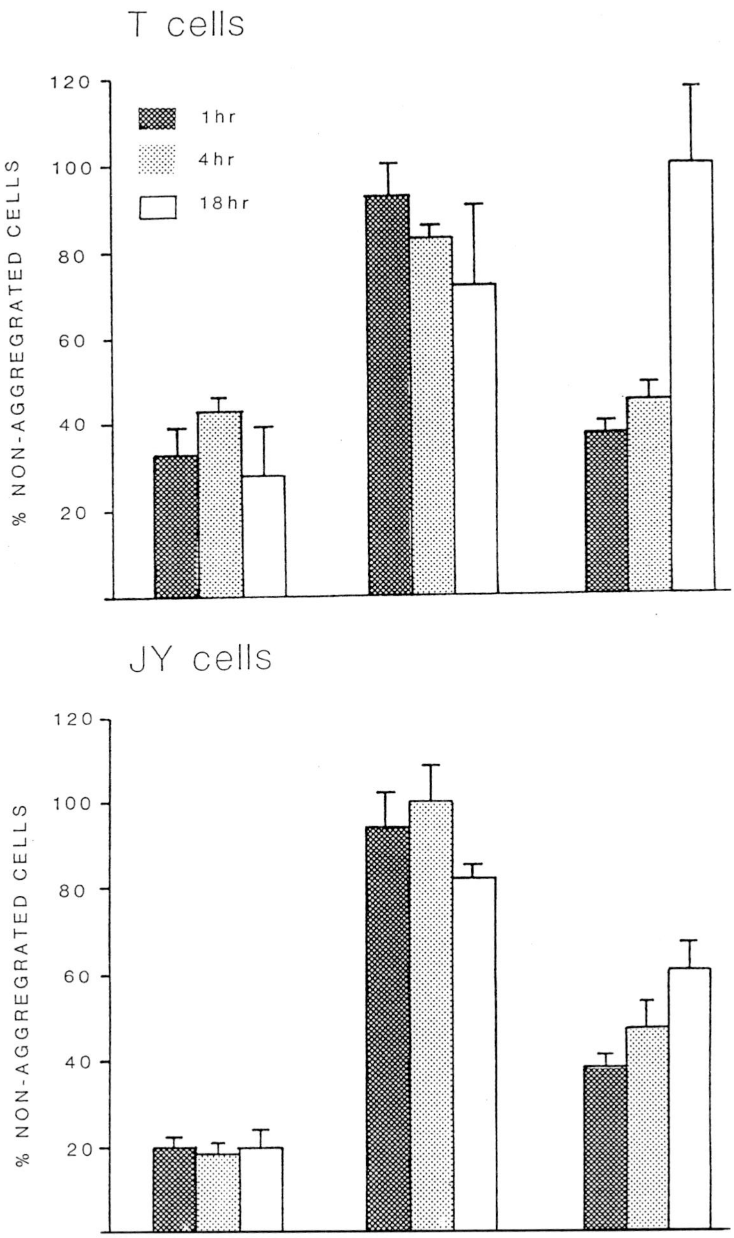

Figure 17.1 The Effect Of JW7 On Leukocyte Adhesion. 1) Tlymphocytes. Non-adherent cells separated from peripheral blood on Ficoll/Hypaque were stimulated with PHA/PdBu and cultured for 2 weeks in the presence of rIL2. The cells were then deprived of IL2 for 3 days and restimulated with 50nM PdBu. Cluster formation was analysed in the presence or absence of antibody. 2) JY Cells. The cells of this B lymphoblastoid line form clusters spontaneously in culture. The clusters were disaggregated by centrifugation and the single cell suspension was then cultured in the presence of 50nM PdBu to induce rapid re-aggregation.

```
GCCTTGGGAAGCCCAGGCCCGGCAGCC ATG GCG GTG GAA GGA GGA ATG
                            MET ALA VAL GLU GLY GLY MET

AAA TGT GTG AAG TTC TTG CTC TAC GTC CTC CTG CTG GCC TTT
LYS CYS VAL LYS PHE LEU LEU TYR VAL LEU LEU LEU ALA PHE

TGC GCC TGT GCA GTG GGA CTG ATT GCC GTG GGT GTC GGG GCA
CYS ALA CYS ALA VAL GLY LEU ILE ALA VAL GLY VAL GLY ALA

CAG CTT GTC CTG AGT CAG ACC ATA ATC CAG GGG GCT ACC CCT
GLN LEU VAL LEU SER GLN THR ILE ILE GLN GLY ALA THR PRO

GGC TCT CTG TTG CCA GTG GTC ATC ATC GCA GTG GGT GTC TTC
GLY SER LEU LEU PRO VAL VAL ILE ILE ALA VAL GLY VAL PHE

CTC TTC CTG GTG GCT TTT GTG GGC TGC TGC GGG GCC TGC AAG
LEU PHE LEU VAL ALA PHE VAL GLY CYS CYS GLY ALA CYS LYS

GAG AAC TAT TGT CTT ATG ATC ACG TTT GCC ATC TTT CTG TCT
GLU ASN TYR CYS LEU MET ILE THR PHE ALA ILE PHE LEU SER

CTT ATC ATG TTG GTG GAG GTG GCC GCA GCC ATT GCT GGC TAT
LEU ILE MET LEU VAL GLU VAL ALA ALA ALA ILE ALA GLY TYR

GTG TTT AGA GAT AAG GTG ATG TCA GAG TTT AAT AAC AAC TTC
VAL PHE ARG ASP LYS VAL MET SER GLU PHE ASN ASN ASN PHE

CGG CAG CAG ATG GAG AAT TAC CCG AAA AAC AAC CAC ACT GCT
ARG GLN GLN MET GLU ASN TYR PRO LYS ASN ASN HIS THR ALA

TCG ATC CTG GAC AGG ATG CAG GCA GAT TTT AAG TGC TGT GGG
SER ILE LEU ASP ARG MET GLN ALA ASP PHE LYS CYS CYS GLY

GCT GCT AAC TAC ACA GAT TGG GAG AAA ATC CCT TCC ATG TCG
ALA ALA ASN TYR THR ASP TRP GLU LYS ILE PRO SER MET SER

AAG AAC CGA GTC CCC GAC TCC TGC TGC ATT AAT GTT ACT GTG
LYS ASN ARG VAL PRO ASP SER CYS CYS ILE ASN VAL THR VAL

GGC TGT GGG ATT AAT TTC AAC GAG AAG GCG ATC CAT AAG GAG
GLY CYS GLY ILE ASN PHE ASN GLU LYS ALA ILE HIS LYS GLU

GGC TGT GTG GAG AAG ATT GGG GGC TGG CTG AGG AAA AAT GTG
GLY CYS VAL GLU LYS ILE GLY GLY TRP LEU ARG LYS ASN VAL

CTG GTG GTA GCT GCA GCA GCC CTT GGA ATT GCT TTT GTC GAG
LEU VAL VAL ALA ALA ALA ALA LEU GLY ILE ALA PHE VAL GLU

GTT TTG GGA ATT GTC TTT GCC TGC TGC CTC GTG AAG AGT ATC
VAL LEU GLY ILE VAL PHE ALA CYS CYS LEU VAL LYS SER ILE

AGA AGT GGC TAC GAG GTG ATG TAG GGGTCTGGTCTCCTCAGCCTCCT
ARG SER GLY TYR GLU VAL MET END

CATCTGGGGGAGTGGAATAGTATCCTCCAGGTTTTTCAATTAAACGGATTATTTT

TTCA...
```

Figure 17.2 Complete Nucleotide and Amino Acid Sequence Of CD63.

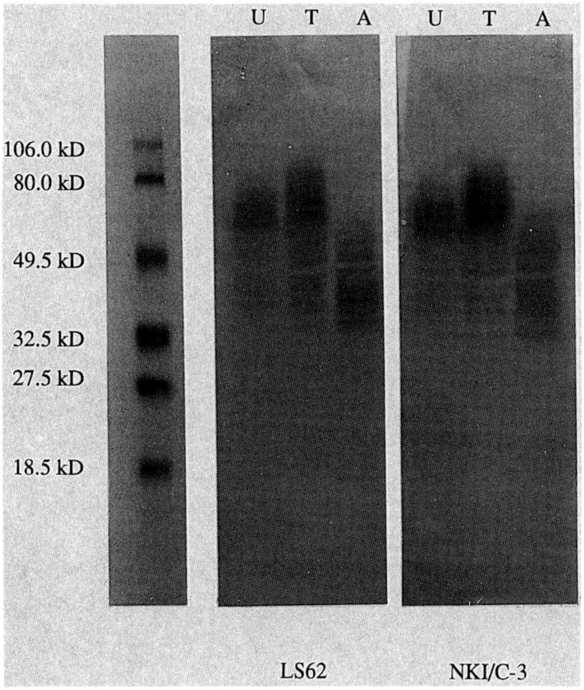

Figure 17.3 Western Blot of CD63 Expressed by The Monocytic Cell Lines U937 (U) and Thp-1 (T), and by A375 (A) Melanoma Cells. Obtained By using antibodies LS62 (Sikora *et al.* 1987) NKI/C-3 (Vennegoor *et al.* 1985).

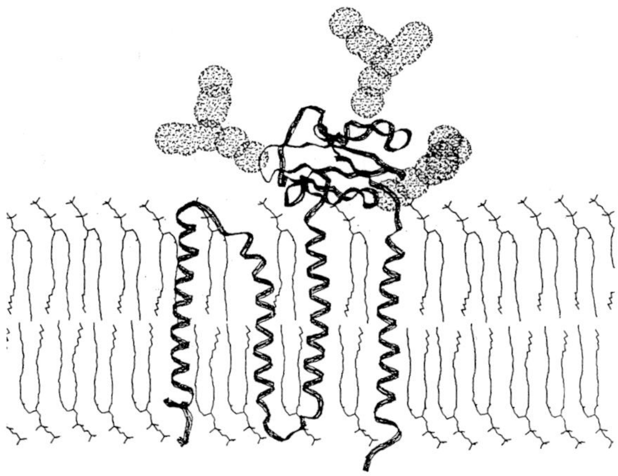

Figure 17.4 Model Of CD63. Based on the hydrophobicity plot of the predicted amino acid sequence. Courtesy of Dr P. Artymiuk.

advanced, it has been suggested that the protein is involved in tumour progression (Hotta *et al.* 1988). Our most recent investigations indicate that CD63 is a signalling molecule involved in regulating cell-cell interactions. Further research on this glycoprotein could give important insights into its role in both the immune system and tumour metastasis.

Acknowledgements

This work was supported by the Yorkshire Cancer Research Campaign.

REFERENCES

1. Dransfield, I. (1987) Molecular studies on human monocytes. Ph.D. Thesis, University of Sheffield.
2. Hotta, H., Ross, A. H., Huebner, K., Isobe, M., Wendeborn, S., Chao, M. V., Ricciardi, R. P., Tsujimoto, Y., Croce, C. M. and Koprowski, H. (1988) Molecular cloning and characterization of an antigen associated with early stages of melanoma tumor progression. *Cancer Res.*, **48**, 2955–2962.
3. Metzelaar, M. J., Wijngaard, P. L. J., Peters, P. J., Sixma, J. J., Nieuwenhius, H. K. and Clevers, H. C. (1991) CD63 antigen. *J. Biol. Chem.*, **266**, 3239–3245.
4. Partridge, L. J., Dransfield, I. and Burton, D. R. (1989) A novel leukocyte adhesion molecule recognised by antibody JW7. *Tissue Antigens*, **33**, 285.
5. Partridge, L. J., Dransfield, I. and Burton, D. R. (1990) Characterisation of a monoclonal antibody recognizing a novel leucocyte adhesion molecule. *Biochem. Soc. Trans.*, **18**, 274–275.
6. Seed, B. and Aruffo, A. (1987) Molecular cloning of the CD2 antigen, the T-cell erythrocyte receptor, by a rapid immunoselection procedure. *Proc. Natl. Acad. Sci. USA*, **84**, 3365–3369.
7. Sikora, L. K. J., Pinto, A., Demetrick, D. J., Dixon, W. T., Urbanski, S. J., Temple, W. and Jerry, L. M. (1987) Characterization of a novel neuroglandular antigen (NGA) expressed on abnormal human melanocytes. *Int. J. Cancer*, **39**, 138–145.
8. Vennegoor, C., Calafat, J., Hageman, P.H., van Buitenen, F., Janssen, H., Kolk, A. and Rümke, P.H. (1985) Biochemical characterization and cellular localization of a formalin-resistant melanoma-associated antigen reacting with monoclonal antibody NKI/C-3. *Cancer Immunol. Immunother.*, **23**, 93–100.

18. STRUCTURAL-FUNCTIONAL CHARACTERIZATION OF A MONOCLONAL ANTIBODY DIRECTED AGAINST P-SELECTIN (CD62/GMP-14/PADGEM): USE IN STUDYING CELL–CELL INTERACTIONS

J. L. MCGREGOR, E. CHIGNIER and L. MCGREGOR

INSERM Unité 331, Institut Pasteur de Lyon , Faculté de Médecine Alexis Carrel, Lyon, France

Cell-cell and cell-matrix interactions play a key role in the immunological defense of the body against infections, prevention of haemorrhage following blood vessel injury, wound healing and growth. Extensive evidence is now available to show that these cellular interactions are largely mediated by families of adhesion molecules such as the integrins, cadherins, leucine-rich glycoproteins and immunolobulins superfamily. These adhesion molecules are also known to play a vital role in tumour growth and metastasis. A new family of adhesion molecules or glycoprotein receptors, known as selectins, has recently been described. One of its members, P-selectin (also known as GMP-140, PADGEM or CD62) is rapidly expressed by platelets or endothelial cells following their activation. P-selectin in is thought to act as the first bridge between activated endothelial cells, or platelets, and leukocytes. This work reviews the structural and functional characterization of a monoclonal antibody (LYP20) directed against P-selectin.

ADHESION MOLECULES AND CELL FUNCTIONS

Increasing evidence is now available to show that cell-cell and cell-matrix interactions is mediated by several families of adhesion molecules or glycoprotein receptors.[1,2,3,4] These families of receptors include the integrins, cadherins, leucine-rich glycoproteins, selectins and immunoglobulins superfamily.[1–4] Additional members to established receptor families, or the identification of new receptor families that have yet to be characterized, are constantly being found. These cellular interactions play a key role in the immunological defense of the body against infections, prevention of haemorrhage following blood vessel injury, wound healing and growth.[5,6] For example, circulating platelets will rapidly adhere (in seconds) to denuded vascular subendothelium, following injury, via their GPIb-IX complex (member of the leucine-rich glycoprotein receptor family) and GPIa-11a (VLA-2), GPIc-IIa (VLA-5), GPIc*-IIa (VLA-6) (member of the integrin receptor family) to respectively von Willebrand factor, collagen, fibronectin and laminin.[7] Platelets adhering to the subendothelium wall will induce circulating platelets to interact with them by secreting aggregating substances.[8] Interaction between platelets will occur via GPIIb-IIIa (member

Correspondence. John L. McGregor Ph.D, INSERM U331, Faculté de Médecine Alexis Carrel, rue Guillaume Paradin, F-69372 Lyon Cedex 08, France. Tel: 33 78013162 Fax: 33 78778612 of the integrin receptor family) and adhesive ligands such as fibrinogen, von Willebrand factor.[9,10] Certain glycoproteins present on the activated platelet surface will favour leukocytes-platelet interactions or act as catalytic sites for the coagulation cascade.[11,12] The thrombus or haemostatic plug, needed to arrest bleeding in injured vessels, is poorly formed or is absent in patients with diseased platelets showing a congenital absence, reduction or disfunction of certain platelet glycoprotein receptors (GPIb-IX, GPIIb-IIIa, GPIa-IIa).[13] Leukocytes will adhere to activated endothelial cells, via their CD11a/CD18, CD11b/CD18, CD11c/CD18 receptors, to extravasate to sites of infection. Absence of the CD11/CD18 complex in leukocytes of some patients will lead to severe life threatening infections.[14] Some of the above described receptors, present for example on platelets or endothelial cells, are implicated in the metastatic process allowing cancer cells to spread from a tumour site to different parts of the human body.[15]

A new family of glycoprotein receptors, known as selectins, has recently been described.[16] The selectin family is currently composed of 3 members (P-selectin, E-selectin, L-selectin). This work reviews structural and functional characterization of a monoclonal antibody (LYP20) directed against P-selectin.

P-SELECTIN, A NEW ADHESION MOLECULE

P-selectin, also known as GMP-140 (Granule Membrane Protein)[17,18] PADGEM (platelet activation-dependent granule-eternal membrane)[19] or CD62, is an integral membrane glycoprotein that is expressed on the activated platelet surface following release of α-granules. Moreover, P-selectin is also rapidly expressed by vascular endothelial cells activated by agonists such as thrombin, histamine and complement protein C5b-9. In endothelial cells, P-selectin is colocalized with von Willebrand factor in Weibel-Palade bodies.[16–19] The isolation and sequencing of the P-selectin cDNA shows this 140 kD cysteine rich glycoprotein to have a lectin domain, an epidermal growth factor (EGF) region, nine repeats related to complement-binding proteins, a transmembrane segment and a short cytoplasmic region.[20] The P-selectin cDNA has strong homologies to 2 other members of the selectin family, E-selectin (also known as endothelial leukocyte adhesion molecule-1 or ELAM-1) and L-selectin (also known as the lymphocyte homing receptor or Leu8/TQ1 in man or gp90MEL-14 in mouse).[16,20] Increasing evidence is available to show that P-selectin mediates the interaction between activated platelets and monocytes or neutrophils. Moreover, at very early stages of endothelial cell activation, by thrombin or histamine, P-selectin together with platelet activating factor (PAF) are rapidly expressed and appear to play a key role in tethering leukocytes.[11,16]

MONOCLONAL ANTIBODY LYP20

Mice were immunized with chymotrypsin-treated human platelets.[21] Hybridomas were obtained by the fusion of spleen cells of immunized mice with sp2/0-Ag14 as previously described.[22] Supernatants of secreting hybridomas were screened against resting or activated formaldehyde-treated platelets, or P-selectin enriched platelet glycoprotein lysates using an enzyme-linked immunosorbent assay (ELISA) system. Selected hybridomas were further characterized by Western blots, immunoprecipitation and platelet aggregation studies. The selected monoclonal antibody (LYP20) was found to be an IgG_1 as determined by an isotyping kit (Amersham International, UK). LYP20 supernatants or ascites was isolated on a protein A-Sepharose column (Pharmacia, Uppsala, Sweden). The number of ^{125}I-labelled LYP20 molecules bound to thrombin-activated platelets (1U/ml) was significantly higher (12,200 ± 1,184 molecules/platelet, kd = 5 ± 0.61 nmol/L) than that observed on resting platelets (2,400 ± 266 molecules/platelet).[23] Moreover, LYP20 was very sensitive in detecting thrombin-activated human platelets by flow cytometry techniques. LYP20 recognized on a Western blot a band migrating at 128 Kd under non-reducing conditions but was unable to bind to its epitope on reduced samples.[23] The band of 128 Kd was not immunoprecipitated by LYP20 if the P-selectin content in samples was depleted by a monoclonal antibody (S12, a generous gift from Dr. R. P. McEver) directed against this antigen. Similarly, if LYP20, but not monoclonal antibodies directed against GPIIb-IIIa or GPIIa, immunodepleted its antigen from the platelet lysate, then P-selectin was not immunoprecipitated by S12.[23] The antigen immunoprecipitated by LYP20 has identical apparent molecular weights, under non-reducing or reducing conditions, as P-selectin. These results show that LYP20 is directed against P-selectin. Moreover, competitive binding experiments showed that LYP20 and S12, are directed against different epitopes on P-selectin.[23]

EFFECT OF LYP20 ON PLATELET-PLATELET AND PLATELET-LEUKOCYTE INTERACTIONS

Monoclonal antibody LYP20 or its F(ab)′2 fragments inhibited platelet-platelet interaction or aggregation of platelets induced by agonists such as collagen (by 60%) or thrombin (by 50%). This inhibitory effect of LYP20 observed on washed platelets or platelets in rich plasma (PRP) had no effect on the binding of fibrinogen to the GPIIb-IIIa complex.[23] The present data indicates that P-selectin may play a certain role in the mechanism leading to irreversible platelet aggregation. LYP20 by rapidly binding to P-selectin molecules, just expressed by activated platelets, appears to interfere with the activity of this adhesion molecule. It is important to note that the work of

W. M. Isenberg, R. P. McEver, Y. V. Jacques and D. F. Bainton (unpublished observations) shows that P-selectin is the only adhesion molecule found, 15 min after thrombin-stimulation, in areas of contact of platelets having irreversibly aggregated. Activation of platelets leads to the activation of the GPIIb-IIIa complex and the binding of fibrinogen allowing platelet-platelet interaction.[10] It remains to be seen at what stage of platelet activation, P-selectin and other surface or α-granule glycoproteins, come in to play in cementing platelet aggregation.

Rosetting of thrombin-activated platelets to U937 cells, a monocytic cell line, was inhibited by LYP20 (by 40%).[23] Monoclonal antibodies directed against GPIIb-IIIa, tPA and CD9 had no effect on platelet rosetting to U937 cells. LYP20 is the first monoclonal antibody to inhibit rosetting of activated platelets to U937 cells or monocytes. However, LYP20 had a less inhibitory effect on rosetting of platelets to monocytes than an anti-P-selectin polyclonal antibody.[24] A polyclonal antibody directed against CD36 (also known as GPIIIb or GPIV) was observed to inhibit the rosetting of activated platelets to U937 cells. The possible role of CD36 and thrombospondin in mediating platelet-monocyte interaction was previously reported by Silverstein *et al.*[25] At this stage, it would appear that platelet-leukocyte interaction may be mediated by P-selectin, CD36 and possibly other antigens. It is of interest to note that monoclonal antibody LYP20 also bound to P-selectin expressed by thrombin-activated cultured human umbilical endothelial cells (HUVEC). Such binding of LYP20 inhibited the adhesion of leukocytes to the activated HUVEC (Murphy J., Bordet J. C., Wyler B., Defrance T., McGregor J. L., unpublished results). The site on P-selectin, bearing the LYP20 epitope, may be of importance in the function of this adhesion molecule in cell-cell interactions. As previously indicated, the LYP20 epitope appears to be disulphide bridge dependent and is not recognized by S12. LYP20, but not S12, was observed to cross-react with rat platelets and intensely bind to thrombin-activated rat platelets (3875 ± 750 molecules/platelet) compared to resting ones (645 ± 240 molecules/platelet).[26] The glycoprotein band immunoprecipitated by LYP20 from labelled rat platelet lysate had the same apparent moleculer weight as that observed for human platelets. Studies are currently performed to locate the epitope of LYP20 on P-selectin using proteolytic degradation, affinity chromatography and gas phase sequencing techniques. It is conceivable that the complement determining regions (CDR) of the heavy and light variable regions of LYP20 may bear protein sequences similar to that expressed by a ligand or a glycoprotein receptor binding to P-selectin. To identify such sequences, the heavy and light variable regions of LYP20 are currently being amplified by PCR using LYP20 hybridoma RNA and appropriate primers. Amplified PCR product will then be sequenced and run on a data bank to identify possible homologies with unknown ligands or receptors binding P-selectin.

Acknowledgements

Supported by Association pour la Recherche contre le Cancer (subvention 6586), Ligue Nationale Française contre le cancer and MGEN.

REFERENCES

1. Pardi, R., Inverardi, L. and Bender, J. R. (1992) Regulatory mechanisms in leukocyte adhesion: flexible receptors for sophisticated travelers. *Immunology Today*, **13(3)**, 100–105.
2. Hynes, R. O. (1992) Integrins: Versatility, modulation, and signaling in cell adhesion. *Cell*, **69**, 11–25.
3. Butcher, E. C. (1991) Leukocyte-endothelial cell recognition: Three (or more) steps to specificity and diversity. *Cell*, **67**, 1033–1036.
4. Springer, T. A. (1990) Adhesion receptors of the immune system. *Nature*, **346**, 425–434.
5. Shimizu, Y., Newman, W., Tanaka, Y. and Shaw, S. (1992) Lymphocyte interactions with endothelial cells. *Immunology Today*, **13(3)**, 106–112.
6. McEver, R. P. (1991) GMP-140: A receptor for neutrophils and monocytes on activated platelets and endothelium. *J. of Cellular Biochem.*, **45**, 156–161.
7. Roth, G. J. (1992) Platelets and blood vessels: the adhesion event. *Immunology Today*, **13(6)**, 224–230.
8. Kinlough-Rathbone, R. L. and Mustard, J. F. (1987) Endogenous mediators of platelet activation, In *Platelets in biology and pathology III*, edited by MacIntyre, D. E. and Gordon, J. L. Elsevier, Amsterdam, pp. 262–268.
9. McGregor, J. L. and Clemetson, K. J. (1988) Biochemistry of the platelet membrane, In *Platelet Immunology*, edited by Kaplan-Gouet, C. and Salmon, C.H. Karger, Basel, pp. 5–31.
10. Marguerie, G. A., Ginsberg, M. H. and Plow, E. F. (1987) The platelet fibrinogen receptor. In *Platelets in biology and pathology III* edited by MacIntyre, D. E. and Gordon, J. L. Elsevier, Amsterdam, pp. 95–126.
11. Zimmerman, G. A., Prescott, S. M. and McIntyre, T. M. (1991) Endothelial cell interactions with granulocytes: tethering and signaling molecules. *Immunology Today*, **13(3)**, 93–100.
12. Bevers, E. M., Rosing, J. and Zwall, R. F. A. (1987) Platelets and coagulation. In *Platelets in biology and pathology III*, edited by MacIntyre, D. E. and Gordon, J. L. Elsevier, Amsterdam, pp. 127–160.
13. Nurden, A. T. (1987) Abnormalities of platelet glycoproteins in inherited disorders of platelet function. In *Platelets in biology and pathology III*, edited by MacIntyre, D. E. and Gordon, J. L. Elsevier, Amsterdam, pp. 37–94.
14. Wardlaw, A. (1990) Leukocyte adhesion to endothelium. *Clin. Exp. Allerg.*, **20**, 619–626.
15. Belloni, P. N. and Tressler, R. J. (1990) Microvascular endothelial cell heterogeneity: Interactions with leukocytes and tumor cells. *Cancer and Metastasis Reviews*, **8**, 353–389.
16. McEver, R. P. (1991) Selectins: Novel receptors that mediate leukocyte adhesion during inflammation. *Thromb. Haemost.*, **65**, 223–228.

17. McEver, R. P. and Martin, N. N. (1984) A monoclonal antibody to a membrane glycoprotien binds only to activated platelets. *J. Biol. Chem.*, **259**, 9799–9804.
18. McEver, R. P., Beckstead, J. H. and Moore, K. L., *et al.* (1989) GMP-140, a platelet alpha-granule membrane protein, is also synthesized by vascular endothelial cells and is localized in Weibel Palade bodies. *J. Clin. Invest.*, **84**, 92–99.
19. Bonfanti, R., Furie, B. C., Furie, B. and Wagner, D. D. (1989) PADGEM (GMP-140) is a component of Weibel-Palade bodies in human endothelial cells. *Blood*, **73**, 1109–1112.
20. Johnston, G. I., Cook, R. G. and McEver, R. P. (1989) Cloning of GMP-140, a granule membrane protein of platelets and endothelium: Sequence similarity to proteins involved in cell adhesion and inflammation. *Cell*, **56**, 1033–1044.
21. McGregor, J. L., McGregor, L. J., Lyons, M., Bauer, A. S. and Huppert, J. (1985) Identification of the fibrinogen binding sites on chymotrypsin-treated human blood platelets using a monoclonal antibody. *Thromb. Haemost.*, **54**, 236a.
22. McGregor, J. L., McGregor, L., Bauer, A. S., Catimel, B., Brochier, J., Dechavanne, M. and Clemetson, K. J. (1986) Identification of two distinct regions within the binding sites for fibrinogen and fibronectin on the IIb-IIIa human platelet membrane glycoprotein complex by monoclonal antibodies P2 and P4. *Eur. J. Biochem.*, **159**, 443–449.
23. Parmentier, S., McGregor, L., Catimal, B., Leung, L. L. K. and McGregor, J. L. (1991) Inhibition of platelet functions by a monoclonal antibody (LY20) directed against a granule membrane glycoprotein (GMP-140/PADGEM). *Blood*, **77(8)**, 1743–1739.
24. Larsen, E., Celi, A., Gilbert, G. E., Furie, B. C., Erban, J. K., Bonfanti, R., Wagner, D. D. and Furie, B. (1989) PADGEM protein: A receptor that mediates the interaction of activated platelets with neutrophils and monocytes. *Cell*, **59**, 305–312.
25. Silverstein, R. L. and Nachman, R. L. (1987) Thrombospondin binds to monocytes-macrophages and mediates platelet-monocyte adhesion. *J. Clin. Invest.*, **79**, 867–874.
26. Winocour, P. D., Chignier, E., Parmentier, S. and McGregor, J. L. (1992) A member of the selectin family (GMP140/PADGEM) is expressed on thrombin-stimulated rat platelets *in vitro*. *Comp. Biochem. Physiol.*, **102A(2)**, 265–271.

19. EXPRESSION OF ADHESION MOLECULES, HLA-DR AND CD68 BY CERVICOVAGINAL EPITHELIAL CELLS IN PATIENTS WITH CHRONIC VAGINAL CANDIDOSIS

G. MARIA MÜLLER and IRENE TAUSCH

Department of Dermatology, Medical School, Humboldt University, Berlin, Germany

The interaction of Candida albicans and host defence cells and the immunomodulating properties which have been reported for mannoproteins of Candida albicans are very important for the local immune response in infected patients. A lot of investigations carried out lately emphasize the assumption that glycosylated surface molecules are of special importance in the adhesion process of yeast cells. The aim of the study is to analyze possible modulation in the expression of these molecules in patients suffering from vulvovaginal candidosis. Therefore, cytospins of cervovaginal secretions taken from 15 patients and from 12 healthy asymptomatic women were investigated. The cell count was checked routinely and the expression of CD 11a, b, c, CD 18, ICAM-1, HLA-DR and CD 68 was established by APAAP technique. The evaluation (number of positive cells, intensity of staining) was made by 500 cells of the equivalent subset. Recurrent infections with Candida albicans cause changes on some cell surface determinants by epithelial cells:

- Increased expression of HLA-DR molecules (number and intensity) depending on the activity of the disease.
- Spotlike development of CD 68 on the normally negative cells.
- CD 11a, b, c, CD 18 and ICAM-1 were not seen on the epithelial cells. The chronic inflammatory process accompanying the infection would have induced these molecules.

KEYWORDS: Adhesion molecules, HLA-DR, CD 68, vaginal candidosis

INTRODUCTION

A precondition for fungal infections is the ability of pathogens to adhere to and penetrate into the epidermal structure. The actual spreading of the disease is dependent on interaction between Candida albicans, epithelial cells and host defence cells. For immunocompetent cells in the skin and mucous membranes adhesion molecules play a very important role in the defence of pathogens.

The cell surface of Candida albicans is composed of a variety of polysaccharides such as glucan, chitin, and mannan. The first two components are primarily elements of the cell structure, while mannan, often covalently linked to protein, constitutes the major antigen of the yeast cells. In addition mannoproteins have enzymatic activity (acid protease) and ligand-receptor function.[1] The immunomodulating properties reported for

Correspondence. Dr. G. Maria Müller, Hautklinik der Medizinischen Fakultät der Humboldt-Universität (Charité), Schumannstr. 20/21, Berlin 10098, Germany.
Tel: 2802 4951 Telefax: 2802 2035

mannoproteins are of special interest in this context.[2,3–5] Eigentler *et al.* and Gustafson *et al.* describe a protein on the surface of yeast-phase Candida albicans that is related to CD 11b/CD 18 with an affinity for C3bi.[6,7]

With these similarities and interconnections in view, we have tried to investigate two problems:

- Are there any qualitative and/or quantitative differences between healthy controls and patients suffering from recurrent candidiasis in expression of adhesion molecules, HLA-DR and CD 68?
- Which conclusions with regard to local defence mechanism can be drawn from these results?

MATERIALS AND METHODS

Subjects

Fifteen females median age 28 (range 19–41) years with recurrent vulvovaginal candidiasis were enrolled in this study. Eleven healthy female volunteers median age 30 (range 26–39) years served as controls. All persons were not receiving any medication.

Cell smears

The cervicovaginal secretions were obtained by washing with a warmed phosphate buffered salt solution (PBS) and kept on ice until centrifugation in a cytocentrifuge (Cytospin 2, SHANDON, Great Britain). The air-dried smears were frozen, wrapped in plastic foil and stored at −20°C until staining was performed.

Staining method

Staining was performed by alkaline phosphatase anti-alkaline phosphatase procedure (APAAP) as described in DAKO-Diagnostika references.[8] In principle, the slides were first incubated with murine monoclonal antibody (mAb), followed by incubation with rabbit-antimouse (bridging) antibody. Subsequently, a murine mAb against alkaline phosphatase preincubated with alkaline phosphatase was applied. In order to enhance the sensitivity of this procedure, the number of phosphatase molecules on the surface was increased by using one or two layers of bridging antibody and antiphosphatase antibody. Bound phosphatase molecules were visualised using naphthol AS-MX phosphate and fast-red TR salt (Sigma Chemie GmbH, Germany). Antigen-positive cells were easily identified, since they showed a vivid red reaction which contrasted clearly with the haematoxylin counterstain. Cellular details were clearly seen in labelled preparations, so that simultaneous morphological identification became possible.

Evaluation

The number of positive cells of 500 subset cells were counted. The staining intensity was subdivided in 0, +, + + and + + +.

Antibodies

Details of the monoclonal antibodies used in the study are given in the table below. The APAAP complex was purchased from DAKO-Diagnostika GmbH (Germany).

Table 19.1 Antibodies used

Antibody	*CD*	*Specificity*	*Source*	
MHM 24	11a	LFA-1, -chain	DAKO	782
2 LPM 19c	11b	C3bi receptor		741
KB 90	11c	Protein 150, 95		732
MHM 783	18	LFA-1, -chain		783
KP1 (9)	68	Macroph./Monoc.		814
CR3/43	—	HLA-DR		775
DK 25	8	T Cytotox./Supp.		707
MT 310	4	T Helper/Inducer		716
UCHT 1	3	Pan T		835
IOB 4	19	Pan B	DIANOVA	0533
IOB	22	Pan B		0779
IOL S4	54	ICAM-1		0544

RESULTS

Comparisons of the material obtained from the patients with that of the healthy controls revealed some interesting differences:

HLA-DR

In all cases of candidiasis the labelling intensity of macrophages/monocytes was stronger compared with the healthy controls. The pattern of labelling was dotlike and surface-membrane-bound. Lymphocytes were rarely found in the secretions investigated and in cases of candidiasis a number of cells showed a moderate marginal staining. The frequent occurence of HLA-DR antigen suggests the presence of a large number of activated cells. Epithelial cells show these antigens weakly and in small areas. The frequency and intensity of expression was dependent on the activity of the disease. After successful treatment the epithelial cells became slowly negative. The pattern of reactivity is illustrated in Figs 19.1 and 19.2.

CD 11a, b, c, CD 18

CD 11a, b, 11c and CD 18 were observed on granulocytes and monocytes and most of the lymphocytes were CD 11b positive and CD 18 positive. The epithelial cells were negative in all cases.

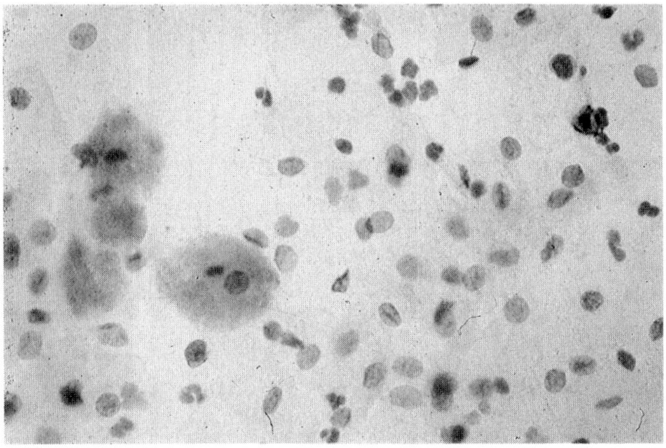

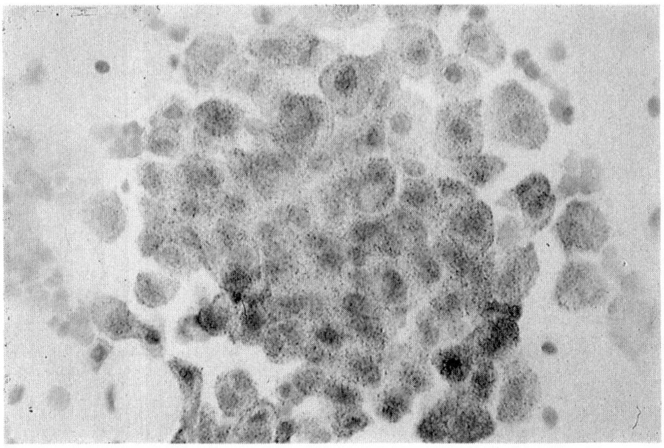

Figures 19.1 and 19.2 Cell smears from cervovaginal secretions in candidosis. Pattern of reactivity with monoclonal antibody against HLA-DR (CR3/43). APAAP technique counterstained with hemalaun. Scale bar 1:683. Fig. 1: before treatment. Fig. 2: after treatment.(See Colour Plate XVI at the back.)

CD 68

Granulocytes and monocytes/macrophages showed a strong, diffuse cytoplasmatic staining and were seen in a great number in patients. Additionally, the epithelial cells of these persons showed spot-like stained areas (Fig. 19.3), whereas the same cells from the healthy persons were constantly negative.

CD 54 (ICAM-1)

All cells in all the secretions investigated, were found to be completely negative.

DISCUSSION

Women with chronic vaginal candidiasis do not develop adhesion molecules (CD 11a, b, c, CD 18, ICAM-1) on the surface of epithelial cells. These molecules are involved in the adhesion of neutrophilic granulocytes and in the presentation of MHC-associated antigen to the antigen specific receptor in T-cells. In human skin and mucous membranes these molecules were physiologically only found in low amounts (Langerhans cells, melanocytes). Although it is known they can be induced *e.g.* during inflammatory processes, in certain diseases (bullous pemphigoid, lupus erythematosus), by certain cytokines (interferon-gamma, tumour necrosis factor, interleukin 6 and 7), by UVB irradiation and are downregulated in other cases (Epstein-Barr virus positive Burkitt's lymphomas, Sézary syndrome). Chronic inflammatory

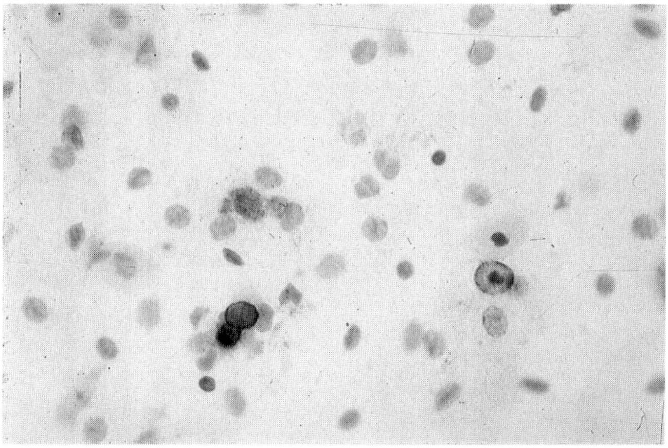

Figure 19.3 Cell smears from cervovaginal secretions in candidosis. Pattern of reactivity with monoclonal antibody against CD 68 (KP1). APAAP technique counterstained with hemalaun. Scale bar 1:683. (See Colour Plate XVII at the back.)

processes accompanying candida infections should be expected to induce these molecules. Downregulation of adhesion molecules seems to be one mechanism of immune modulation which is found to occur only in a defined host environment; another process might be linked with antigen CD 68. KP1 (CD 68) stains macrophages in a wide variety of human tissues and reacts with myeloid precursors and peripheral blood granulocytes.[9] Similarly, mast cells consistently exhibit a strong cytoplasmatic immunoreactivity. Langerhans cells or interdigitating reticulum cells are either negative or show weak or restricted areas of reactivity.[10–12] Of epithelial malignancies tested (47 cases) CD 68 (Ki-M6) was positive only in renal cell carcinoma (11/14).[13] Areas of spotlike staining in epithelial cells are a remarkable phenomenon considering that the same cells from the healthy persons were constantly negative. CD 68 detects a highly glycosylated intracellular protein with a molecular weight of approximately 110 kD.[10] In many investigations carried out lately the assumption is made that glycosylated surface molecules are of particular importance for the adhesion process in yeast cells.[1,3,4,14] Tronchin *et al.* and Macura *et al.* established that preincubation of buccal epithelial cells with Concanavalin A inhibits the adhesion of yeast cells, as does suppression of mannoprotein production by Tunicamycin.[15–17] Methyl-alpha-D-mannoside, L-Fucose, N-acetyl-D-glucosamin, or D-mannose have the same effect and they may also function as epithelial cell receptors.[16,18] Unfortunately, the exact molecular structure and biological function of the CD 68 molecule still remains elusive[19] and a discussion of the CD 68 molecule as a target structure for the yeast cells would be premature. In addition to these effects polysaccharide antigens of Candida albicans have been shown to generate a complex series of interactions resulting in the suppression of T and B lymphocyte response to Candida antigens *in vitro.*[2,20,21]

Finally, it should be said, that we could not find any indications of a defect in the process of antigen processing and/or presentation. Macrophages/monocytes were often found in the secretions of patients and invariably showed a very intense colouring when they were stained with HLA-DR antibodies. Epithelial cells contain these antigens too, but most weakly and in small areas as this is known from other skin diseases. The frequency of occurence and the intensity of staining was dependent on the activity of the disease. Similar findings were described by Jontell *et al.* They noticed class II transplantation antigens (HLA-DR and HLA-DQ) in epithelial cells in cases of oral candidiasis.[22] In this context, it is important to note that human candida-specific T cells require HLA-DR compatible macrophages for their activation.[23,24]

CONCLUSIONS

Susceptibility to recurrent mucocutaneous infections may be due to complex immunoregulatory mechanisms activated by antigens of candida and no one specific defect has been consistently identified as being the major predisposing factor for chronic infection. The downregulation of adhesion molecules appears to be one mechanism for the local immune modulation. Another possibility could be the expression of CD 68 in epithelial cells.

REFERENCES

1. Calderone, R. A. and Braun, P. C. (1991) Adherence and receptor relationships of Candida albicans. *Microbiological Reviews*, **55**, 1–50.
2. Cuff, C. F., Taub, D. D. and Roggers, T. J. (1989) The induction of T suppressor cells with a soluble extract of Candida albicans. *Cellular Immunology*, **122**, 71–82.
3. Domer, J. E., Elkins, K., Ennist, D. and Baker, P. (1988) Modulation of immune response by surface polysaccharides of Candida albicans. *Reviews of Infection Diseases*, **54**, 419–422.
4. Domer, J. E., Garner, R. E. and Befidi-Mengùe, R. N. (1989) Mannan as an antigen in cell-mediated immunity (CMI) assay and as a modulator of mannan-specific (CMI). *Infection and Immunity*, **57**, 693–700.
5. Witkins, S. S., Hirsch, J. and Ledger, W. J. (1986) A macrophage defect in women with recurrent candida vaginitis and its reversal *in vitro* by prostaglandin inhibitors. *American Journal of Obstetrics and Gynecology*, **155**, 790–786.
6. Eigentler, A., Schulz, T. F. and Larcher, C. *et al.* (1989) C3bi-binding protein on Candida albicans: temperature dependent expression and relationship to human complement receptor type 3. *Infections and Immunology*, **57**, 616–622.
7. Gustafson, K. S., Vercellotti, G. M. and Bendel, C. M. *et al.* (1991) Molecular mimicry in Candida albicans. Role of an integrin analogue in adhesion of the yeast to human endothelium. *Journal of Clinical Investigation*, **87**, 1896–1902.
8. Dakopatts (1991) APAAP-Färbemethode für Blut- und Knochen-mark-Ausstkriche sowie Gewebeschnitte. Company information from DAKO-Diagnostika GmbH, Hamburg.
9. Dakopatts (1991) DAKO-Kits für Immunenzymatische Färbemethoden, allgemeine Testbeschreibung. Company information from DAKO-Diagnostika GmbH, Hamburg.
10. Dakopatts (1991) Specification sheet, Code. -No. M 814, LOT- No. 099, DAKO Diagnostika GmbH, Hamburg.
11. Knapp, W., Dörken, B., Gilles, W. R., Rieber, E. P., Schmidt, R. E., Stein, H. and von dem Borne, A. E. G. (1989) White cell differentiation antigens. In *Leukocyte Typing* IV, Appendix A, Oxford/London/Tokyo, Oxford University Press, pp. 1090–1092.
12. Knapp, W., Rieber, P., Dörken, B., Schmidt, R. E., Stein, H. and von dem Borne, A. E. G. (1989) Towards a better definition of human leucocyte surface molecules. *Immunology Today*, **10**, 253–259.
13. Horny, H. P., Schaumburg-Lever, G., Bolz, S., Geerts, M. L. and Kaiserling, E. (1990) Use of monoclonal antibody KP 1 for identifying normal and neoplastic human mast cells. *Journal of Clinical Pathology*, **43**, 719–722.
14. Gloghini, A., Volpe, R., Canzonieri, V. and Carbone, A. (1991) Immunohistochemical characterization of Ki-M6 monoclonal antibody in Bouin-fixed,

paraffin-embedded sections of normal and neoplastic human tissues. *Virchows Archiv für Pathologie, Anatomie und Histopathologie*, **418**, 355–360.
15. Abu-el-Teen, K., Ghannum, M. and Stretton, R. J. (1989) Effects of subinhibitory concentrations of antifungal agents on adherence of Candida spp. to buccal epithelial cells *in vitro*. *Mycoses*, **32**, 551–562.
16. Tronchin, B. Bonchara, J. P. and Robert, R. (1989) Dynamic changes of the cell wall surface of Candida albicans associated with germination and adherence. *European Journal of Cell Biology*, **50**, 285–290.
17. Macura, A. B. and Tondyra, E. (1989) Influence of some carbohydrates and concanavalin A on the adherence of Candida albicans *in vitro* to buccal epithelial cells. *International Journal of Medical Microbiology*, **272**, 196–201.
18. Micklem, K., Rigney, E., Cordell, J., Simmons, D., Stross, P., Turley, H., Seed, B. and Mason, D. (1989) A human macrophage-associated antigen (CD 68) detected by six different monoclonal antibodies. *British Journal of Haematology*, **73**, 6–11.
19. Olsen, I. (1990) Oral adhesion of yeasts. *Acta odontologica Scandinavica*, **48**, 45–53.
20. Pulford, K. A., Sipos, A., Cordell, J. L., Stross, W. P. and Mason, D. Y. (1990) Distribution of the CD 68 macrophage/myeloid associated antigen. *International Immunology*, **2**, 973–980.
21. Ashman, R. B. and Papadimitriou, J. M. (1989) Genetic regulation of pathogenesis and host responses in fungal infections. In *Immunology of Fungal Infections*, edited by Kurstak, E. Marcel Dekker, New York/Basel, pp. 347–371.
22. Durandy, A., Fischer, A., LeDeist, F., Drouhet, E. and Criscelli, C. (1987) Mannan-specific and mannan-induced T cell suppressive activity in patients with chronic mucocutaneous candidiasis. *Journal of Clinical Immunology*, **7**, 400–409.
23. Jontell, M., Scheynius, A., Ohman, S. C. and Magusson, B. (1986) Expression of class II transplantation antigens by epithelial cells in oral candidosis, oral lichen planus and gingivitis. *Journal of Oral Pathology*, **15**, 484–488.
24. Ashman, R. B. and Papadimitriou, J. M. (1990) What's new in the mechanisms of host resistance to Candida albicans infection? *Pathology, Research and Practice*, **186**, 527–534.
25. Jose, Y., Komori, K., Inoye, Nomura, K., Yamamura, M. and Tsuji, K. (1981) Role of macrophages in T lymphocyte response to candida allergen in man with special reference to HLA-D and DR. *Clin. Exp. Immunol.*, **45**, 152–157.

INDEX

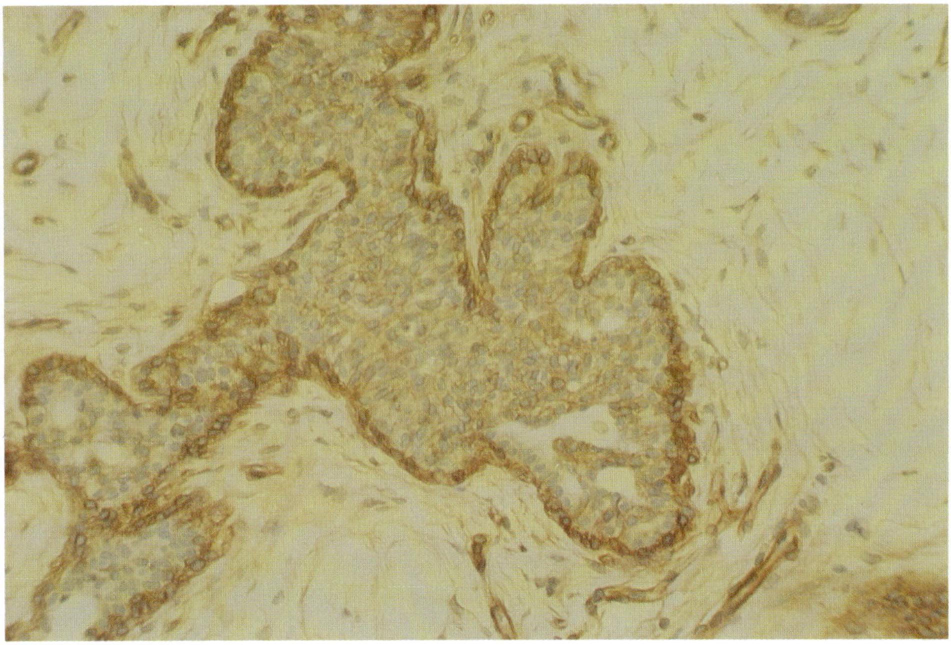

COLOUR PLATE I. See G. W. H. Stamp *et al.* Figure 2.1, page 17

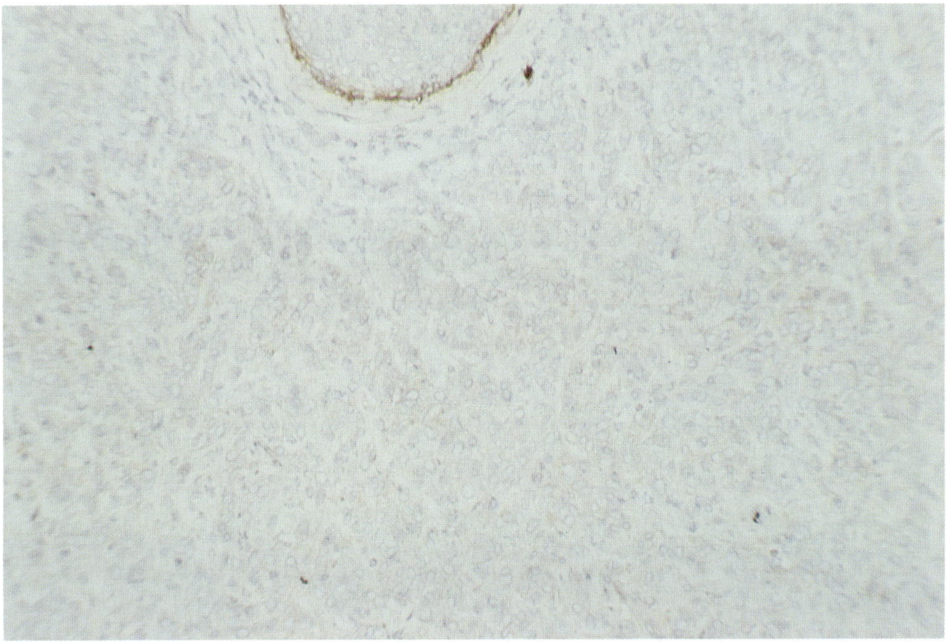

COLOUR PLATE II. See G. W. H. Stamp *et al.* Figure 2.2, page 20

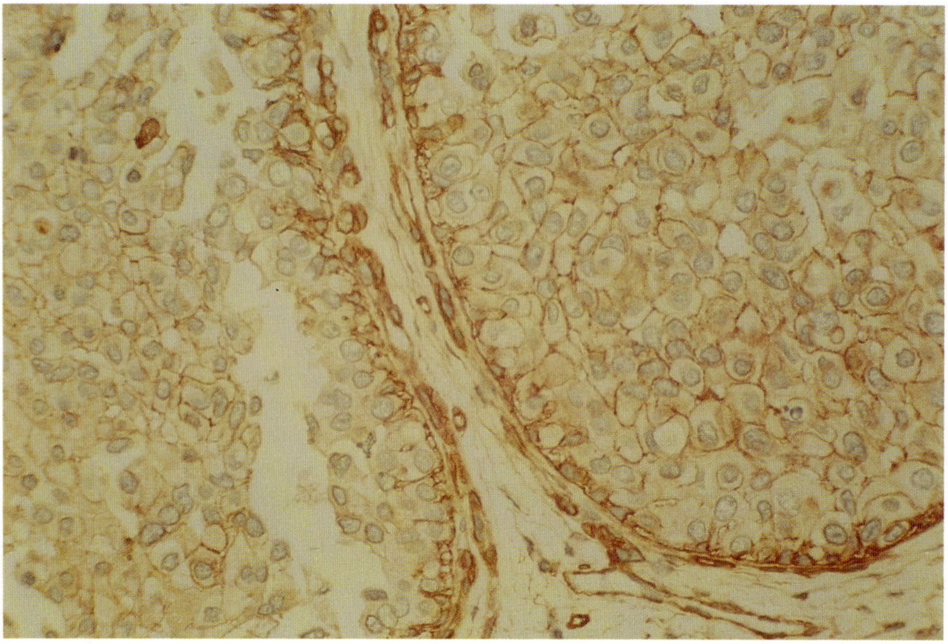

COLOUR PLATE III. See G. W. H. Stamp *et al.* Figure 2.3, page 21

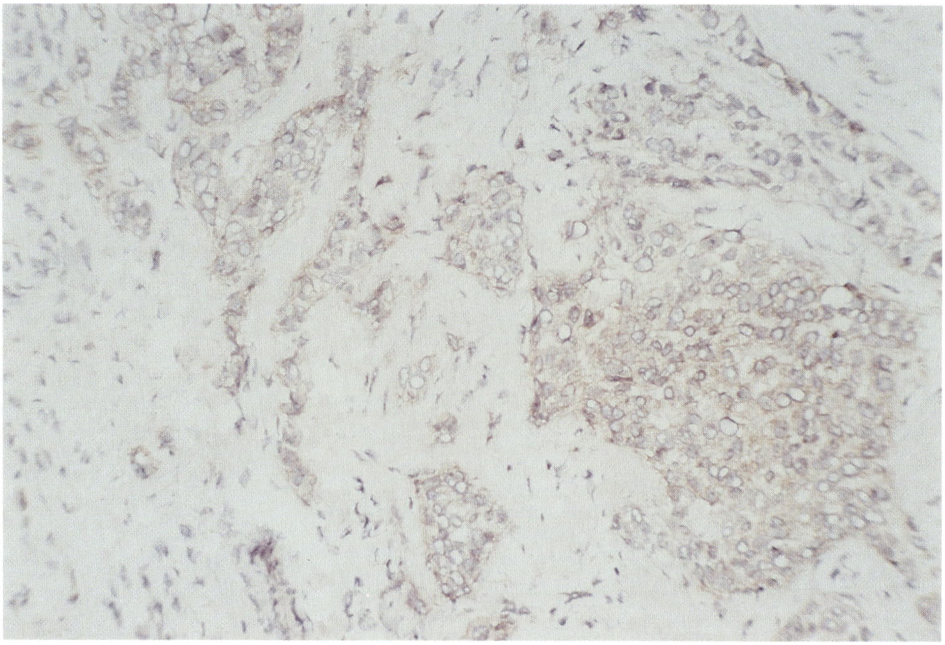

COLOUR PLATE IV. See G. W. H. Stamp *et al.* Figure 2.4, page 22

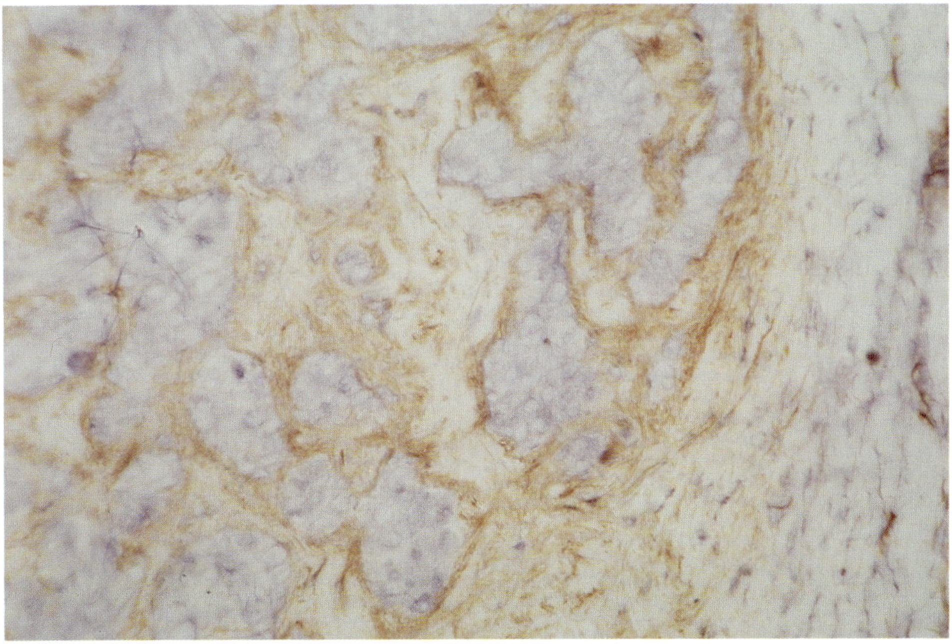

COLOUR PLATE V. See G. W. H. Stamp *et al.* Figure 2.5, page 25

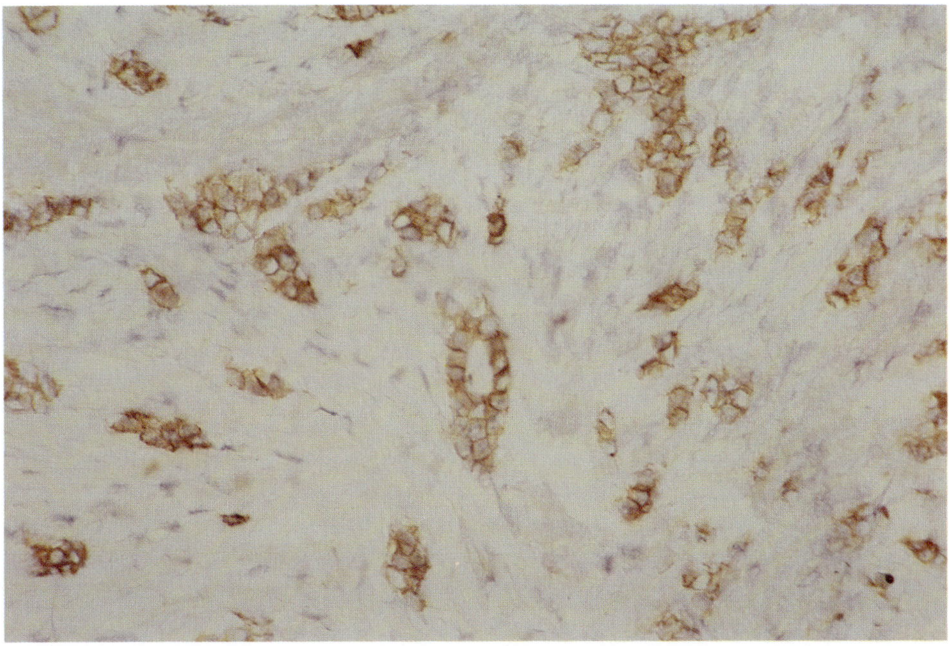

COLOUR PLATE VI. See G. W. H. Stamp *et al.* Figure 2.6, page 26

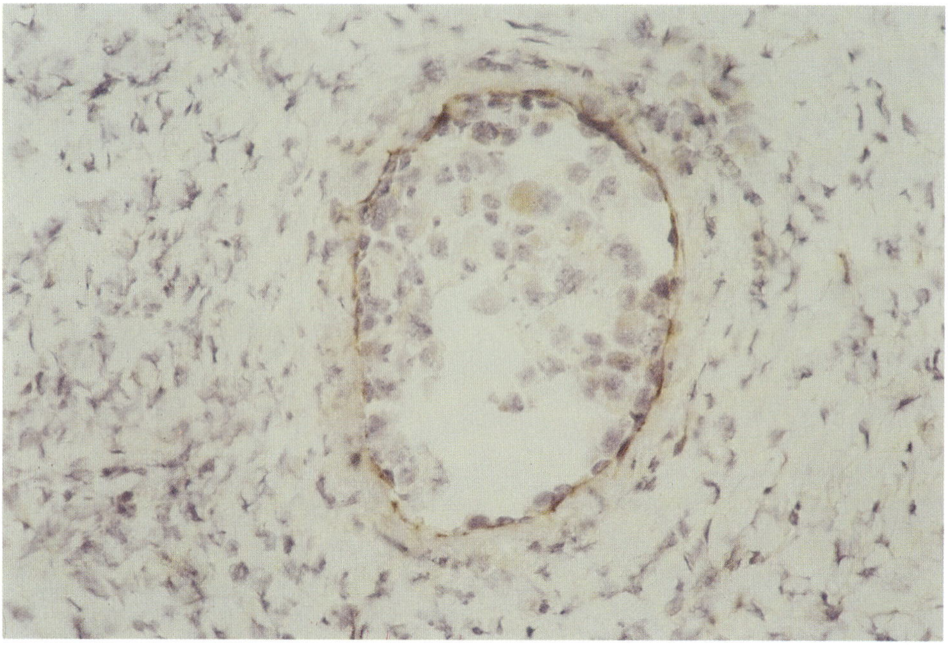

COLOUR PLATE VII. See G. W. H. Stamp *et al.* Figure 2.7, page 27

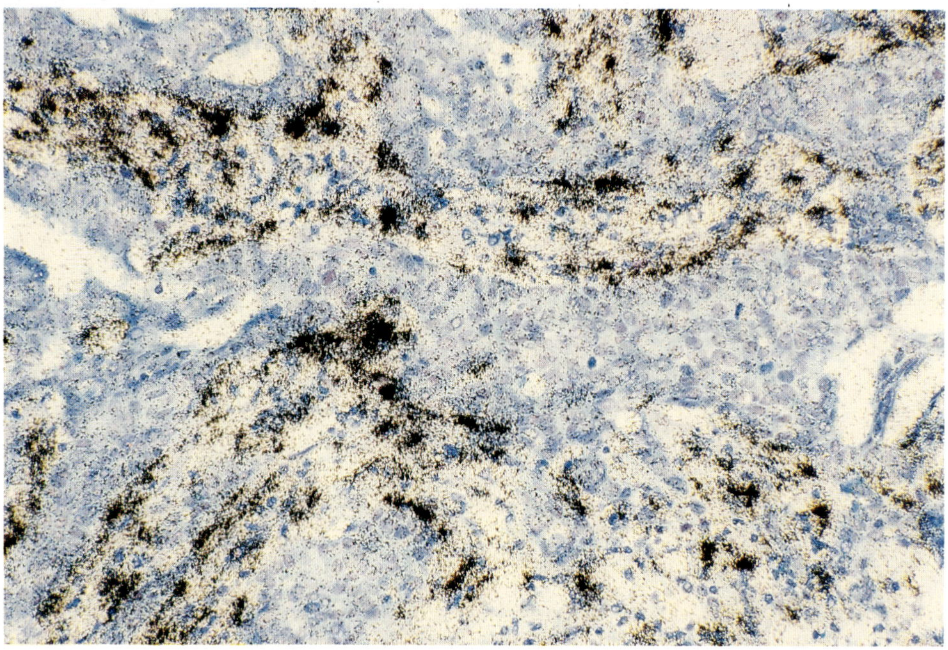

COLOUR PLATE VIII. See G. W. H. Stamp *et al.* Figure 2.8, page 28

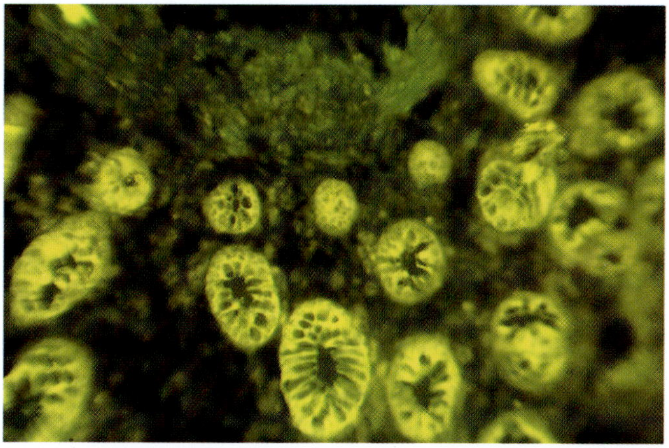

COLOUR PLATE IX. See Frank S. Steven *et al.* Figure 11.1, page 123

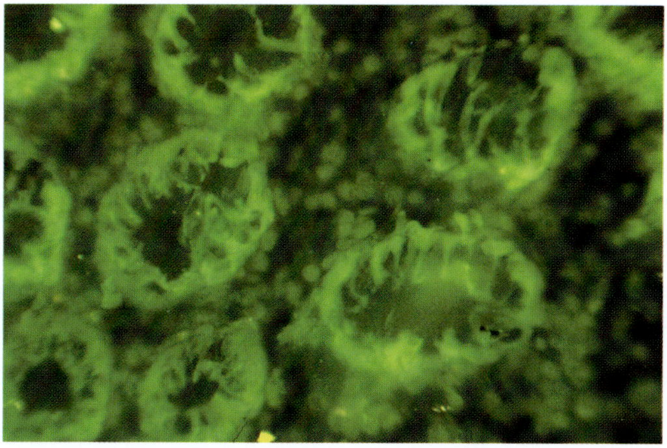

COLOUR PLATE X. See Frank S. Steven *et al.* Figure 11.2, page 124

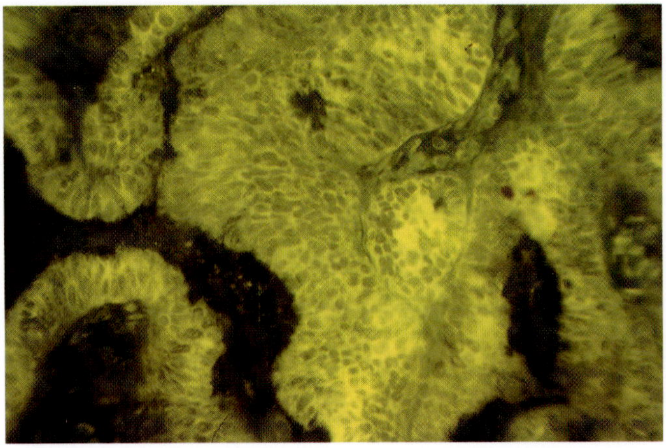

COLOUR PLATE XI. See Frank S. Steven *et al.* Figure 11.3, page 125

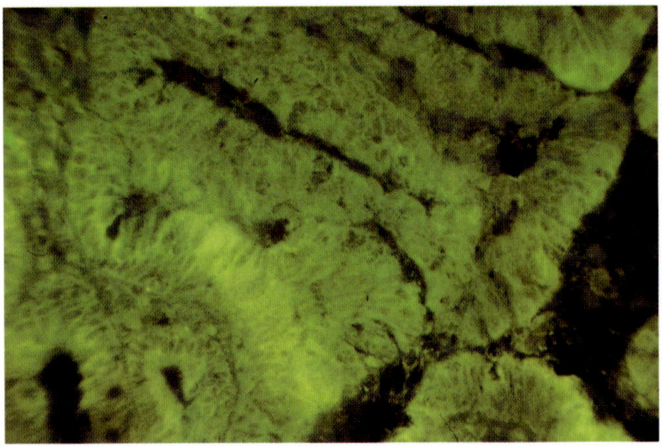

COLOUR PLATE XII. See Frank S. Steven *et al.* Figure 11.4, page 125

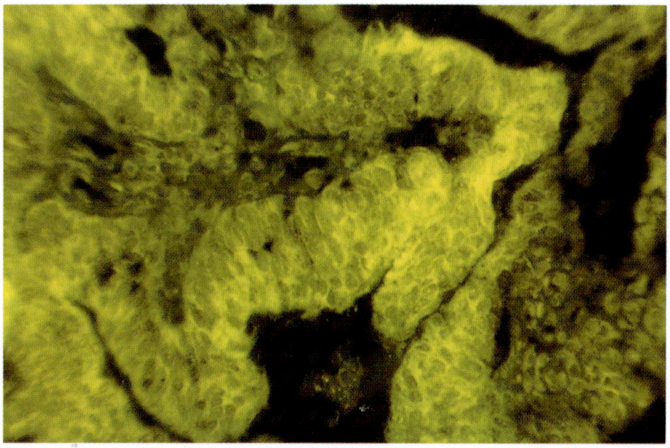

COLOUR PLATE XIII. See Frank S. Steven *et al.* Figure 11.5, page 126

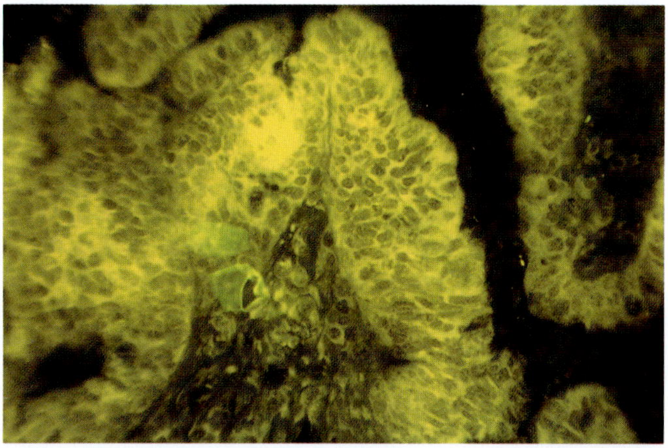

COLOUR PLATE XIV. See Frank S. Steven *et al.* Figure 11.6, page 127

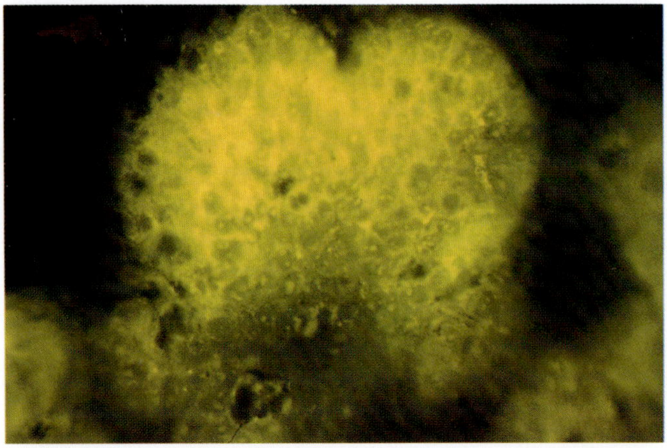

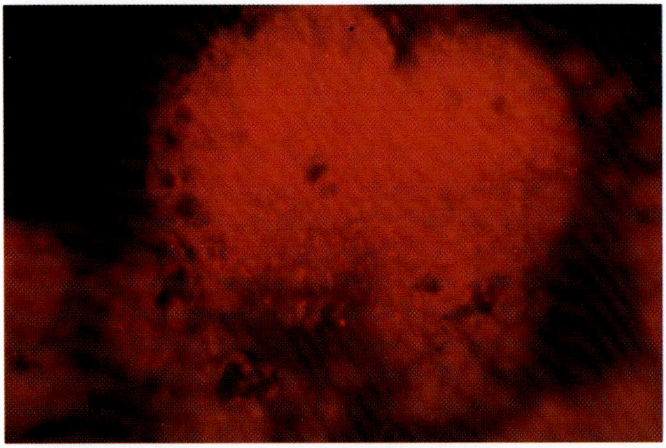

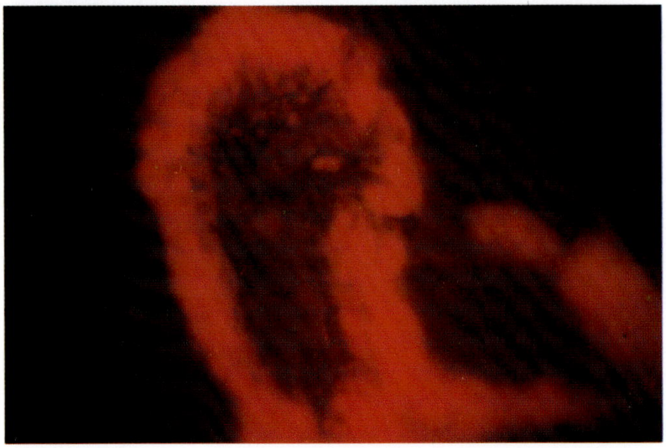

COLOUR PLATE XV. See Frank S. Steven *et al.* Figure 11.7,8,9, page 129

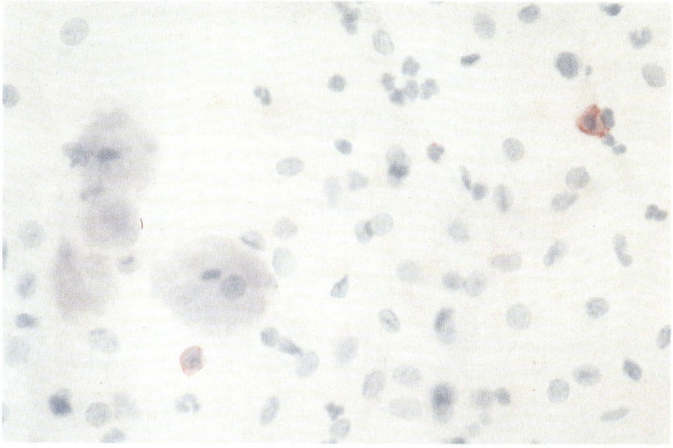

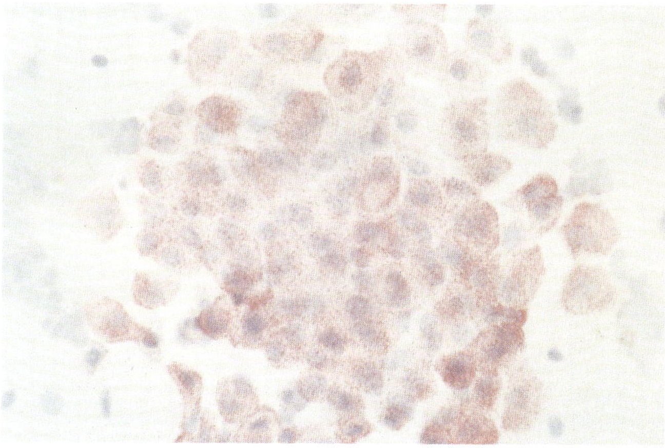

COLOUR PLATE XVI. G. Maria Müller and Irene Tausch Figure 19.1 and 19.2, page 226

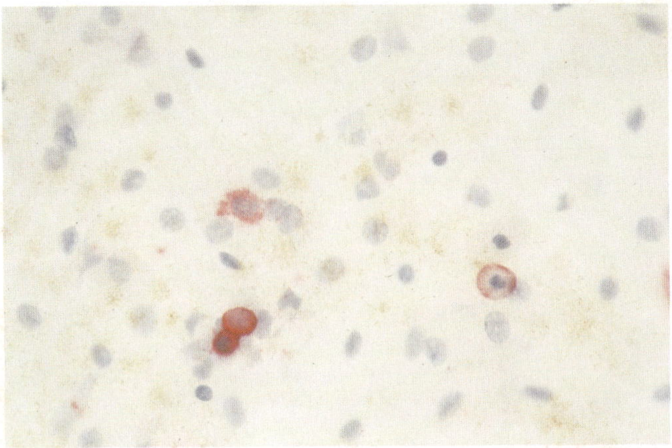

COLOUR PLATE XVII. G. Maria Müller and Irene Tausch Figure 19.3, page 227